Ecosystem Engineers
Plants to Protists

Academic Press is an imprint of Elsevier
30 Corporate Drive, Suite 400, Burlington, MA 01803, USA
525 B Street, Suite 1900, San Diego, California 92101-4495, USA
84 Theobald's Road, London WC1X 8RR, UK

This book is printed on acid-free paper. ∞

Library of Congress Cataloging-in-Publication Data
Ecosystem engineers : plsnts to protists / Kim Cuddington . . . [et al.].
      p.   cm.
   Includes bibliographical references and index.
   ISBN-13: 978-0-12-373857-8 (hard cover : alk. paper)   1. Habitat (Ecology)—Modification.   2. Ecology.   I. Cuddington, Kim.
   QH541.E3197   2007
   577—dc22

                                                        2007017144

British Library Cataloguing-in-Publication Data
A catalogue record for this book is available from the British Library.

978-0-12-373857-8

For information on all Academic Press publications
visit our Web site at www.books.elsevier.com

Printed and bound in the United Kingdom
Transferred to Digital Printing, 2011

Ecosystem Engineers

Plants to Protists

Kim Cuddington
Ohio University

James E. Byers
University of New Hampshire

William G. Wilson
Duke University

Alan Hastings
University of California

ELSEVIER

AMSTERDAM • BOSTON • HEIDELBERG • LONDON
NEW YORK • OXFORD • PARIS • SAN DIEGO
SAN FRANCISCO • SINGAPORE • SYDNEY • TOKYO

Academic Press is an imprint of Elsevier

CONTENTS

7 ☞ *CARPOBROTUS* AS A CASE STUDY OF THE COMPLEXITIES OF SPECIES IMPACTS   139

Nicole Molinari, Carla D'Antonio, and George Thomson

8 ☞ ECOSYSTEM ENGINEERING IN THE FOSSIL RECORD: EARLY EXAMPLES FROM THE CAMBRIAN PERIOD   163

Katherine N. Marenco and David J. Bottjer

9 ☞ HABITAT CONVERSION ASSOCIATED WITH BIOERODING MARINE ISOPODS   185

Theresa Sinicrope Talley and Jeffrey A. Crooks

14 ➤ SYNTHESIS OF ECOSYSTEM ENGINEERING THEORY   275

William G. Wilson

Section IV   SOCIO-ECONOMIC ISSUES AND MANAGEMENT SOLUTIONS

15 ➤ RESTORING OYSTER REEFS TO RECOVER ECOSYSTEM SERVICES   281

Jonathan H. Grabowski and Charles H. Peterson

16 ➤ MANAGING INVASIVE ECOSYSTEM ENGINEERS: THE CASE OF *SPARTINA* IN PACIFIC ESTUARIES   299

John G. Lambrinos

PREFACE

The purpose of this collection is to present some of the diversity of ideas and studies about species that can be classified as "ecosystem engineers." As with any developing concept, we find disagreement about the meaning and usefulness of this term in the literature and among ourselves. The idea for the book arose in a National Center for Ecological Analysis and Synthesis (NCEAS) working group designed to develop models of ecosystem engineering species. Our meetings could be characterized as lively, punctuated as they were by vigorous debates regarding definitions and arguments over whether a particular species' actions were appropriately characterized as engineering. Given that a small group of eight people with an active interest in the concept could not reach an agreement about definition, it is even less likely that the larger scientific community will do so in the immediate future. Notably, though, all eight found utility in the concept. In these pages, we invite other authors to contribute to this diversity of opinion in the hope that the variety of ideas and applications will engender further research in this area and a concomitant refinement of the concept. Given the breadth of the topic, only an edited book like this one, which draws on a wide range of authors, could hope to provide even the semblance of a balanced overview.

To begin, what is an ecosystem engineer? In proposing the concept, Jones et al. (1994, 1997) describe species which physically modify, maintain, or create habitats. They give as one canonical example, beavers, which create pond habitats by building dams that modify water flow regimes. One key characteristic of this activity is that it is not directly linked to the processes of consumption. That is, while beavers consume the living tissue of the trees, it is not this consumption that leads directly to the creation of a pond. It is here that most debates regarding ecosystem engineering seem to originate. That is, do we focus on the process of ecosystem engineering (i.e., the modification of the abiotic environment through nontrophic interactions) and label those actions ecosystem engineering, or do we instead focus on the outcome of species activities (i.e., the creation of habitat regardless of means). This type of distinction can lead to fierce discussions over whether the seastar *Pisaster ochraceus* in Bob Paine's classic (1961) study acts as an ecosystem engineer, and consequently, whether there is any difference between keystone species and ecosystem engineers.

Similarly, there has been much discussion regarding the utility of the ecosystem engineer concept. After all, all species to some extent modify their environment simply by existing. Opponents argue that ecosystem engineering cannot be used to distinguish one group of species from another, or one type of activities from another, and as a result, they suggest that the concept has no utility at all. Clearly, this claim is too extreme, as illustrated by the contributions in this collection and in the ecological literature that use the concept to increase the explanatory power of their studies. It is only by considering the effects of habitat modification that the importance of some species and their actions can be discovered. For example, the salt marsh grass *Spartina alterniflora* on the West coast of the United States invades and modifies the coastal mudflats into a thickly vegetated tidal plane with much reduced wave action and increased sedimentation rates, greatly influencing community composition. The community-level effects of this invasive species cannot be understood from a food web diagram, or even an ecosystem model of energy flows. The effects of such habitat modification entirely change the ecosystem.

Apart from ecological understanding, the management of such impacts may also require the incorporation of ecosystem engineering. One key to developing management plans is the understanding that the habitat modification effects of a given species may not subside with the demise of the species. Many ecosystem engineering effects are characterized by legacy effects which persist after death. The quintessential example may be coral, whose reef structures often persist for centuries and provide an engineered substratum for a community unparalleled in its diversity. Although coral are engaged in trophic interactions, the habitat provisioning by the coral is the attribute managers clearly seek to protect or restore. The number of artificial reef restoration programs seeking to introduce objects that replicate the coral's structure is testament to this.

We have organized the book into four sections. The first lays out the historical origins and broad concepts of ecosystem engineering. Additionally, it presents some of the contrasting viewpoints on definitions mentioned above. Section 2 presents some in-depth examples of ecosystem engineers. A major aim of this section is to provide tangible, highly varied examples to apply to conceptual and theoretical developments in other sections. Chapters in Section 3 develop the mathematical theory of ecosystem engineers and review the very brief ecosystem engineer theoretical literature. Finally, the authors of Section 4 address applied examples where ecosystem engineers have been important to the success or failure of resource management, restoration, or conservation. Each section has a concluding chapter that brings together the contributions in that section into a more unified framework.

We hope that the biggest contributions of our book are to stimulate discussion of ecosystem engineering, and perhaps spur further development of viable tools to aid its study, particularly to practical applications. As exemplified here, ecosystem engineering has many indications of being a powerful way of categorizing an important subset of ecological interactions. Assessing its history and merits, presenting solid examples, recapping and developing relevant theory, and examining successful applications are thus all important and timely aspects to present in our collected volume.

Kim Cuddington
James E. Byers
William G. Wilson
Alan Hastings

CONTRIBUTORS

Numbers in parentheses indicate the chapter to which the author has contributed.

Sebastien Barot (5): BIOSOL, Centre IRD Bondy, France

Manuel Blouin (5): BIOSOL, Université de Paris 12, France

David J. Bottjer (8): Department of Earth Sciences, University of Southern California, Los Angeles, California

Sol Brand (17): Mitrani Department of Desert Ecology, Ben-Gurion University, Beer-Sheva, Israel

Natalie Buchman (2): Department of Biological Sciences, Ohio University, Athens, Ohio

James E. Byers (10): Department of Zoology, University of New Hampshire, Durham, New Hampshire

Jeffrey A. Crooks (9): Tijuana River National Estuarine Research Reserve, Imperial Beach, California

Kim Cuddington (2, 4, 13): Department of Biological Sciences, Ohio University, Athens, Ohio

Carla D'Antonio (7): Ecology, Evolution, & Marine Biology, University of California, Santa Barbara, Santa Barbara, California

Thibaud Decaëns (5): ECODIV, Université de Rouen, France

Erez Gilad (12): Department of Solar Energy and Environmental Physics, Ben-Gurion University, Beer-Sheva, Israel

Jonathan H. Grabowski (15): Gulf of Maine Research Institute, Portland, Maine

Jorge L. Gutierrez (1): Institute of Ecosystem Studies, Millbrook, New York

Alan Hastings (13, 20): University of California, Davis, California

Juan José Jimenez (5): FAO, Rome, Italy

Clive G. Jones (1): Institute of Ecosystem Studies, Millbrook, New York

Pascal Jouquet (5): BIOSOL, Centre IRD Bondy, France

Rob Klinger (18): Section of Evolution & Ecology, University of California, Davis, California

John G. Lambrinos (16): Department of Horticulture, Oregon State University, Corvallis, Oregon

Patrick Lavelle (5): BIOSOL, Université Pierre et Marie Curie (Paris 6) and IRD, France

John T. Lill (6): Department of Biological Sciences, George Washington University, Washington, D.C.

Katherine N. Marenco (8): Department of Earth Sciences, University of Southern California, Los Angeles, California

Robert J. Marquis (6): University of Missouri-St. Louis, St. Louis, Missouri

Ehud Meron (12): Department of Solar Energy and Environmental Physics, Ben-Gurion University, Beer-Sheva, Israel

Nicole Molinari (7): Ecology, Evolution, & Marine Biology, University of California, Santa Barbara, Santa Barbara, California

Yarden Oren (17): Mitrani Department of Desert Ecology, Ben-Gurion University, Beer-Sheva, Israel

Avi Perevolotsky (17): Agricultural Research Center, The Volcani Center, Bet Dagan, Israel

Ivette Perfecto (19): University of Michigan, Ann Arbor, Michigan

Charles H. Peterson (15): Insititute of Marine Sciences, University of North Carolina at Chapel Hill, Morehead City, North Carolina

Antonello Provenzale (12): Universita di Genova e della Basilicata, Savona, Italy

Moshe Shachak (12, 17): Mitrani Department of Desert Ecology, Ben-Gurion University, Beer-Sheva, Israel

Theresa Sinicrope Talley (9): Department of Environmental Science and Policy, University of California, Davis, California

George Thomson (7): Ecology, Evolution, & Marine Biology, University of California, Santa Barbara, Santa Barbara, California

John Vandermeer (19): University of Michigan, Ann Arbor, Michigan

Jost von Hardenberg (12): Universita di Genova e della Basilicata, Savona, Italy

William G. Wilson (3, 11, 14): Biology Department, Duke University, Durham, North Carolina

Justin P. Wright (11): Biology Department, Duke University, Durham, North Carolina

I

HISTORY AND DEFINITIONS OF ECOSYSTEM ENGINEERING

We begin with contributions discussing the history of the ecosystem engineer concept, its definition, and its utility. As with other terms in the ecological literature (e.g., *keystone species*), ecosystem engineering has been met with debate about its usefulness and precise definition. In this section, authors attempt to bring clarity to this discussion by outlining the historical antecedents of the idea, discussing its potential usefulness, and providing more nuanced definitions. It should be recognized that the controversy regarding this idea is reflected in the different definitions provided by different authors. However, the differences of opinion expressed here do advance this debate by moving past somewhat trivial difficulties and striking at some of the key issues, such as the inclusion of both positive and negative interactions, and the value of a process-based vs. an outcome-based definition.

1

ON THE PURPOSE, MEANING, AND USAGE OF THE PHYSICAL ECOSYSTEM ENGINEERING CONCEPT

Clive G. Jones and Jorge L. Gutiérrez

1.1 ⬤ INTRODUCTION

There has been substantial growth of interest in the concept of physical ecosystem engineering by organisms since the publication of Jones et al. 1994 and 1997a. The concept has certainly catalyzed new case studies, methods, modeling, generalization, and synthesis (see reviews by Lavelle et al. 1997; Crooks 2002; Coleman and Williams 2002; Gutiérrez et al. 2003; Wright and Jones 2004, 2006; Boogert et al. 2006; Caraco et al. 2006; Gutiérrez and Jones 2006; Jouquet et al. 2006; Moore 2006; Hastings et al. 2007; also see Table 1.1). However, the concept has also generated controversy and uncertainty over meaning, usage, and purpose (e.g., Jones et al. 1997b; Power 1997a, 1997b; Reichman and Seabloom 2002a, 2002b; Wilby 2002), reflected in the following questions. Don't all organisms change the environment? Aren't all organisms therefore ecosystem engineers? If so, isn't the concept too broad to be useful? Don't engineers always have large or large-scale impacts? Shouldn't engineers be limited to species with large effects? Aren't engineers and keystone species the same? Isn't engineering equivalent to facilitation or positive influence? Isn't the approach overly reductionist? Why do we need the concept? How can we use it?

TABLE 1.1 Illustrative usage of the physical ecosystem engineering concept.

Conceptual Application	References
Population dynamics	
When survival depends on habitat modification	Gurney and Lawton 1996
Linked to dynamics of patch creation	Wright et al. 2004
Invasion	Cuddington and Hastings 2004
Community organization	
Consequences for community structure	Flecker 1996, Flecker and Taylor 2004, Gutiérrez and Iribarne 1999
Species interactions and altered resource availability or abiotic stress	Gutiérrez and Iribarne 2004, Daleo et al. 2006
Patterns of species distribution	Escapa et al. 2004, Jouquet et al. 2004
Variation in species responses across abiotic gradients	Crain and Bertness 2005, Wright et al. 2006, Badano and Cavieres 2006b
Environmental heterogeneity and species diversity at patch and landscape scales	Wright et al. 2002, 2003, 2006; Lill and Marquis 2003; Badano and Cavieres 2006a, 2006b
Parsing species effects into trophic (assimilatory–dissimilatory) and nontrophic contributions	Crooks and Khim 1999, Wilby et al. 2001
Structural legacies and community organization	Gutiérrez and Iribarne 1999
Species diversity in fossil communities	Parras and Casadío 2006
Assessing effects on community organization	Badano et al. 2006
Predicting patch-level richness effects	Wright and Jones 2004
Ecosystem processes	
Controls on material fluxes between ecosystems	Caraco et al. 2006, del-Val et al. 2006, Gutiérrez et al. 2006
General determinants of biogeochemical heterogeneity	Gutiérrez and Jones 2006
Integration with state factors	Jones et al. 2006
Conservation, restoration, and management	
Global change scenarios for soil	Lavelle et al. 1997
Persistence of endangered species	Pintor and Soluk 2006
Support of species diversity via habitat diversity	Bangert and Slobodchikoff 2006
Conceptual models for management and conservation of threatened species	Goubet et al. 2006
Evaluation of abiotic restoration options	Byers et al. 2006

Uncertainty, misconstrual, and misunderstanding impede scientific progress, but since no concept is ever born fully developed, they also justify clarification. Concepts that cannot eventually be sufficiently unambiguously defined as to be made operational deserve to disappear. Further, while a concept is not a theory, it is a foundation upon which theory is built, and the foundation must be solid if one has any aspiration for theory development (Pickett et al. 1994). The questions outlined in Jones et al. (1994) clearly beg theory development.

Here we present a perspective on selected aspects of the purpose, meaning, and usage of the concept, including some new thoughts, some clarification, and some reification. We briefly describe the domain, general purpose, and components of the concept. We then define the two coupled, direct interactions comprising ecosystem engineering— the *physical ecosystem engineering process* responsible for abiotic change, and *physical ecosystem engineering consequence* that addresses biotic effects of abiotic change. We clarify the meaning of "ecosystem" in *ecosystem engineer*. We address causes of process ubiquity and how they lead to general expectations of consequence. We examine sources of context-dependent variation in engineer effect magnitude and significance and what needs to be known to predict effects. We define conditions for detectable engineering effects and the condition for large effects, all other factors being equal (i.e., *ceteris paribus*). We argue against unspecified conflation of process and consequence. We illustrate where explicit consideration of influential physical ecosystem engineering may or may not be needed, point out what the concept has been used for, and suggest general topics where it might be useful. We end with comments on how conceptual breadth relates to utility, and what perspective on species interactions is reflected in the concept. Our overall intent is conceptual clarification and amplification.

1.2 ● ON THE DEFINITION

ON THE CONCEPTUAL DOMAIN, GENERAL PURPOSE, AND COMPONENTS

Physical ecosystem engineering as defined by Jones et al. (1994, 1997a; Table 1.2) is a particular form of abiotic environmental modification by organisms that often, but not invariably, has effects on biota and their interactions. Abiotic environmental change occurs as a consequence of the physical structure of organisms or via organisms causing changes in the physical structure of the living and nonliving materials. These abiotic changes can then affect biota, including the engineer. Biotic influence

TABLE 1.2 Definitions of physical ecosystem engineering.

Jones et al. 1994: "Ecosystem engineers are organisms that directly or indirectly modulate the availability of resources (other than themselves) to other species by causing physical state changes in biotic or abiotic materials. In so doing they modify, maintain and/or create habitats. The direct provision of resources by an organism to other species, in the form of living or dead tissues is not engineering."

Jones et al. 1997a: "Physical ecosystem engineers are organisms that directly or indirectly control the availability of resources to other organisms by causing physical state changes in biotic or abiotic materials. Physical ecosystem engineering by organisms is the physical modification, maintenance or creation of habitats. Ecological effects of engineers on many other species occur in virtually all ecosystems because the physical state changes directly create non-food resources such as living space, directly control abiotic resources, and indirectly modulate abiotic forces that, in turn, affect resource use by other organisms. Trophic interactions, i.e., consumption, decomposition and resource competition are not engineering."

Physical ecosystem engineering process: Organismally caused, structurally mediated changes in the distribution, abundance, and composition of energy and materials in the abiotic environment arising independent or irrespective of changes due to assimilation and dissimilation.

Ecosystem engineering consequence: Influence arising from engineer control on abiotic factors that occurs independent or irrespective of use of or impact of these abiotic factors on the engineer or the participation by the engineer in biotic interactions, despite the fact that these can all affect the engineer and its engineering activities.

"Ecosystem" in Ecosystem Engineering: A place with all the living and nonliving interacting. Hence, *ecosystem* refers to the biotic on abiotic of the engineering process and the abiotic on biotic of engineering consequence.

For discussion see text and cited references.

encompasses organisms, populations, communities, ecosystems, and landscapes and can be integrated by thinking of physical ecosystem engineering as the creation, modification, maintenance, and destruction of habitats. The concept therefore addresses some but not all of the ways organisms can change the abiotic environment and the consequences thereof.

The concept was developed to encompass a variety of disparate and oft-ignored ecological phenomena not addressed by the historical focus of ecology on trophic relations (i.e., predation, resource competition, food webs, energy flow, nutrient cycling, and the like). Ecologists had long been familiar with many examples (see Chapter 2, Buchman). Some specialty areas in ecology and other disciples had emphasized some aspects (e.g., marine sediment bioturbation, mammalian soil disturbance, geomorphology). Nevertheless, as evidenced by omission from ecological textbooks, formal recognition and study of the general process and its consequences were not central to ecological science. So the primary purpose of the papers (Jones et al. 1994, 1997a) was to draw attention to the ubiquity and importance of this process and its consequences, to provide an integrative general framework, to lay out a provisional question-based research agenda, and to give it a name.

The concept addresses the combined influence of two coupled direct interactions. The first is the way organisms change the abiotic environment—the *physical ecosystem engineering process*. The second is how these abiotic changes affect biota—*ecosystem engineering consequence*. The distinction reveals important criteria of demarcation for what is and is not physical ecosystem engineering, exposes context dependency for effects that enhance prediction of effect magnitudes and significance, and helps clarify the purpose of the concept and how one might use it. In the following text we examine these two component interactions before briefly reintegrating them with the overall concept.

ON THE PHYSICAL ECOSYSTEM ENGINEERING PROCESS

The *physical ecosystem engineering process* can be defined as the following: Organismally caused, structurally mediated changes in the distribution, abundance, and composition of energy and materials in the abiotic environment arising independent or irrespective of changes due to assimilation and dissimilation.

"Organismally caused" distinguishes the process from purely abiotic forces (i.e., climatic and geologic processes) that are functional analogs when they change the same abiotic variables. Wind and elephants both uproot trees creating tip-up mounds. Organismal causation also invokes potential for spatial and temporal differences in the resulting abiotic environment compared to purely abiotic forces, even when the mean abiotic change is the same (cf. Reichman and Seabloom 2002a). Elephants and wind both may knock over trees, but different factors are needed to predict when and where such events might occur (Pickett et al. 2000).

"Structurally mediated changes" reflects the requirement for abiotic change to arise via structural change (i.e., physical state changes, Jones et al. 1994, 1997a). This can occur autogenically where the living organism is the structure, or allogenically where the organism makes the structure from living or nonliving materials (Jones et al. 1994). Thus if there is no structural change there is no physical ecosystem engineering process. This requirement distinguishes this process from other ecological processes that may have the same abiotic effect (e.g., increased nitrogen in aquatic invertebrate burrows can result from invertebrate excretion *and* from increased oxygen supply that controls microbial mineralization, Aller 1988), or the same overall biotic response (e.g., increased macrophyte growth in the presence of burrows, Bertness 1985).

Inherent in structural mediation but not explicit in the definition is recognition that structures have some degree of persistence. Dead autogenic engineers and allogenic engineering leave structural legacies with concomitant abiotic effects, with the persistence of legacies being a function of construct durability and the abiotic and biotic forces causing their disappearance (Jones et al. 1994, Hastings et al. 2007).

"Changes in the distribution, abundance, and composition of energy and materials in the abiotic environment" is the most general possible description of abiotic influence. Such effects are not unique to the engineering process. Geomorphic structures can have similar abiotic effects (e.g., rocks and trees both cast shade), and as discussed in following text, organismal uptake and release of materials can bring about comparable abiotic changes. However, within a structural context, ecosystem engineering encompasses organismally changed structure (e.g., a burrow, leaves tied by caterpillars, earthworm litter burial), interactions of structure with various forms of kinetic energy (e.g., hydrological attenuation by beaver dams), abiotic consequences of such kinetic interactions (e.g., sedimentation behind the dam), and interactions of organismally made structures and kinetic energy imparted by organisms (e.g., burrowing polychaetes pumping water by body movement, Evans 1971). For further discussion of some of these relationships, see Gutiérrez and Jones 2006.

Finally, the requirement that abiotic change occur "independent or irrespective of changes due to assimilation and dissimilation" distinguishes the engineering process from changes caused by the universal processes of organismal uptake (light, water, nutrients, other minerals, O_2, CO_2, trace gases, organic compounds) and release (carbon and nutrients in litter, woody debris, feces, urine, and carcasses; water, O_2, CO_2, trace gases, H^+, other organic and inorganic chemicals). Since the physical ecosystem engineering process can result in altered energy and material flows (e.g., water kinetic to potential energy in a beaver impoundment and sedimentation of suspended materials), and these can involve chemical changes (e.g., redox effects on beaver pond sediment geochemistry due to reduced water column oxygen exchange), this part of the definition is a necessary and important qualifier for the non-assimilatory and nondissimilatory (or "nontrophic") basis of any abiotic effects.

It is worth further exploring what we mean by "independent or irrespective," since it informs where the engineering process begins and ends. "Independent," in the context of our definition, means that there are many other life processes unrelated to or only very distally related to assimilation and dissimilation that can result in changes to structure and the abiotic environment—growth, predator and stress avoidance, and movement, to name but a few. Examples include wind attenuation by trees, nests and dens that shelter animals, and the hoofprints and trails made by large animals.

"Irrespective," in the context of our definition, means that many organismal activities associated to varying degrees with assimilatory and dissimilatory transfers also have structural influences whose effects on the abiotic occur regardless of any influence of the transfers. For example, leaf litter affects soil-gas exchange and rain splash impact irrespective of its role as a resource for decomposers (Facelli and Pickett 1991). Trees cast shade, in part because they assimilate photons (uptake) and in part because, like any physical structure, they absorb and reflect photons (engineering). Desert porcupines always dig soil to feed on bulbs (Shachak et al. 1991); soil effects occur irrespective of consumption but are always associated with it. Effects of insect defoliation on the understory physical environment (e.g., Doane and McManus 1981) depend upon consumption amount (along with extant canopy structure and extrinsic abiotic conditions) but occur irrespective of effects on trees or caterpillars or altered nutrient cycling via frass. The central point is not that assimilation–dissimilation must always occur separately from the engineering process, although as noted in preceding text it is often independent, but that any co-occurrence requires the distinction if we are to invoke either engineering or assimilation–dissimilation as a causal explanation for abiotic change.

ON ECOSYSTEM ENGINEERING CONSEQUENCE

Abiotic changes due to the engineering process are the starting point of consequence. While worthy of study alone (e.g., erosion, hydrology, sedimentation, pedogenesis, heat balance, physical gas exchange, etc.), they necessarily underpin all consequences for biota and their interactions on which we now focus. We can broadly define *consequence* as the following: Influence arising from engineer control on abiotic factors that occurs independent or irrespective of use of or impact of these abiotic factors on the engineer or the participation by the engineer in biotic interactions, despite the fact that all these can affect the engineer and its engineering activities.

"Control" (*modulation* is equivalent) is analogous to a faucet on a pipe; flow is regulated independent or irrespective of water use. Thus beaver dams control hydrology and flood and drought impact (Naiman et al. 1988), while dead mollusk shells control living space, enemy-free space, and abiotic stress (Gutiérrez et al. 2003). The term *control* helps distinguish engineering effects on biota and their interactions from any other influence of the engineer via other types of ecological interactions (e.g., abiotic resource uptake and direct resource competition; role as predator, prey, pollinator, or disperser).

"Abiotic factors" is shorthand for the large number of abiotic influences on biota and their interactions very familiar to ecologists. All that differs here is recognition that an organism is responsible for abiotic change via structural change, but the kinds of abiotic variables are no different. They are the following: consumable energy and materials (e.g., light, nutrients, water); nonconsumable resources (e.g., living space, enemy- or competitor-free space); and abiotic constraint or enablement including direct abiotic influences on organisms (e.g., temperature, salinity, wind, redox) and influences on information exchange or cues used by organisms (e.g., sound attenuation or amplification, temperature, light quality).

This first part of the definition ("influence arising from engineer control on abiotic factors") contains an important, unstated but implicit recognition that since species and their interactions vary in their sensitivity to the abiotic, engineer effects will be context dependent on the degree of abiotic change caused by the engineering process *and* the degree of abiotic limitation, constraint, or enablement experienced by species. Such context dependency applies to direct abiotic effects on species (e.g., trapped runoff water on plant growth, Eldridge et al. 2002) and abiotic influences on species interactions (e.g., how engineer-altered resources influence plant competition, Shachak et al. 1991; how refugia may affect predator–prey interactions, e.g., Usio and Townsend 2002).

The latter part of the definition ("that occurs . . . engineering activities") recognizes the potential importance of engineering feedbacks to the engineer and effects of other biotic interactions on engineering activities. It also emphasizes that the relationship between the engineer and its engineering effects is fundamentally no different from the effect of the engineer on other species, i.e., effects arise via control on abiotic factors. Again, it excludes any other types of ecological interactions that the engineer may have with other biota, while recognizing that if these other interactions affect engineer density, engineering activities, and structural change, they can then affect the degree and type of abiotic change.

ON COMBINING ENGINEERING PROCESS AND CONSEQUENCE

Given a suitably broad construal of habitat encompassing all relevant abiotic aspects of place along with some biotic effect, process and consequence can be usefully combined into the recognition that physical ecosystem engineering is organismal, structurally mediated habitat change, conforming to the definition of Jones et al. (1994, 1997a).

We think the definitions of ecosystem engineering process and consequence enhance the overall definition of physical ecosystem engineering, helping provide clear criteria of demarcation as to what it is and what it is not. There is no fundamental change in either the intent or meaning of the concept, hopefully just illumination. As we show later, this collectively informs expectations for effect magnitude and significance, and how to use the concept.

ON "ECOSYSTEM" IN *ECOSYSTEM ENGINEERING*

We will not go into the meaning of the word *engineer.* It is certainly neither defined nor treated tautologically in the concept, and this issue has been adequately discussed (Power 1997a, 1997b; Jones et al. 1997b; Wright and Jones 2006). However, we will make a brief comment on "ecosystem" in *ecosystem engineer.* Some have construed the meaning as large scale or extensive. However, the meaning derives from Tansley (1935). His definition of ecosystem was size independent. An ecosystem can be large or small, but it is always a place with all the living and non-living interacting (Likens 1992, Pickett and Cadenasso 2002). Thus here "ecosystem" refers to the biotic–abiotic–biotic interactions representing the engineering process (biotic on abiotic) and consequence (abiotic on biotic). Certainly, some engineers can affect the functioning of large areas (e.g., oyster reef influences on estuarine flows and sedimentation, Ruesink et al. 2005; tsunami attenuation by mangrove forests, Kathiresan and Rajendran 2005), but they often have local effects (e.g., animal burrow, woodpecker hole, phytotelmata, birds nest). So, although the spatial scale of engineering is an interesting and important topic (e.g., see Hastings et al. 2007), it is neither a defining feature of the concept, nor the meaning of the word *ecosystem* in the concept.

1.3 ➤ ON PROCESS UBIQUITY

Are all organisms capable of the physical ecosystem engineering process? Based on the definition of the process and first principles of physics, the answer is almost certainly yes for all free-living organisms, although this clearly cannot be empirically proven. All physical structures interact with kinetic energy (i.e., radiant as light, heat, sound; energized fluids as water, air, and other gases). The inanimate and animate do not fundamentally differ in this regard. All free-living organisms have physical structures (autogenic). Many alter the physical structure of their surroundings (allogenic). Some, such as bioturbators, also generate kinetic energy in their structurally modified surroundings (allogenic). All these structures are inserted into abiotic kinetic energy flows. Physics tell us that these structures must affect and be affected by those flows, resulting in some degree of energy transformation and the redistribution of energized fluids and the materials they may contain. Given sufficiently accurate and diverse measurement instrumentation, it is a reasonable bet that all structures will result in some detectable change in one or more abiotic variables. A bird's nest affects local turbulent airflow, and mobile animals cast temporary shade, even though these almost certainly have no broader significance. So in this sense the physical ecosystem engineering process is an extended property of life. This should not be a blinding revelation, but then, nor is the fact that all free-living organisms also necessarily change the abiotic environment via the uptake and release of energy and materials.

Organisms therefore cannot be physically engineering unless they directly cause structural change within an abiotic milieu. So, ignoring the obviously trivial (e.g., shade cast by moving animals), it follows that if they are not causing such changes they are not engineering; and if they are not free-living they cannot engineer (cf. Thomas et al. 1998). We might expect greater capacity for influence when organisms are or make persistent rather than ephemeral structures (Jones et al. 1997a). Organismally created structures that are large relative to the abiotic environment experienced by other biota might be more influential (e.g., forests, Holling 1992; impoundments in tree holes or phytotelmata, Fish 1983; leaves tied by caterpillars, Lill and Marquis 2003) than those that are relatively small (e.g., effects of herb shade on large mammals). Small-bodied autogenic engineers likely have to be numerous (e.g., algae, Townsend et al. 1992) or aggregated into larger structures (e.g., microbial biofilms, Battin et al. 2003) to have large abiotic effects. It seems reasonable to suppose that small allogenic engineers will either have to have large per capita effects (e.g., earthworms, Darwin 1890, Lavelle et al. 1997) and/or be numerous (e.g., termites, Dangerfield et al. 1998, Jouquet et al. 2006) to cause substantive abiotic change (Jones et al. 1994, 1997a).

While the preceding is somewhat informative, it is clearly insufficient to predict what abiotic changes will occur, how large they will be, or what the biotic significance may be—issues we turn to next.

1.4 ☙ ON EFFECT MAGNITUDE AND SIGNIFICANCE

Ubiquity of a life process does not equate to universality of importance. We should expect that the physical ecosystem engineering process may often have little consequence, in the same way that energy and material uptake and release by many of the organisms in an ecosystem are not central to understanding energy flow, nutrient cycling, or food web dynamics. Nor for that matter is ubiquity a cause for phenomenological dismissal. Some physical engineering is significant, just as the uptake and release of energy and materials by some organisms is important. The challenge is to determine what makes the difference between the significant and insignificant.

The answer is it depends on context, and we think the separation of physical ecosystem engineering into process and consequence helps address this context dependency. First, from the definition of *process*, there can be no abiotic effect, hence no biotic consequence, without structural change. Second, given structural change, depending on the abiotic variable(s) of interest selected and baseline abiotic conditions (i.e., the structurally unmodified state), measurable abiotic change may or may not occur, depending upon structural form and abiotic milieu. The physical properties of structures and the physics of their interaction with kinetic energy are central to predicting this effect. Third, given some detectable abiotic effect, changes may be the same as, or larger or smaller than, those caused by other forces (i.e., purely abiotic or assimi-latory–dissimilatory). Further and as noted earlier, the spatial or tempo-ral dynamics of such abiotic effects may be the same as or different from those due to other forces. Thus we can judge the importance of the engineering process in terms of abiotic change relative to the effect magnitudes and dynamics due to these other forces acting on the same abiotic variable(s). Fourth, given some abiotic change, we should then expect that whether or not there will be biotic consequence will depend upon the degree of abiotic change (magnitude and direction) *and* the sensitivity of the biota or their interactions to this abiotic variable in terms of limitation, constraint, or enablement. An understanding of species sensitivities relative to baseline abiotic conditions can be used to predict the particular response. Finally, given some abiotic effect on some biotic response variable of interest, we can judge the relative import of the engineering in comparison to other forces (abiotic or other types of biotic interactions) affecting the same biotic response variable.

The preceding dependencies allow for a very precise definition of when physical ecosystem engineering will have a biotic effect. If an organism causes structural change that results in an abiotic change that is larger than or different from that caused by other abiotic or biotic forces; and if biota are sensitive to that degree or type of abiotic change; and if the biotic responses to these abiotic changes are greater than those due to other biotic forces acting on the same biotic response variable; then there will be a detectable engineering effect. If any one of those conditions does not hold, there will be no detectable effect. It follows that physical engineering by organisms that causes large abiotic changes affecting highly sensitive biota where there is no other influence (i.e., *ceteris paribus*) will have large effects.

While the preceding analysis identifies the primary sources of context dependency and how to address them, it is clear on both theoretical and empirical grounds that we should expect that, overall, physical ecosystem engineering by organisms can have no effect, or positive or negative effects; and that any effects will vary from small to large (Jones et al. 1994, 1997a). Such considerations indicate that it might be unwise to conflate process and consequence without clear accompanying statements of conditionality.

As ecologists we seek to predict and explain the significant. We doubt anyone could get a paper published on the lack of effects of turbulence due to bird's nests on canopy gas exchange, or the lack of effects of shade cast by mobile animals on plant growth. Scientists know how to avoid the trivial, so we are not concerned that the literature will be overwhelmed by such papers. We are, however, very much concerned about the opposite tendency, that of merging engineering process and consequence into statements that are solely about the significant without appropriate statements of conditionality.

We note an unhealthy tendency in the literature for such unspecified conjunction, and we think this a dangerous deviation from the meaning and intent of the concept that seriously weakens its value. Thus we are not at all enamored of statements that can be construed as saying the equivalent of the following: All engineers have large effects; or engineers ought to be restricted to those that have large effects; or keystone species and engineers are the same; or engineers have mostly positive or facilitative effects. Based on the original papers that discussed these issues

(Jones et al. 1994, 1997a), other papers pointing out the same problem (Boogert et al. 2006, Gutiérrez and Jones 2006, Wilby 2002, Wright and Jones 2006), and the preceding considerations, we think such statements are scientifically indefensible on both empirical (e.g., Wright and Jones 2004) and theoretical grounds unless they are accompanied by clear statements of conditionality. Such unconditional statements are epistemologically equivalent to saying that predation always has large effects on prey density; or we will only call it a predator if it has a large effect; or that a predator invariably negatively affects prey density—statements we know not to be universally true (e.g., Adams et al. 1998, Strauss 1991, Wooton 1994).

Physical ecosystem engineering is a process that may have significant consequence given certain conditionalities outlined in preceding text. We are as concerned as anyone with being able to predict which species will be important engineers and what and how big their effects will be; it is *the* central theoretical challenge to which the concept can contribute. We already know that organismal activities that change structure vary, that structures vary, that baseline abiotic environments vary, that resulting abiotic change varies, and that species vary in their sensitivity to abiotic factors. We do not think this challenge can be met by unspecified conflation that thereby eliminates the very sources of variation in cause and effect. Ecological outcomes are often context dependent. Little is to be gained by ignoring this in our quest for general understanding.

1.5 ● ON USAGE

That a concept exists and is used by some should not obligate others to use it, nor should the fact that it is unnecessary in some situations preclude consideration of its utility elsewhere. Nor should we, as authors, attempt to proscribe usage; this is anathema to creativity and assumes omniscience we lack. Instead, we will illustrate some situations when explicit consideration of physical ecosystem engineering may not be needed even though it may be influential, briefly point out what the concept has been used for, and make a few suggestions for general topics where it might be particularly useful.

Many ecological questions about abiotic environmental effects can be answered by taking the abiotic as a given or treating it as stochastic variation. We do not need to consider the engineering if the abiotic is measured as an independent variable, and we make no inference about causation. If the abiotic is not measured, any assumptions about and conclusions based on independence in abiotic state or dynamics, or treatment as stochastic abiotic variation, are violated if it is engineered. This is because the spatial and temporal dynamics of the abiotic environment will, in some way, reflect the factors influencing the engineer and its engineering activities. If the engineering can legitimately be treated as an externality (i.e., no engineer feedback), the abiotic still can be taken as a given, even though it is "made" by the engineer, again provided it is measured and provided no assumptions are made that its dynamics are independent of biota. If the engineer is not an externality to the system, then whether or not the engineering has to be explicitly considered will be a function of the degree to which engineering feedbacks to the engineer and structural legacies alter dynamics. For example, if the abiotic is always changed the same way and to the same degree over the same space and time scales as the presence of the engineer, then the engineering could be collapsed into presence–density of the engineer.

Parsimony suggests that other extant models or concepts may serve as well or better than engineering in some circumstances, even when engineering is responsible for observed effects. For example, plant shade is, in part (see earlier text), an engineering process controlled by canopy architecture, leaf area index and photon absorption, and reflection properties of leaves; however, simple light competition models often suffice (e.g., Canham et al. 2006). Such models are not appropriate for understanding habitat creation for understory plants, since this is not competition; either nonmechanistic facilitation models or engineering models could be used. If we are interested in how variation in light quantity and quality within a forest creates habitat diversity for understory species, we may need to measure some of the preceding physical engineering variables across species. But perhaps we might also collapse this into light quality neighborhoods associated with certain tree species, taking the underlying engineering processes as given.

One might imagine that consideration of engineering would be *de rigueur* in studies on the population dynamics of obvious, significant ecosystem engineers. However, we may not have to explicitly expose the engineering under all circumstances. To date, modeling and theoretical studies indicate that explicit consideration is required under five basic circumstances: When engineering feedbacks affect density-dependent regulation (Gurney and Lawton 1996, Wright et al. 2004; also see Chapter 3, Wilson); when structural legacies created by engineers introduce lagged environmental decay (Gurney and Lawton 1996, Wright et al. 2004, Hastings et al. 2007); when mobile engineers exhibit differential preference for various engineered environmental states (Wright et al. 2004); when engineering is optional and dependent on environment state (Wright et al. 2004); and when the engineering has spatial dimensions that do not simply relate to the presence of the engineer (e.g., extensive influence, Hastings et al. 2007).

So, in general, if we seek causal explanation of abiotic change, including its dynamics, we may often, but not invariably, invoke physical ecosystem engineering, but this does not mean that all the underlying details always require exposure. Clearly, understanding when explicit consideration is *de rigueur* would be of considerable value, and modeling can do much to help answer this question. Perhaps the easiest answer to the usage question is just to point out where the concept seems to have been useful over the last 12 years. Table 1.1 illustrates some of the diversity of ecological questions that have substantively made use of the concept in population, community, and ecosystem ecology, and in conservation, restoration, and management.

We end this section with some eclectic suggestions of general topic areas where we think consideration of the ecosystem engineering dimensions may be particularly worthwhile: abiotic heterogeneity, its consequences and context dependency; explanation of indirect, legacy, keystone, foundation, and facilitative species effects; assessing relative contributions of species to multiple processes; understanding species effects at various levels of organization, especially comparative studies; habitat creation, maintenance, and destruction by species; understanding human environmental impacts; and using species to achieve conservation, restoration, and environmental management goals.

1.6 ● ON BREADTH AND UTILITY

We have periodically heard comments that the ecosystem engineering concept is too broad to be useful. Certainly the concept is broad, but we do not understand this reasoning. Many ecological concepts are at least as broad in scope and are very useful (e.g., the ecosystem, predation, competition [as a process], nutrient cycling, energy flow, dispersal). Some concepts are broad and still under debate as to their utility (e.g., keystone species, intermediate disturbance, ecological thresholds, functional groups). Some broad concepts have been abandoned as not being particularly useful (e.g., Clemensian superorganism, balance of nature, phytosocial sintaxa). Breadth is determined by the variety of phenomena encompassed by the central idea. Conceptual value is judged by the degree to which it affords better scientific understanding, given sufficient time for a community of investigators to further develop and assess it. We leave it to the community to judge whether the concept has been useful and still can be useful based on the literature and our preceding discussion.

If ecosystem engineering encompassed only beaver or only gophers, it would be so narrow that it would be just a species description and neither interesting nor useful. If, based on its definition, the concept attempted to encompass all types of abiotic change by all organisms, then it would be incorrect, an impediment, and too broad. The conceptual domain is, however, very specific. It refers only to organismally caused, structurally mediated abiotic change and its biotic effects. The breadth arises from the fact that many organisms do this to some degree. While we can recognize subclasses within (e.g., autogenic, allogenic), we cannot arbitrarily include some organisms that fit the definition, while excluding others that also fit the definition. This is another reason why we consider that defining an ecosystem engineer as such only when it has a large effect is a fundamental deviation from the purpose of the concept. Such a deviation would force us into confronting the same insoluble problem facing the keystone species concept: how to universally define species importance in a context-dependent world with variable outcomes.

1.7 ‹ ON THE UNDERLYING PERSPECTIVE

The ecosystem engineering concept has certainly led to a wider appreciation of the ubiquity of organismally caused, structurally mediated abiotic change and its effects on organisms, populations, communities, ecosystems, and landscapes. We think it helps provide a broader view of nature, one extending beyond the dominant trophic perspective. Nevertheless, it is also a perspective. It is just a way of looking at certain things organisms do that affect the way they interact with the abiotic environment and hence each other.

It is a mechanistic rather than a phenomenological view. The ecosystem engineering process and organismal abiotic sensitivity both must be considered to predict outcomes. To some who consider outcomes the Holy Grail, in ecology—we agree that predicting outcomes is a Grail— such a mechanistic, context-dependent perspective may seem insufficiently phenomenological. On the other hand, as pointed out by Wright and Jones (2006), many process-based concepts have ultimately turned out to be more useful than outcome-based ones, perhaps reflecting their greater suitability for addressing context dependency.

To others, the abstraction of organismal features relevant to engineering may seem like reductionism or atomization. Yet the focus on relevant organismal features has been of great value in other areas of ecology (e.g., predation, direct resource competition, vectoring). It does not preclude recognition of multiple roles of species, nor their integrated total effect. So akin to these other areas, identifying organismal attributes relevant to engineering can contribute to our understanding of context-dependent species effects, while facilitating cross-species and cross-system comparisons (for excellent examples, see Crooks and Khim 1999, Wilby et al. 2001).

1.8 ● A CONCLUDING REMARK ON CONCEPT AND THEORY

As pointed out in this chapter's introduction, a concept is not a theory, but it is a foundation upon which theory is built. This foundation must be solid. We hope that our discussion helps provide some solidification with a concomitant reduction in uncertainty, misconstrual, and misunderstanding. We do think the concept can be built into more fully developed theory. Indeed, we see clear signs that this is happening. Many of the examples of use of the concept (Table 1.1) involve general hypotheses, frameworks, methodologies, models, and applications that all contribute to theory development. There is, however, much to be done before we would call physical ecosystem engineering a developed theory; not least, demonstrating that the concept can help predict which species will have what magnitude of engineering effects, on which abiotic variables, with what biotic consequence, in which types of abiotic environments.

ACKNOWLEDGMENTS

We thank Bob Holt for pointing out abiotic information use by organisms and that engineering also can control signals made by them; Brian Silliman for urging us to address the question "Are all organisms ecosystem engineers?"; colleagues in the NCEAS working group, "Habitat modification in conservation problems: Modeling invasive ecosystem engineers" (Jeb Byers, Jeff Crooks, Kim Cuddington, Alan Hastings, John Lambrinos, Theresa Talley, and Will Wilson) for valuable discussion; and countless authors, colleagues, and students for their writings and comments on the concept over the last 12 years. We thank the Andrew W. Mellon Foundation and the Institute of Ecosystem Studies for financial support. CGJ thanks the state and region of the Île de France for a Blaise Pascal International Research Chair via the Fondation de École Normale Supérieure. This chapter resulted from a working group at the National Center for Ecological Analysis and Synthesis, a center funded by NSF (Grant No. DEB-94–21535), the University of California at Santa Barbara, and the State of California. This is a contribution to the program of the Institute of Ecosystem Studies.

REFERENCES

Adams, P.A., Holt, R.D., and Roth, J.D. (1998). Apparent competition or apparent mutualism? Shared predation when populations cycle. *Ecology* 79:201–212.

Aller, R.C. (1988). Benthic fauna and biogeochemical processes in marine sediments: The role of burrow structures. In *Nitrogen Cycling in Coastal Marine Environments*, T.H. Blackburn and J. Sorenson, Eds. New York: John Wiley and Sons, pp. 301–338.

Badano, E.I., and Cavieres, L.A. (2006a). Ecosystem engineering across ecosystems: Do engineer species sharing common features have generalized or idiosyncratic effects on species diversity? *Journal of Biogeography* 33:304–313.

Badano, E.I., and Cavieres, L.A. (2006b). Impacts of ecosystem engineers on community attributes: Effects of cushion plants at different elevations of the Chilean Andes. *Diversity and Distributions* 12:388–396.

Badano E.I., Jones, C.G., Cavieres, L.A., and Wright, J.P. (2006). Assessing impacts of ecosystem engineers on community organization: A general approach illustrated by effects of a high-Andean cushion plant. *Oikos* 115:369–385.

Bangert, R.K., and Slobodchikoff, C.N. (2006). Conservation of prairie dog ecosystem engineering may support arthropod beta and gamma diversity. *Journal of Arid Environments* 67:100–115.

Battin, T.J., Kaplan, L.A., Newbold, J.D., and Hansen, C. (2003). Contributions of microbial biofilms to ecosystem processes in stream mesocosms. *Nature* 426:439–442.

Bertness, M.D. (1985). Fiddler crab regulation of *Spartina alterniflora* production on a New England salt marsh. *Ecology* 66:1042–1055.

Boogert, N.J., Paterson, D.M., and Laland, K.N. (2006). The implications of niche construction and ecosystem engineering for conservation biology. *BioScience* 56:570–578.

Byers, J.E., Cuddington, K., Jones, C.G., Talley, T.S., Hastings, A., Lambrinos, J.G., Crooks, J.A., and Wilson, W.G. (2006). Using ecosystem engineers to restore ecological systems. *Trends in Ecology and Evolution* 21:493–500.

Canham, C.D., Papaik, M.J., Uriarte, M., McWilliams, W.H., Jenkins, J.C., and Twery, M.J. (2006). Neighborhood analyses of canopy tree competition along environmental gradients in New England forests. *Ecological Applications* 16:540–554.

Caraco N., Cole, J., Findlay, S., and Wigand, C. (2006). Vascular plants as engineers of oxygen in aquatic systems. *BioScience* 56:219–225.

Coleman, F.C., and Williams, S.L. (2002). Overexploiting marine ecosystem engineers: Potential consequences for biodiversity. *Trends in Ecology and Evolution* 17:40–44.

Crain, C.M., and Bertness, M.D. (2005). Community impacts of a tussock sedge: Is ecosystem engineering important in benign habitats? *Ecology* 86:2695–2704.

Crooks, J.A. (2002). Characterizing ecosystem-level consequences of biological invasions: The role of ecosystem engineers. *Oikos* 97:153–166.

Crooks, J.A., and Khim, H.S. (1999). Architectural vs. biological effects of a habitat-altering, exotic mussel, *Musculista senhousia. Journal of Experimental Marine Biology and Ecology* 240:53–75.

Cuddington, K., and Hastings, A. (2004). Invasive engineers. *Ecological Modelling* 178:335–347.

Daleo, P., Escapa, M., Alberti, J., and Iribarne, O.O. (2006). Negative effects of an autogenic ecosystem engineer: Interactions between coralline turf and an ephemeral green alga. *Marine Ecology-Progress Series* 315:67–73.

Dangerfield, J.M., McCarthy, T.S., and Ellery, W.N. (1998). The mound-building termite *Macrotermes michaelseni* as an ecosystem engineer. *Journal of Tropical Ecology* 14:507–520.

Darwin, C. (1890). *The Formation of Vegetable Mould, Through the Action of Worms, with Observations on Their Habits.* New York: D. Appleton and Co.

del-Val, E., Armesto, J.J., Barbosa, O., Christie, D.A., Gutiérrez, A.G., Jones, C.G., Marquet P.A., and Weathers, K.C. (2006). Rain forest islands in the Chilean semiarid region: Fog-dependency, ecosystem persistence and tree regeneration. *Ecosystems* 9:598–608.

Doane, C.D., and McManus, M.L. (Eds). (1981). *The Gypsy Moth: Research Towards Integrated Pest Management.* Washington, D.C.: U.S. Department of Agriculture.

Eldridge, J.D., Zaady, E., and Shachak, M. (2002). Microphytic crusts, shrub patches and water harvesting in the Negev Desert: The Shikim system. *Landscape Ecology* 17:587–597.

Escapa, M., Iribarne, O.O., and Navarro, D. (2004). Effects of the intertidal burrowing crab *Chasmagnathus granulatus* on infaunal zonation patterns, tidal behavior, and risk of mortality. *Estuaries* 27:120–131.

Evans, S.M. (1971). Behavior in polychaetes. *Quarterly Review of Biology* 46:379–405.

Facelli, J.M., and Pickett, S.T.A. (1991). Plant litter: Its dynamics and effects on plant community structure. *Botanical Review* 57:1–32.

Fish, D. (1983). Phytotelmata: Flora and fauna. In *Phytotelmata: Terrestrial Plants as Hosts for Aquatic Communities,* J.H. Frank and L.P. Lounibos, Eds. Medford, NJ: Plexus, pp. 1–27.

Flecker, A.S. (1996). Ecosystem engineering by a dominant detritivore in a diverse tropical stream. *Ecology* 77:1845–1854.

Flecker, A.S., and Taylor, B.W. (2004). Tropical fishes as biological bulldozers: Density effects on resource heterogeneity and species diversity. *Ecology* 85:2267–2278.

Goubet, P., Thebaud, G., and Pettel, G. (2006). Ecological constraints on *Sphagnum* bog development: A conceptual model for conservation. *Revue d'Ecologie-La Terre et la Vie* 61:101–116.

Gurney, W.S.C., and Lawton, J.H. (1996). The population dynamics of ecosystem engineers. *Oikos* 76:273–283.

Gutiérrez, J.L., and Iribarne, O.O. (1999). Role of Holocene beds of the stout razor clam *Tagelus plebeius* in structuring present benthic communities. *Marine Ecology-Progress Series* 185:213–228.

Gutiérrez, J.L., and Iribarne, O.O. (2004). Conditional responses of organisms to habitat structure: An example from intertidal mudflats. *Oecologia* 139:572–582.

Gutiérrez, J.L., and Jones, C.G. (2006). Physical ecosystem engineers as agents of biogeochemical heterogeneity. *BioScience* 56:227–236.

Gutiérrez, J.L., Jones, C.G., Groffman, P.M, Findlay, S.E.G., Iribarne, O.O, Ribeiro, P. D., and Bruschetti, C.M. (2006). The contribution of crab burrow excavation to carbon availability in surficial salt-marsh sediments. *Ecosystems* 9:647–658.

Gutiérrez, J.L., Jones, C.G., Strayer, D.L., and Iribarne, O.O. (2003). Mollusks as ecosystem engineers: The role of shell production in aquatic habitats. *Oikos* 101:79–90.

Hastings, A., Byers, J.E., Crooks, J.A., Cuddington, K., Jones, C.G., Lambrinos, J.G., Talley, T.S., and Wilson, W.G. (2007). Ecosystem engineering in space and time. *Ecology Letters* 10:153–164.

Holling, C.S. 1992. Cross-scale morphology, geometry, and dynamics of ecosystems. *Ecological Monographs* 62:447–502.

Jones, C.G., Gutiérrez, J.L., Groffman, P.M., and Shachak, M. (2006). Linking ecosystem engineers to soil processes: A framework using the Jenny State Factor Equation. *European Journal of Soil Biology* 42:S39–S53.

Jones, C.G., Lawton, J.H., and Shachak, M. (1994). Organisms as ecosystem engineers. *Oikos* 69:373–386.

Jones, C.G., Lawton, J.H., and Shachak, M. (1997a). Positive and negative effects of organisms as ecosystem engineers. *Ecology* 78:1946–1957.

Jones, C.G., Lawton, J.H., and Shachak, M. (1997b). Ecosystem engineering by organisms: Why semantics matters. *Trends in Ecology and Evolution* 12:275.

Jouquet, P., Boulain, N., Gignoux, J., and Lepage, M. (2004). Association between subterranean termites and grasses in a West African savanna: Spatial pattern analysis shows a significant role for *Odontotermes* n. *pauperans. Applied Soil Ecology* 27:99–107.

Jouquet, P., Dauber, J., Lagerlof, J., Lavelle, P., and Lepage, M. (2006). Soil invertebrates as ecosystem engineers: Intended and accidental effects on soil and feedback loops. *Applied Soil Ecology* 32:153–164.

Kathiresan, K., and Rajendran, N. (2005). Coastal mangrove forests mitigated tsunami. *Estuarine Coastal and Shelf Science* 65:601–606.

Lavelle, P., Bignell, D., Lepage, M., Wolters, V., Roger, P., Ineson, P., Heal, O.W., and Dhillion, S. (1997). Soil function in a changing world: The role of invertebrate ecosystem engineers. *European Journal of Soil Biology* 33:159–193.

Likens, G.E. (1992). *The Ecosystem Approach: Its Use and Abuse.* Oldendorf/Luhe, Germany: Ecology Institute.

Lill, J.T., and Marquis, R.J. (2003). Ecosystem engineering by caterpillars increases insect herbivore diversity on white oak. *Ecology* 84:682–690.

Moore, J.W. (2006). Animal ecosystem engineers in streams. *BioScience* 56:237–246.

Naiman, R.J., Johnston, C.A., and Kelley, J.C. (1988). Alteration of North American streams by beaver. *BioScience* 38:753–762.

Parras, A., and Casadío, S. (2006). The oyster *Crassostrea? hatcheri* (Ortmann, 1897), a physical ecosystem engineer from the upper Oligocene–lower Miocene of Patagonia, southern Argentina. *Palaios* 21:168–186.

Pickett, S.T.A., and Cadenasso, M.L. (2002). The ecosystem as a multidimensional concept: Meaning, model, and metaphor. *Ecosystems* 5:1–10.

Pickett, S.T.A., Cadenasso, M.L., and Jones, C.G. (2000). Generation of heterogeneity by organisms: Creation, maintenance, and transformation. In *The Ecological Consequences of Environmental Heterogeneity: The 40th Symposium of the British Ecological Society*, M.J. Hutchings, E.A. John, and A.J.A. Stewart, Eds. Oxford: Blackwell Science, pp. 33–52.

Pickett, S.T.A., Kolasa, J., and Jones, C.G. (1994). *Ecological Understanding: The Nature of Theory and the Theory of Nature.* San Diego: Academic Press.

Pintor, L.M., and Soluk, D.A. (2006). Evaluating the non-consumptive, positive effects of a predator in the persistence of an endangered species. *Biological Conservation* 130:584–591.

Power, M.E. (1997a). Estimating impacts of a dominant detritivore in a neotropical stream. *Trends in Ecology and Evolution* 12:47–49.

Power, M.E. (1997b). Ecosystem engineering by organisms: Why semantics matters— Reply. *Trends in Ecology and Evolution* 12:275–276.

Reichman, O.J., and Seabloom, E.W. (2002a). The role of pocket gophers as subterranean ecosystem engineers. *Trends in Ecology and Evolution* 17:44–49.

Reichmann, O.J., and Seabloom, E.W. (2002b). Ecosystem engineering: A trivialized concept? Response. *Trends in Ecology and Evolution* 17:308.

Ruesink, J.L., Lenihan, H.S., Trimble, A.C., Heiman, K.W., Micheli, F., Byers, J.E., and Kay, M.C. (2005). Introduction of non-native oysters: Ecosystem effects and restoration implications. *Annual Review of Ecology Evolution and Systematics* 36:643–689.

Shachak, M., Brand, S., and Gutterman, Y. (1991). Patch dynamics along a resource gradient: Porcupine disturbance and vegetation pattern in a desert. *Oecologia* 88:141–147.

Strauss, S.Y. (1991). Indirect effects in community ecology: Their definition, study and importance. *Trends in Ecology and Evolution* 6:206–210.

Tansley, A.G. (1935). The use and abuse of vegetational concepts and terms. *Ecology* 16, 284–307.

Thomas, F., Renaud, F., de Meeus, T., and Poulin, R. (1998). Manipulation of host behaviour by parasites: Ecosystem engineering in the intertidal zone? *Proceedings of the Royal Society of London Series B-Biological Sciences* 265:1091–1096.

Townsend, D.W., Keller, M.D., Sieracki, M.E., and Ackelson, S.G. (1992). Spring phytoplankton blooms in the absence of vertical water column stratification. *Nature* 360:59–62.

Usio, N., and Townsend, C.R. (2002). Functional significance of crayfish in stream food webs: Roles of omnivory, substrate heterogeneity and sex. *Oikos* 98:512–522.

Wilby, A. (2002). Ecosystem engineering: A trivialized concept? *Trends in Ecology and Evolution* 17:307.

Wilby, A., Shachak, M., and Boeken, B. (2001). Integration of ecosystem engineering and trophic effects of herbivores. *Oikos* 92:436–444.

Wooton, J.T. (1994). The nature and consequences of indirect effects in ecological communities. *Annual Review of Ecology and Systematics* 25:443–466.

Wright, J.P., Flecker, A.S., and Jones, C.G. (2003). Local versus landscape controls on plant species richness in beaver meadows. *Ecology* 84:3162–3173.

Wright, J.P., Gurney, W.S.C., and Jones, C.G. (2004). Patch dynamics in a landscape modified by ecosystem engineers. *Oikos* 105:336–348.

Wright, J.P., and Jones, C.G. (2004). Predicting effects of ecosystem engineers on patch-scale species richness from primary productivity. *Ecology* 85:2071–2081.

Wright, J.P., and Jones, C.G. (2006). The concept of organisms as ecosystem engineers ten years on: Progress, limitations, and challenges. *BioScience* 56:203–209.

Wright, J.P., Jones, C.G., Boeken, B., and Shachak, M. (2006). Predictability of ecosystem engineering effects on species richness across environmental variability and spatial scales. *Journal of Ecology* 94:815–824.

Wright, J.P., Jones, C.G., and Flecker, A.S. (2002). An ecosystem engineer, the beaver, increases species richness at the landscape scale. *Oecologia* 132:96–101.

2

A HISTORICAL PERSPECTIVE ON ECOSYSTEM ENGINEERING

Natalie Buchman, Kim Cuddington, and John Lambrinos

2.1 ⬤ INTRODUCTION

Ecologists have been aware that animals and plants modify the physical environment for at least 150 years, even though the term *ecosystem engineer* was not coined until 1994 (Jones et al. 1994). As we have argued elsewhere (Beisner and Cuddington 2005), awareness of the historical development of any ecological topic can lead to deeper understanding and more rapid scientific progress. In this spirit, we outline some of the major areas of research on ecosystem engineering that have been important historically, focusing on early studies in the late 1800s to the more recent contributions of the early 1990s (Figure 2.1).

Before we begin, it is worth noting that there is some controversy about the appropriate use of the term *ecosystem engineering*. Jones et al. (1994) originally defined an ecosystem engineer as an organism that creates, modifies, or maintains a habitat by altering the availability of resources to other organisms. More recent definitions emphasize the alteration of the physical environment by these species (Jones et al. 1997, Guttiérez and Jones 2006). Some authors have argued that these definitions include all organisms, and claim that the term should be restricted to those species that have large impacts on the environment and

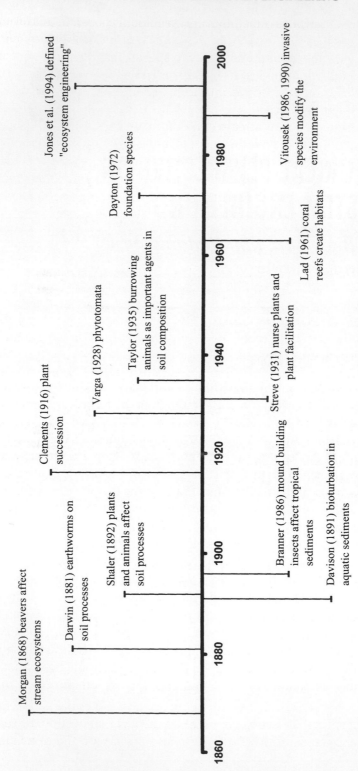

FIGURE 2.1 Timeline of selected important studies related to the ecosystem engineering concept.

associated communities (Reichman and Seabloom 2000a). In this review, we focus on studies of organisms that alter the physical environment without attempting to limit our survey to species that have large effects, however defined. In practice, most published literature will emphasize those species thought to have important effects, and our review will certainly share this bias.

There has been a steady stream of literature on species-specific modification of the physical environment since the late 1800s. For example, Morgan (1868) claimed that beaver impact the hydrology and geomorphology of stream ecosystems. Soon after, Lyell (1873) noted that organisms could locally and superficially alter geomorphology. Darwin described the actions of earthworms on soil and sediment processes (Darwin 1881), while Shaler (1882) reviewed the effects of many other species on soils. The observation that organisms interact with each other indirectly through physical habitat modification has also played an important, although often implicit, role in the development of key ecological concepts such as succession and facilitation (see Cowles 1911, and Bertness and Callaway 1994).

The number of studies that are encompassed in this topic is, however, vast, and our review should be considered selective rather than exhaustive. Following major divisions in the literature, we have organized these studies into four categories: soil processes, plant succession, microclimate modification, and habitat creation. We note that this topic categorization helped early scientists to draw analogies between seemingly disparate organisms. For example, Elton (1927) suggests that land crabs on coral islands played a similar role to that of earthworms in continental regions. This type of analogy seems to us to be a precursor of the more overarching category of ecosystem engineer.

2.2 ⬤ SOIL AND SEDIMENT PROCESSES

The study of soil processes is an area where the impact of ecosystem engineering has been, and continues to be, widely recognized. Lavelle (2002) recently argued ecosystem engineering is more important for the regulation of soil processes than trophic interactions. Historically, many studies in the past 150 years have addressed the large effect of animals and plants on soil and sediment processes at local scales. For example, in 1892, Shaler provided an early overview of the impact of microorganisms, animals, and plants on soil processes. In addition, early ecologists recognized that these soil engineers, by altering the composition and structure of soils, may also affect other organisms that occupy the soils. Moreover, some early workers were also aware of the functional analogies of different organisms in different soil and sediment environments. Most early investigators, however, focused on particular groups of organisms and their effects soil processes. We provide an overview of this area of research using the most commonly studied groups such as earthworms, mound-building insects, burrowing rodents, benthic fauna, and terrestrial plants.

Darwin's (1881) famous book described the enormous effect that earthworms have on physical soil structure through their burrowing, mixing, and casting activities. It has been known for some time that earthworms grind soil particles in their digestive tract and so cause increased aeration of soil (Shaler 1892). In a later work, Hopp (1946) observed that the action of casting aggregates soil particles and allows soil to drain readily, which aids in the prevention of erosion. Since then, many ecologists have found that the addition of earthworms to soil causes increased growth and vitality of vegetation because of improved soil structure and moisture penetration (Hopp and Slater 1948, van Rhee 1965, Stockdill 1966). The mixing and incorporation of organic matter in soils by earthworms also have a long history of study. Early researchers observed earthworms pulling leaves and other organic matter into their burrows, thereby introducing organic matter into lower levels of the soil (Shaler 1892). Later studies found that earthworm casts increase microbial activity (Barley and Jennings 1959, Parle 1963, Jeanson 1960), which causes dead material to be incorporated into the soil surface at a much faster rate (Stockdill 1966). As a result, nitrogen and phosphorus in soils increase with activities of earthworms (Lunt and Jacobson 1944, Barley and Jennings 1959, Aldag and Graff 1975, Sharpley and Syers 1976).

Early ecologists realized that the action of mound-building insects in tropical regions was analogous to the effects of earthworms in temperate climates (Branner 1896). Many species of ants and termites profoundly modify physical soil properties by selecting and redistributing soil particles during the construction of nests, mounds, and foraging galleries. These activities have a direct influence on soil characteristics such as bulk density, turnover rates, profile development, and water infiltration. These physical changes can secondarily influence a number of important soil processes such as redox and nutrient cycling (Lobry de Bruyn and Conacher 1990).

Early studies catalogued some of these effects. Shaler (1892) claimed that in some areas, ants are so numerous that they transfer 1/5 inch of new soil to the surface each year. In early work in the tropics, Branner (1896) observed ant and termite mounds covering large areas and consisting of many tons of soil. He also noted that mound-building ants brought up soil from lower levels to the surface, which causes large soil particles to break up and also promotes the incorporation of organic material. Bell (1883) argued that organic debris such as leaves and dead plant tissue would not be as readily incorporated into soil if it were not for the burrowing action of insects. It was also observed that ants drag their food and leaves of plants into underground channels, which causes increases in organic matter content (Branner 1896). Later studies, in various systems (Baxter and Hole 1967, Wiken et al. 1976, Mandel and Sorenson 1982, Levan and Stone 1983, Carlson and Whitford 1991), found that ant mounds have increased levels of inorganic and organic material, lowered bulk density, and altered soil structure as compared to surrounding nonmound areas. Finally, recent work suggests that the action of mound building, subterranean gallery construction, and redistribution of soil particles by ants also increases the water content, pH (Rogers and Lavigne 1974, Briese 1982, Beattie and Culver 1983), and the rate of colonization by microorganisms because of the increase in decomposable material and access due to the underground tunnels (Czerwinski et al. 1971, Lockaby and Adams 1985).

In addition to these direct effects on soil processes, some researchers concluded that vegetation patterns are influenced by ant (Czerwinski et al. 1971, King 1977, Beattie and Culver 1983) and termite mounds (Glover et al. 1965). Salem and Hole (1968) found that ant activities of depositing subsoil on the surface and excavating chambers caused a reduction in bulk density and an increase in available nutrients to plants. Black seed harvester ants (*Messor andrei*) create large nest mounds of excavated soil. In California grasslands these mounds support plant assemblages that are distinct from nonmound vegetation. Hobbs (1985) hypothesized that selective seed harvesting by *M. andrei* caused the unique mound vegetation. Brown and Human (1997), however, used ant exclusions to demonstrate that granivory had little influence on vegetation dynamics. Instead, physical differences between mound and nonmound patches such as soil temperature seem to be driving vegetation patterns. Studies such as this that explicitly test alternative mechanisms have been rare. Most studies examining the influence of ant and termites on vegetation or ecosystem traits have been correlative. In general, however, research into the mechanisms by which soil invertebrates act as ecosystem engineers is more advanced than any other area (see Jouquet et al. 2006 for a recent review).

Burrowing rodents have figured prominently in the recent literature on ecosystem engineering (e.g., Reichman and Seabloom 2002a, 2002b). Early researchers also observed that burrowing rodents, like soil invertebrates, cause a mixing of the soils and the addition of vegetation and other organic compounds into the burrows (Shaler 1891, Green and

Reynold 1932). Taylor and McGinnie (1928) found that the burrowing action of the kangaroo rats, pocket gophers, ground squirrels, and prairie dogs has an enormous impact on soil properties and vegetation growth. Taylor (1935) observed that the sublayers of soil are brought to the surface by this action, which increases the fertility of soils. Early work also illustrated that digging of burrows increases water infiltration and retention (Green and Reynold 1932, Taylor 1935), which could have a positive effect on plant growth or influence community structure (Reynolds 1958), and also affect soil erosion and stability (Arthur et al. 1929, Meadows 1991). Early and more recent analysis of soils worked over by rodents shows an increase in calcium, magnesium, bicarbonate, nitrate, and phosphorus (Green and Reynold 1932) and increased microbial activity (Meadows 1991).

Analogous studies in aquatic systems also have a long history. Davison (1891) was one of the first to investigate the bioturbation of aquatic sediments in a study of a polychaete feeding on tidal flats. Miller (1961) suggested that 2–3 cm of marsh sediment is reworked due to the deposit feeding action of fiddler crabs. Rhoads and Young (1970) found that deposit feeders affect grain size, increase water content at sediment surface, change sediment stability, and affect species diversity in aquatic sediments. Both early and later studies demonstrated that the production of fecal pellets and burrowing action by various aquatic species have large effects on aquatic sediments (Moore 1931; Brinkman 1932; Rhoads 1963, 1967; McMaster 1967; Rhoads and Young 1970; Brenchley 1981; Posey et al. 1991). With few exceptions (e.g., Aller 1982), the burrowing action of macrofauna increases organic matter, solute transport, oxygen content, sulphate and nitrate reduction, and metabolic activity in aquatic sediments (e.g., Anderson and Kristersen 1991). For example, Bertness (1985, 1991) found that in marsh habitats, the burrowing of fiddler crabs caused an increase in soil drainage, soil oxidation-reduction potential, decomposition of below-ground plant debris, and cord grass production. Reichelt (1991) found that the construction of burrows by meiofauna redistributes sediment, which affects the physical, chemical, and biological properties of the system. Other studies catalogue the effects of various species on sediment composition, such as herbivorous snails (Bertness 1984), atyid shrimp (Brenchley 1981, Pringle et al. 1993), crayfish (Soluk and Craig 1990, Wallace et al. 1981, Pringle and Blake 1994), and fish (Flecker 1996).

Of course, plant species also have dramatic effects on soils and sediments. As early as 1892, Shaler described the profound effect plant roots have on soils. They cause movement of soil, breakup of rocks, addition of organic compounds, and creation of channels from decaying roots. He also noted that the overturning of trees will bring nutrient-rich subsoil to the surface. Other early studies reported that forest soils had increased air, water, and organic matter content compared to bare unforested regions (Ramann 1897, Hoppe 1898, Albert 1912, Engler 1919), and various mechanisms mediated by root growth were invoked to explain such differences. The accumulation of dead leaves as litter also changes the microenvironment of soils by altering surface structure, drainage, and heat and gas exchange (Facelli and Pickett 1991). Various types of vegetation employ similar engineering mechanisms that affect the pattern of soil properties in many different environments (oak tree in heath land: Muller 1887; orange trees in Florida orchards: Jamison 1942; desert shrubs: Fireman and Hayward 1952, Muller and Muller 1956).

This disparate literature describing how organisms affect soil processes at the local scale has historically had only a minor influence on studies examining large-scale geomorphological processes. Lyell's (1873) early observations and Charles Darwin's (1842) theory of coral atoll formation notwithstanding, for most of its early history the discipline of geomorphology focused mainly on understanding how physical processes influence the evolution of landforms (Chorley and Beckinsale 1991). In the last several decades, however, researchers have increasingly recognized the important influence that organisms can have on landform development (Viles 1988, Butler 1988, Stallins 2006). The incorporation of biological feedbacks into physical process models has been especially useful in understanding highly dynamic processes operating over relatively short time scales, such as coastal erosion and desertification (Costanza et al. 1990, von Hardenberg et al. 2001)

2.3 ➡ SUCCESSION

The investigation of successional processes has been tightly related to the studies of organisms' effects on soil processes. Cowles (1911) provided a detailed overview of early succession studies. One of the first observations of this process was made by William King in *Philosophical Transactions* (1685), in which he described bog formation due to the production of peat. Biberg (1749) initiated the idea that moss and lichen establishment on unoccupied rocks causes the production of soil, and subsequently the establishment of vegetation. Early appreciation of engineering mechanisms is also found in early successional studies conducted in various climates and habitats (islands: Reissek 1856; sand dunes: Warming 1891; Rhone Delta: Flechault and Combry 1894; German hearth: Graebner 1895).

Beginning with von Humboldt's (1805) pioneering work on plant geography, the concept of abiotic determinism had a strong influence on the development of the concept of succession. Steenstrup (1842) was one of the first to argue that vegetation changes preserved in the fossil record reflected changes in European climate since the last glacial period. In early succession models climate was the primary mechanism regulating climax community composition, either as integrated units (Clements 1916) or as individually reacting species (Gleason 1939). These early succession models did implicitly assume that species–environment interactions partly drove successional changes towards the eventual climax community. Empirical studies documented the stabilization of sand dunes by pioneer plants allowing the colonization of other vegetation (Cowles 1889, Olson 1958), and the ultimate impact of microclimate modification (see following text) was noted in the action of nurse plants in providing shade for the colonization and growth of vegetation (Streve 1931, Niering et al. 1963).

Early examples of the dynamic interaction between the biotic and abiotic environments greatly influenced the development of the ecosystem concept. In introducing the term, Tansley (1935) argued that plant communities were in dynamic equilibrium with the abiotic environment such that "the biome is determined by climate and soil and in its turn reacts, sometimes and to some extent on climate, always on soil." In contrast to the early conceptual models of succession, later models described successional changes explicitly as the outcome of direct species interactions. These models, however, focused on the ultimate population impacts of the species interactions rather than on the underlying mechanisms (e.g., Horn 1974, Connell and Slatyer 1977). Recently, it has been found that invasive plant species can have a large impact on community change (Vitousek 1986, 1990).

2.4 ⬥ MICROCLIMATE MODIFICATION, FACILITATION, AND INHIBITION

The influence of plants on the local microclimate was first noted by Jozef Paczoski in the mid 1800s (Maycock 1967). The alteration of local conditions can either facilitate or inhibit the growth of new plants. Various studies have described the positive effects that plants can have on the environment and the colonization of other species. In an early study, Streve (1931) found that there was a greater amount of herbaceous plants, perennials, and seedlings under bushes and desert trees compared to the surrounding barren areas because of the increased shading and moisture. Ellison (1949) observed that seedling establishment and survival in a depleted alpine range are higher under plant canopies due to less extreme temperature and increased soil moisture. Chapin et al. (1979) found that *Eriophorum vaginatum* dominate the Alaskan tundra because of its tussock growth form, which increases soil temperature and moisture via insulation. More recently, it has been suggested that nurse plants in desert habitats can reduce surface temperatures due to increased shading and enhance the survival and distribution of seedlings (Steenberg and Lowe 1969, Turner et al. 1969, Franco and Nobel 1989). These microhabitats also affect other species in the community. On the forest floor the microenvironment that is created due to shading has profound effects on the microbial community (Williams and Gray 1974).

Plant litter can also affect the microclimate of an area by changing the physical and chemical environment (Facelli and Pickett 1991). McKinney (1929) found that litter aids in the prevention of soil freezing by providing insulation. Litter also intercepts sunlight, reduces thermal amplitude of soils, and can affect the germination and growth of seeds (Bliss and Smith 1985, Fowler and Knauer 1986, Facelli and Pickett 1991). It also can reduce evaporation (Hollard and Coleman 1987, Facelli and Pickett 1991). Similarly, peat also insulates soils, affecting microclimate and increasing soil respiration (Petrone et al. 2001).

Of course, not all habitat modification is beneficial. The inhibition of new vegetation growth by previously established vegetation has long been an area of research and, under some definitions, can be considered ecosystem engineering. Early studies on bogs reported that plant roots give off excretions causing bog water and soil to become toxic to other plant growth (Livingston et al. 1905, Schreiner and Reed 1907, Dachnowski 1908, Tansley 1949). Salisbury (1922) noticed that in woodlands in England the soil was becoming more acidic due to the change in vegetation. The acidification of soils by plants can have a negative impact on the growth of new vegetation (Grubb et al. 1969, Nihlgard 1972). Muller (1953) reported that the toxins produced by desert shrubs significantly impacted the distribution and abundance of other plant species. Several other earlier investigators found salt accumulation (Litwak 1957, Sharma and Tongway 1973) and a change in pH (Fireman and Hayward 1952) in soils beneath plants.

Sometimes such modification occurs in the context of invasion. Exotic species of plants can alter the microclimate to an extent that causes an unfavorable environment for native species. The salt accumulation in an exotic ice plant causes an increase in soil salinity and reduces soil fertility inhibiting the growth of nontolerant plant species allowing it to dominate (Vivrette and Muller 1977, Kloot 1983). The invasion of *Myrica faya*, an actiorrihozal nitrogen fixer, causes an increase of nitrogen in the area surrounding this plant (D'Antonio and Vitousek 1992). Exotic grasses in semi-arid scrublands have caused the increase in fire because of the increased production of litter (Parsons 1972).

Similarly, a wide variety of other species alter microenvironments through diverse mechanisms. For example, porcupines dig holes that can become filled with water, which then become favorable sites of plant colonization (Yair and Rutin 1981). In aquatic environments, plankton biomass and distribution can affect heat content and thermal structure of lakes due to the light interception and reflection off of these particles (Mazumber et al. 1990).

2.5 ● HABITAT CREATION

Both plants and animals create habitats for themselves and other organisms. This creation of habitat can contribute to species diversity and distribution. Early on, Müller (1879) noted that plant physical structures are habitats for animals and plants, and in 1928, Varga coined the term *phytotomata* to describe the small aquatic habitats created by plants. Möbius (1877) discussed the community of organisms inhabiting oyster beds, "which find everything necessary for their growth and continuance such as suitable soil, sufficient food, the requisite percentage of salt and a temperature favorable to their development." Since these early studies, many species have been identified as habitat creators. Debris dams are created by fallen forest trees, which alter the morphology and stability of streams and so create habitats for various organisms (Heede 1972, Keller and Swanson 1979, Likens and Bilby 1982). Kelp forests (Round 1981) and sea grass prairies (Jones et al. 1994) also support a diverse abundance of plant and animal communities. In more recent literature, it has been noted that leaf shelters serve as homes for other species after they have been abandoned by their arthropod creators (Fukui 2001). Even organisms such as small algae have a large impact on the creation of habitats. Coral reefs are formed dominantly by the action of algae overgrowing and cementing accrual together (Ladd 1961, Womersley and Bailey 1969, Round 1981, Anderson 1992), which provides a habitat for many aquatic organisms. In 1972, Dayton collectively defined these organisms as *foundation species* that build the structure of the environment.

One of the earliest (Morgan 1868) and most intensely studied species that create habitats are beavers (Naiman 1988, Wright et al. 2002). Early and later studies have determined that beaver dams play an important role on stream ecosystem dynamics by changing hydrology (Gard 1961, Smith et al. 1991), nutrient cycling (Francis et al. 1985, Naiman et al. 1991, Yavitt et al. 1992), decomposition dynamics (Hodgkinson 1975, Naiman et al. 1986), nutrient availability (Wilde et al. 1950, Johnston and Naiman 1990), and biogeochemical cycles (Naiman et al. 1994). The activities of beaver affect wildlife (Bradt 1947, Swank 1949, Grasse and Putnam 1950, Rutherford 1955), stream invertebrates (Hanson and Campbell 1963, McDowell and Naiman 1986), fish (Gard 1961, Hansen and Campbell 1963, Snodgrass and Meffe 1998), and vegetation (Johnston and Naiman 1990, Feldman 1995, Barnes and Dibble 1988, Wright et al. 2002). Such consequences are long lasting (Rudeman and Schoonmaker 1938, Ives 1942, Naimen et al. 1994), spatially extensive, and result in legacy effects after the dam has been abandoned (Neff 1957). In a similar fashion, it has long been noted that alligators also play a very important role as habitat creators in wetland ecosystems. Beard (1938) claimed that wallow digging by alligators had a great impact on organisms in wetland ecosystems. Wallows provide refuge for aquatic and terrestrial vertebrates, invertebrates, and microorganisms (Allen and Neil 1952, Loveless 1959, Finlayson and Moser 1991). The creation of these holes allows the survival of many organisms, and thereby increases local species richness and diversity (Kushlan 1974). Wallows also play a role in shaping plant community structure (Craighead 1968, Palmer and Mazzotti 2004). In addition, alligators also create nest mounds that are used by turtles (Dietz and Jackson 1979) and other reptiles (Kushlan and Kushlan 1980). More recent studies have found that the wallows of crocodiles in Australian swamps are analogous in function to Florida wetland alligator wallows (Magnusson and Taylor 1982).

Mollusks are another group whose importance in the creation of habitats was noted early on (e.g. Möbius 1877). The production of mollusk shells in aquatic environments serves many purposes, such as provision of hard substrate, protection from predation and from physical and physiological stress, and modulation of solute and particulate transport (Gutiérrez et al. 2003).

Early studies found that a variety of organisms live in these habitats, including fish (Breder 1942), octopus (Voss 1956), hermit crabs (Reese 1969), and many other organisms. More recent ecologists have found that shell-producing species can have a large impact on aquatic ecosystems because of the abundance (Russell-Hunter 1983), durability

(Kidwell 1985, Powell et al. 1989), and diverse species occupancy (McLean 1983) of the shell structure.

2.6 ☙ CONCLUSION

Far from being newly recognized phenomena, this historical review makes it clear that ecologists have been actively engaged in studying the myriad ways in which species alter their physical environment for the entire history of ecology. Is there any benefit then in grouping these phenomena under a common term of *ecosystem engineering* (Jones et al. 1994)? One of the principal challenges facing the science of ecology is the immensely complex and contingent nature of its units of study (Strong 1980, Simberloff 2004). One important tool for ordering this complexity has been to identify key functional traits that have important influence on community or ecosystem processes. Early on, authors like Shaler (1881) and Elton (1927) perceived that even very different species often can share similar functional roles within ecosystems.

Yet, there have been two notable problems in the implementation of this realization over the preceding years. First, many definitions of function have been phenomenological and nonmechanistic. Second, there has been a near obsession with the contribution of trophic mechanisms to functional roles (e.g., Paine 1969), to the neglect of nontrophic mechanisms. We suggest that the ecosystem engineering concept helps remedy both of these difficulties. As the studies in this review illustrate, engineering mechanisms are ubiquitous and play diverse functional roles across a range of ecosystems. The ecosystem engineering concept helps unify under common mechanistic functions a diverse array of processes that previously had been treated as idiosyncratic species–environment interactions. Moreover, the overarching grouping of *ecosystem engineer* may now move us to draw parallels between species whose effects on the physical environment are quite different, and whose ecosystem functions may also seem quite different (e.g., crabs that affect soil processes vs. plants that form phytotomata). This categorization hopefully will help facilitate the integration of these processes into ecological models that historically have focused exclusively on trophic mechanisms. From a more applied point of view, species that provide important engineering-based functions within ecosystems are being targeted for conservation (Crain and Bertness 2006). We are also beginning to appreciate that invasive species exert many of their most pernicious impacts through ecosystem engineering (Crooks 2002), but also that ecosystem engineers can be important tools for the management and restoration of ecosystems (e.g., Byers et al. 2006).

REFERENCES

Albert, R. (1912). Bodenuntersuchungen im Gebiete der Lüneburger Heide. *Zeitschrift fuer Forst und Jagdwesen.* 44:2–10.

Aldag, R., and Graff, O. (1975). N-Fraktionen in Regenwurmlosung und deren Ursprungsboden. *Pedobiologia* 15:151–153.

Allen, R., and Neil, W.T. (1952). The American alligator. *Florida Wildlife* 6:8–9,44.

Aller, R.C. (1982). Carbonate dissolution in nearshore terrigenous muds: The role of physical and biological reworking. *Journal of Geology* 90:79–95.

Anderson, F.O., and Kristersen, E. (1991). Effects of burrowing macrofauna on organic matter decomposition in coastal marine sediments. *Symposium of the Zoological Society of Land* 63:69–88.

Anderson, R.A. (1992). Diversity of eukaryotic algae. *Biodiversity and Conservation* 1:267–292.

Arthur, H., Hype, M., and Redington, P.G. (1929). *Report of the Chief of the Bureau of Biological Survey.* Washington, D.C.: United States Department of Agriculture Bureau of Biological Survey, pp. 1–54.

Barley, K.P., and Jennings, A.C. (1959). Earthworms and soil fertility III: The influence of earthworms on the availability of nitrogen. *Australian Journal of Agricultural Resources* 10:364–370.

Barnes, W.J., and Dibble, E. (1988). The effects of beaver in riverbank forest succession. *Canadian Journal of Botany* 66:40–44.

Baxter, F.P., and Hole, F.D. (1967). Ant (*Formica cinerea*) pedoturbation in a prairie soil. *Soil Science Society of America Proceedings* 31:425–428.

Beard, D.B. (1938). *Wildlife Reconnaissance.* Washington, D.C.: U.S. Department of the Interior, National Park Service, Everglades National Park Project, p. 106.

Beattie, A.J., and Culver, D.C. (1983). The nest chemistry of two, seed-dispersing ant species. *Oecologia* 56:99–103.

Beisner, B., and Cuddington, K. (2005). Why a history of ecology? An introduction. In *Ecological Paradigms Lost: Routes to Theory Change.* K. Cuddington and B. Beisner, Eds. Burlington, MA: Elsevier Academic Press, pp. 1–6.

Bell, R. (1883). The causes of fertility of the land in the Canadian north-west territories. *Trans. Royal Society of Canada* 8:157–162.

Bertness, M.D. (1984). Habitat and community modification by an introduced herbivorous snail. *Ecology* 65:370–381.

——. (1985). Fibbler crab regulation of *Spartina alterniflora* production on a New England salt marsh. *Ecology* 66:1042–1055.

——. (1991). Zonation of *Spartina* spp. in a New England salt marsh. *Ecology* 72:138–148.

Bertness, M.D., and Callaway, R.M. (1994). Positive interactions in communities. *Trends in Ecology and Evolution* 9:191–193.

Biberg, I.J. (1749). Oeconomia naturae. *Amoenitates Academicae* 2:1–52.

Bliss, D., and Smith, H. (1985). Penetration of light into soil and its role in the control of seed germination. *Plant, Cell and Environment* 8:475–483.

Bradt, G.W. (1947). *Michigan Beaver Management*. Lancing, MI: Michigan Department of Conservation.

Branner, J.C. (1896). Ants as geologic agents in the tropics. *Bulletin of the Geological Society of America* VII:295–300.

Breder, C.M. (1942). On the reproduction of *Gobiosoma robustum*. *Zoologica* 27:61–65.

Brenchley, G.A. (1981). Disturbance and community structure: An experimental study of bioturbation in marine soft-bottom environments. *Journal of Marine Research* 39:767–790.

Briese, D.T. (1982). Partitioning of resources amongst seed-harvesters in an ant community on semi-arid Australia. *Australian Journal of Ecology* 7:299–307.

Brinkman, R. (1932). Uver die Schichtung und ihre Bedingungen. *Fortschritte der Geologia und Palaentolgie* 2:187–219.

Brown, J.F., and Human, K.G. (1997). Effects of harvester ants on plant species distribution and abundance in a serpentine grassland. *Oecologia* 112:237–243.

Butler, D.R. (1995). *Zoogeomorphology: Animals as Geomorphic Agents*. New York: Cambridge University Press.

Byers, J. E., Cuddington, K., Jones, C., Talley, T., Hastings, A., Lambrinos, J., Crooks, J., and Wilson, W. (2006). Using ecosystem engineers to restore ecological systems. *Trends in Ecology & Evolution* 21:493–500.

Carlson, S.R., and Whitford, W.G. (1991). Ant mound influence on vegetation and soils in a semiarid mountain ecosystem. *Am. Midl. Nat.* 126:125–139.

Chapin, F.S. III., Van Cleve, K., and Chapin, M.C. (1979). Soil temperature and nutrient cycling in the tussock growth form of *Eriophorum vaginatum*. *Journal of Ecology* 67:169–189.

Chorley, R.J., and Beckinsale, R.P. (1991). *The History of the Study of Landforms or the Development of Geomorphology, Volume 3: Historical and Regional Geomorphology 1890–1950*. New York: Routledge.

Clements, F.E. (1916). *Plant Succession*. Publication 242. Washington, D.C.: Carnegie Institute of Washington.

Connell, J.H., and Slatyer, R.O. (1977). Mechanisms of succession in natural communities and their role in community stability and organization. *The American Naturalist* 111:1119–1144.

Costanza, R., Sklar, F.R., and White, M.L. (1990). Modeling coastal landscape dynamics *BioScience* 40:91–107.

Cowles, H.C. (1889). The ecological relations of the vegetation on the sand dunes in Lake Michigan. *Botanical Gazette* 27:95–117.

——. (1911). The causes of vegetative cycles. *Botanical Gazette* 51:161–183.

Craighead, F.C., Sr. (1968). The role of the alligator in shaping plant communities and maintaining wildlife in the southern everglades. *The Florida Naturalist* 41:69–74.

Crain, C. M., and Bertness, M. D. (2006). Ecosystem engineering across environmental gradients: Implications for conservation and management. *Bioscience* 56:211–218.

Crooks, J.A. (2002). Characterizing ecosystem-level consequences of biological invasions: The role of ecosystem engineers. *Oikos* 97:153–166.

Czerwinski, Z., Jakubczyk, H., and Petal, J. (1971). Influence of ant hills on meadow soils. *Pedobiologia* 11:277–285.

Dachnowski, A. (1908). The toxic property of bog water and bog soil. *Botanical Gazette* 46:130–143.

D'Antonio, M., and Vitousek, P.M. (1992). Biological invasions by exotic grasses, the grass/fire cycle, and global change. *Annual Review of Ecology and Systematics* 23:63–87.

Darwin, C. (1842). *The Structure and Distribution of Coral Reefs. Being the First Part of the Geology of the Voyage of the 'Beagle.'* London: Smith, Elder & Co.

——. (1881). *The Formation of Vegetable Mould Through the Action of Worms with Observations of Their Habits.* London: Murray.

Davison, C. (1891). On the amount of sand brought up by Lob worms to the surface. *Geology Magazine* (Great Britain) 8:489.

Dayton, P.K. (1972). Toward an understanding of community resilience and the potential effects of enrichments to the benthos at McMurdo Sound, Antartica. In *Proceedings of the Colloquium on Conservation Problems in Antarctica*, B.C. Parker, Ed. Lawrence, KS: Allen Press, pp. 81–96.

Dietz, D.C., and Jackson, D.R. (1979). Use of alligator nests by nesting turtles. *Journal of Herpetology* 13:510–512.

Ellison, L. (1949). Establishment of vegetation on depleted subalpine range as influence by microenvironment. *Ecol. Monograph* 19:95–121.

Elton, C. (1927). *Animal Ecology.* Chicago: University of Chicago Press.

Engler, A. (1919). Untersuchungen über den Einfluss des Waldes auf den Stand der Gewässer. *Mitteilungen Der Schweizerschen Anstalt Fuer Das Forstilche Versuchswesen* 12:1–626.

Facelli, J.M., and Pickett, S.T.A. (1991). Plant litter: Its dynamics and effects on plant community structure. *Botanical Review* 57:1–32.

Feldman, A.L. (1995). The effects of beaver (*Castor canadensis*) impoundment on plant diversity and community composition in the coastal plain of South Carolina. Thesis. University of Georgia, Athens.

Finlayson, M., and Moser, M. (Eds.). (1991). Wetlands. Oxford: International Waterfowl and Wetlands Research Bureau.

Fireman, M., and Hayword, H.E. (1952). Indicator significance of some scrubs in the Escalante Desert, Utah. *Botanical Gazette* 114:143–155.

Flechault, C., and Combry, P. (1894). Sur la flore de la Camargue et des alluvions du Rhone. *Bulletin of the Society of Botany France* 41:37–58.

Flecker, A.S. (1996). Ecosystem engineering by dominant detritivore in a diverse tropical stream. *Ecology* 77:1845–1854.

Fowler, S.W., and Knauer, G.A. (1986). Role of large particles in the transport of elements and organic compounds through the oceanic water column. *Progress in Oceanography* 16:147–194.

Francis, M.M., Naiman, R.J., and Melillo, J.M. (1985). Nitrogen fixation in subarctic streams influenced by beaver (*Castor canadensis*). *Hydrobiologia* 121:193–203.

Franco, A.C., and Nobel, P.S. (1989). Effect of nurse plants on the microhabitat and growth of cacti. *Journal of Ecology* 77:870–886.

Fukui, A. (2001). Indirect interactions mediated by leaf shelters in animal-plant communities. *Population Ecology* 43:31–40.

Gard, R. (1961). Effects of beaver on trout in Sagehen Creek, California. *Journal of Wildlife Management* 25:221–242.

Gleason, H.A. (1917). The structure and development of plant associations. *Bulletin of the Torrey Botanical Club.* 43:463–481.

Glover, P.E., Trump, E.C., and Wateridge, L.E.D. (1965). Termitaria and vegetation on the Loita Plains of Kenya. *Journal of Ecology* 52:367–377.

Graebner, P. (1895). Studien ueber die norddeutsche Heide. *Botanische Jahrbücher für Systematik, Pflanzengeschichte, und Pflanzengeographie* 20:500–654.

Grasse, J.E., and Putnam, E.F. (1950). Beaver management and ecology in Wyoming. *Wyoming Game and Fish Communication Bulletin* No. 6, 52 pp.

Green, R.A., and Reynold, C. (1932). The influence of two burrowing rodents, *Dipodomys spectabilis spectabilis* (kangaroo rat) and *Neotome albigula albigula* (pack rat) on desert soils in Arizona. *Ecology* 12:73–80.

Grubb, P.J., Green, H.E., and Merrifield, R.C.J. (1969). The ecology of chalk heath: Its relevance to the calcicole-calcifuge and soil acidification problems. *Journal of Ecology* 57:175–212.

Guttiérez, J., and Jones, C. (2006). Physical ecosystem engineers as agents of biogeochemical heterogeneity. *Bioscience* 56:227–236.

Gutiérrez, J.L., Jones, C.G., Strayer, D.L., and Iribarne, O.O. (2003). Mollusks as ecosystem engineers: The role of shell production in aquatic habitats. *Oikos* 101:79–90.

Hanson, W. D., and Campbell, R.S. (1963). The effects of pool size and beaver activity on distribution and abundance of warm-water fishes in a north Missouri stream. *American Midland Naturalist* 69:136–149.

Heede, B.H. (1972). Influences of a forest on the hydraulic geometry of two mountain streams. *Water Resource Bulletin* 8:523–530.

Hobbs, R.J. (1985). Harvester ant foraging and plant species distribution in annual grassland. *Oecologia* 67:519–523.

Hodgkinson, I.D. (1975). Energy flow and organic matter decomposition in an abandoned beaver pond ecosystem. *Oecologia* (Berlin) 21:131–139.

Hollard, E.A., and Coleman, D.C. (1987). Litter placement effects on microbial and organic matter dynamics in an agroecosystem. *Ecology* 68:425–433.

Hopp, H. (1946). Soil conservation: Earthworms fight erosion too. *Soil Conservation* 11:252–254.

Hopp, H., and Slater, C.S. (1948). Influence of earthworms on soil productivity. *U.S. Department of Agriculture Bulletin*, pp. 421–428.

Hoppe, E. (1898). Uber Veränderung des Waldbodens durch Abholzung. *Centralblatt Fuer Das Gesamte Forstwesen* 24:51–64.

Horn, H.S. (1974). The ecology of secondary succession. *Annual Review of Ecol. Systems* 5:25–37.

Ives, R.L. (1942). The beaver-meadow complex. *Journal of Geomorphology* 5:191–203.

Jamison, V.C. (1942). The slow reversible drying of sandy soils beneath citrus trees in central Florida. *Proceedings of the Soil Science Society of America* 7:36–41.

Jeanson, C. (1960). Evolution de la matier organique du sol sous l'action de *Lumbricus herculeus*. *Comptes Rendus Academie des Sciences* 250:3041–3043.

Johnston, C.A., and Naiman, R.J. (1990). The use of geographical information systems to analyze long-term landscape alteration by beaver. *Landscape Ecology* 1:41–57.

Jones, C.G., Lawton, J.H., and Shachak, M. (1994). Organisms as ecosystem engineers. *Oikos* 69:373–386.

———. (1997). Positive and negative effects of organisms as ecosystem engineers. *Ecology* 78:1946–1957.

Jouquet, P., Dauber, J., Lagerlöf, J., Lavelle, P., and Lepage, M. (2006). Soil invertebrates as ecosystem engineers: Intended and accidental effects on soil and feedback loops. *Applied Soil Ecology* 32:153–164.

Keller, E.A., and Swanson, F.J. (1979). Effects of large organic material on channel form and fluvial processes. *Earth Surface Processes* 4:361–380.

Kidwell, S.M. (1985). Palaeobiological and sedimentological implications of fossil concentrations. *Nature* 318:457–460.

King, W. (1685). Of the bogs and loughs of Ireland. *Philosophical Transactions of the Royal Society of London* 15:948–960.

King, T.J. (1977). The plant ecology of ant-hills in calcareous grasslands. I. Patterns of species in relation to ant-hills in southern England. *Journal of Ecology* 65:235–256.

Kloot, P.M. (1983). The role common iceplant (*Mesembryanthemum crystallinum*) in the deterioration of medic pastures. *Australia Journal of Ecology* 8:301–306.

Kushlan, J.A. (1974). Observations on the role of the American alligator (*Alligator mississippiensis*) in the southern Florida wetlands. *Copiea* 31:993–996.

Kushlan, J.A., and Kushlan, M.S. (1980). Everglades alligator nests: Nesting sites for marsh reptiles. *Copeia* 1980:930–932.

Ladd, H.S. (1961). Reef building. *Science* 134:703–715.

Lavelle, P. (2002). Functional domains in soils. *Ecological Research* 17:441–450.

Levan, M.A., and Stone, E.L. (1983). Soil modification by colonies of black meadow ants in a New York old field. *Soil Science Society of America Journal* 47:1192–1195.

Likens, G.E., and Bilby, R.E. (1982). Development, maintenance, and role of organic debris dams in New England streams. In *Sediment Budgets and Routing in Forest Drainage Basins*, F.J. Swanson, R.J. Janda, T. Dunne, and D.N. Swanston, Eds., USDA Forest Service General Technical Report PNW141. USDA Forest Service, Pacific Northwest Forest, and Range Experimental Station, pp. 122–128.

Litwak, M. (1957). The influence of *Tamarix aphylla* on soil composition in the northern Negev of Israel. *Bull. Res. Coun. Israel*, Section D, 6:39–45.

Livingston, B.E., Britton, J.C., and Reid, F.R. (1905). Studies on the properties of an unproductive soil. *U.S. Department of Agriculture Bulletin on Bur. Soils*, 28.

Lobry de Bruyn, L.A., and Conacher, A.J. (1990). The role of termites and ants in soil modification: A review. *Australian Journal of Soil Research* 28:55–93.

Lockaby, B.G., and Adams, J.C. (1985). Pedoturbation of a forest by fire ants. *Journal of the Soil Science Society of America* 99:220–223.

Loveless, C.M. (1959). A study of the vegetation in the Florida Everglades. *Ecology* 40:1–9.

Lunt, H.A., and Jacobson, G.M. (1944). The chemical composition of earthworm casts. *Soil Science* 58:367–375.

Lyell, C. (1873). *Principles of Geology*. New York: D. Appleton Company.

Magnusson, W.E., and Taylor, J.A. (1982). Wallows of *Crocodylus porosus* as dry season refuges in swamps. *Copeia* 2:470–480.

Mandel, R.D., and Sorenson, C.J. (1982). The role of the western harvester ant (*Pogonomyrmex occidentalis*) in soil formation. *Soil Science Society of America Journal* 46:785–788.

Maycock, P.F. (1967). Jozef Paczoski: Founder of the science of phytosociology. *Ecology* 48:1031–1034.

Mazumber, A., Taylor, W.D., McQueen, D.J., and Lean, D.R.S. (1990). Effects of fish and plankton on lake temperature and mixing depth. *Science* 247:312–315.

McDowell, D., and Naiman, R.J. (1986). Structure and function of a benthic invertebrate stream community as influenced by beaver (*Castor canadensis*). *Oecologia* (Berlin) 68:481–489.

McKinney, A.L. (1929). Effect of forest litter on soil temperature and soil freezing in autumn and winter. *Ecology* 10:312–321.

McLean, R. (1983). Gastropod shells: A dynamic resource that helps shape benthic community structure. *Journal of Experimental Marine Biology and Ecology* 69:151–174.

McMaster, R.L. (1967). Compactness variability of estuarine sediments: An in situ study. In *Estuaries*, G.H. Lauff, Ed. Washington, D.C.: American Association for the Advancement of Science, 83:261–267.

Meadows, A. (1991). Burrows and burrowing animals: An overview. *Symposium of the Zoological Society of London* 63:1–13.

Meadows, P.S., and Meadows, A. (Eds.). (1991). *The environmental impact of burrowing animals and animal burrows*. Oxford: Clarendon Press.

Miller, D.C. (1961). The feeding mechanism of fiddler crabs, with ecological considerations of feeding adaptations. *Zoologica* 46:89–101.

Möbius, K. (1877). Reprinted from *Die Auster und die Autserwirthschaft*. Berlin, Wiegnundt, Hempel and Parey. In *Report of the U.S. Commission of Fisheries*, translated by H.J. Rice, 1880, pp. 683–751.

Moore, H.B. (1931). The muds of the Clyde Sea Area. III. Chemical and physical conditions, rate of sedimentation and fauna. *Journal of the Marine Biology Association U.K.* 17:325–358.

Morgan, L.H. (1868). The *American Beaver and His Works*. Philadelphia: J.B. Lippincott.

Muller, C.H. (1953). The association of desert annuals with shrubs. *American Journal of Bot.* 40:53–60.

Müller, F. (1879). Notes on the cases of some south Brazilian Trichoptera. *Transactions of the Entomological Society of London* 1879:131–144.

Muller, P.E. (1887). *Studien über die natürlichen Humusformen and deren Einwirkungen auf Vegetation und Boden.* Berlin: Julius Springer, p. 324.

Muller, W.H., and Muller, C.H. (1956). Association patterns involving desert plants that contain toxic products. *American Journal of Botany* 43:354–355.

Naiman, R.J. (1988). Animal influences on ecosystem dynamics. *BioScience* 38:750–752.

Naiman, R.J., Manning, T., and Johnston, C.A. (1991). Beaver population fluctuations and tropospheric methane emission in boreal wetlands. *Biogeochemistry* 12:1–15.

Naiman, R.J., Melillo, J.M., and Hobbie, J.E. (1986). Ecosystem alteration of boreal forest streams by beaver (*Castor canadensis*). *Ecology* 67:1254–1269.

Naiman, R.J., Pinay, G., Johnston, C.A., and Pastor, J. (1994). Beaver influences on the long-term biogeochemical characteristics of boreal forest drainage networks. *Ecology* 75:905–921.

Neff, D.J. (1957). Ecological effects of beaver habitat abandonment in the Colorado Rockies. *Journal of Wildlife Management* 21:80–84.

Niering, W.A., Whittaker, R.H., and Lowe, C.H. (1963). The saguaro: A population in relation to environment. *Science* 142:15–23.

Nihlgard, B. (1972). Plant biomass, primary production and distribution of chemical elements in a beech and planted spruce forest in south Sweden. *Oikos* 23:69–81.

Olson, J.S. (1958). Rates of succession and soil changes on southern Lake Michigan sand dunes. *Botanical Gazette* 119:125–170.

Paine, R.T. (1969). A note on trophic complexity and community stability. *American Naturalist* 103:91–93.

Palmer, M.L., and Mazzotti, F.J. (2004). Structure of Everglades alligator holes. *Wetlands* 24:115–122.

Parle, J.N. (1963). A microbial study of earthworm casts. *Journal of General Microbiology* 31:13–22.

Parsons, J.J. (1972). Spread of African pasture grasses to the American tropics. *Journal of Range Management* 25:12–17.

Petrone, R.M., Waddington, J.M., and Price, J.S. (2001). Ecosystem scale evapotranspiration and CO_2 exchange from a restored peatland. *Hydrological Processes* 15:283–284.

Pollock, M.M., Naiman, R.J., and Hanley, T.A. (1998). Plant species richness in riparian wetlands—A test of biodiversity theory. *Ecology* 79:94–105.

Posey, M.H., Bumbauld, B.R., and Armstrong, D.A. (1991). Effects of a burrowing mud shrimp, *Upogebia pugettensis* (Dana), on the abundances of macro-infauna. *Journal of Experimental Marine Biology and Ecology* 148:283–294.

Powell, E.N., Staff, G.M., Davies, D.J., and Callender, W.R. (1989). Macrobenthic death assemblages in modern marine environments: Formation, interpretation, and application. *Critical Review of Aquatic Science* 1:555–589.

Pringle, C.M., and Blake, G.A. (1994). Quantitative effects of atyid shrimp (Decapoda: Atyidae) on the depositional environment in a tropical stream: Use of electricity for experimental exclusion. *Canadian Journal of Fishing and Aquatic Science* 51:1443–1450.

Pringle, C.M., Blake, G.A., Covich, A.P., Buzby, K.M., and Finley, A. (1993). Effects of omnivorous shrimp in a montane tropical stream: Sediment removal disturbance of sessile invertebrates and enhancement of understory algal biomass. *Oecologia* 93:1–11.

Ramann, E. (1897). Uver Lochkahlschlä. *Zeitschrift fuer Forst und Jagdwesen* 29:697.

Reese, E. (1969). Behavioral adaptations of intertidal hermit crabs. *American Zool.* 9:343–355.

Reichelt, A.C. (1991). Environmental effects of meiofaunal burrowing. *Symposium of the Zoological Society of London* 63:33–52.

Reichman, O., and Seabloom, E.W. (2002a). Ecosystem engineering: A trivialized concept? *Trends in Ecology and Evolution* 17:308.

——. (2002b). The role of pocket gophers as subterranean ecosystem engineers. *Trends in Ecology and Evolution* 17:44–49.

Reissek, S. (1856). Ueber die Bildungsgeschichte der Donauinseln im mittleren Laufe dieses Stromes. *Flora* 39:622–624.

Reynolds, H.G. (1958). The ecology of the Merriam kangaroo rat (*Dipodomys merriami Mearns*) on grazing land of southern Arizona. *Ecology Monograph* 28:111–127.

Rhoads, D.C. (1963). Rates of sediment reworking by *Yoldia limatula* in Buzzards Bay, Massachusetts, and Long Island Sound. *Journal of Sedimentary Petrology* 33:723–727.

——. (1967). Biogenic reworking of inter and subtidal sediments in Barnstable Harbor and Buzzards Bay, Massachusetts. *Journal of Geology* 75:61–76.

Rhoads, D.C., and Young, D. (1970). The influence of deposit feeding organisms on sediment stability and community trophic structure. *Journal of Marine Research* 28:150–178.

Rogers, L.E., and Lavigne, R.J. (1974). Environmental effects of Western harvester ants on the shortgrass plains ecosystem. *Environ. Ent.* 3:994–997.

Round, F.E. (1981). *The Ecology of Algae.* New York: Cambridge University Press.

Rudeman, R., and Schoonmaker, W.J. (1938). Beaver dams as geologic agents. *Science* 8:523–525.

Russell-Hunter, W.D. (1983). Overview: Planetary distribution and ecological constraints upon the mollusca. In *The Mollusca*, W.D. Russell-Hunter, Ed., Vol. 6 ecology. Orlando: Academic Press, pp. 1–27.

Rutherford, W.H. (1955). Wildlife and environmental relationships of beavers in Colorado forests. *Journal of Forestry* 53:803–806.

Salem, M.Z., and Hole, F.D. (1968). Ant pedoturbation in a forest soil. *Proceedings of Soil Science Society of America* 32:563–567.

Salisbury, E.J. (1922). Stratification and hydrogen-ion concentration of soil in relation to leaching and plant succession with special reference to woodlands. *Journal of Ecology* 4:220–240.

Schreiner, O., and Reed, H.S. (1907). Some factors influencing soil fertility. *U.S. Department of Agriculture Bulletin on Bureau of Soils* 28.

Shaler, N. (1892). Effect of Animals and Plants on Soils. In *Origin and Nature of Soils*. 12th Annual Report, Director U.S. Geol. Survey, Part J. Geology Annual Report, Sector of the Interior. Washington, D.C.: Government Printing Office.

Sharma, M.L., and Tongway, D.J. (1973). Plant induced soil salinity in two brush (*Atriplex* spp.) communities. *Journal of Range Management* 26:121–124.

Sharpley, A.N., and Syers, J.K. (1976). Potential role of earthworm casts for the phosphorus enrichment of run-off waters. *Soil Biology Biochemistry*, 8:341–346.

Simberloff, D. (2004). Community ecology: Is it time to move on? The *American Naturalist* 163:787–799.

Smith, J.N., and Schafer, C.T. (1984). Bioturbation processes in continental slope and rise sediments delineated by Pb-210 microfossil and textural indicators. *Journal of Marine Research* 42:1117–1145.

Smith, M.E., Driscoll, C.T., Wyskowski, B.J., Brooks, C.M., and Cosentini, C.C. (1991). Modification of stream ecosystem structure and function by beaver in the Adirondak Mountains, New York. *Canadian Journal of Zoology* 69:55–61.

Snodgrass, J.W., and Meffe, G.K. (1998). Influence of beavers on stream fish assemblages: Effects of pond age and watershed position. *Ecology* 79:928–942.

Soluk, D.A., and Craig, D.A. (1990). Digging with a vortex: Flow manipulation facilitates prey capture by a predatory stream mayfly. *Limnology and Oceanography* 35:1201–1206.

Stallins, J.A. (2006). Geomorphology and ecology: Unifying themes for complex systems in biogeomorphology. *Geomorphology* 77:207–216.

Steenberg, W.F., and Lowe, C.H. (1969). Critical factors during the first year of life of saguaro (*Cereus giganteus*) at Saguaro National Monument. *Ecology* 50:825–834.

Steenstrup, J.J.S. (1842). Geognostik-geolgisk Undersögelse of Skovmoserne Vidnesdam og Liffemose I det nordlige Sjaelland, ledsaget af sammenlignednde Bemaerkninger, hentede fra Danmarks Skov-, Kjaer- og Lyngmoser I Almindelighed. *Det Kongelige Danske Videnskabernes Selskabs Naturvidenskabe lige og Mathematishev Afhandlingher* 9:17–120.

Stockdill, S.M.J. (1966). The effects of earthworms of pastures. *Proceedings of the New Zealand Ecol. Society* 13:68–75.

Streve, F. (1931). Physical conditions in sun and shade. *Ecology* 12:96–104.

Strong, D.R. (1980). Null hypotheses in ecology. *Synthese* 43:271–285.

Swank, W.G. (1949). *Beaver Ecology and Management in West Virginia*. Conservation Commission of West Virginia, Division of Game Management, Bulletin No. 1.

Tansley, A.G. (1935). The use and abuse of vegetation concepts and terms. *Ecology* 16:284–307.

———. (1949). *Britain's Green Mantle*. London: George Allen and Unwin.

Taylor, W.P. (1935). Some animal relations to soils. *Ecology* 16:127–136.

Taylor, W.P., and McGinnie, W.G. (1928). The bio-ecology of forest and range. *The Scientific Monthly* 27:177–182.

Turner, R.M., Alcorn, S.M., and Olin, G. (1969). Mortality of transplanted saguaro seedlings. *Ecology* 5:835–844.

van Rhee, J.A. (1965). Earthworms activity and plant growth in artificial cultures. *Plant Soil* 22:45–48.

Varga, L. (1928). Ein interessanter Biotop der Bioconöse von Wasserorganismen. *Biologisches Zentralblatt* 48:143–162.

Viles, H. (1988). *Biogeomorphology*. Oxford: Blackwell.

Vitousek, P.M. (1986). Biological invasions and ecosystem properties. In *Biological Invasions of North America and Hawaii*, H.A. Mooney and J. Drake, Eds. New York: Springer, pp. 163–176.

Vitousek, P.M. (1990). Biological invasions and ecosystem processes: Toward an integration of population biology and ecosystem studies. *Oikos* 57:7–13.

Vivrette, N.J., and Muller, C.H. (1977). Mechanism of invasion and dominance of coastal grassland by *Mesembrythemum crystallinum*. *Ecological Monographs* 47:301–318.

Von Hardenberg, J., Meron, E., Shachak, M., and Zarmil, Y. (2001). Diversity of vegetation patterns and desertification. *Physical Review Letters* 87(198101):1–4.

von Humboldt, A. 1807 (1805). *Essai sur la géographie dea plante*s. Paris: Schoell et Tubingue Cotta.

Voss, G.L. (1956). A review of the cephalopods of the Gulf of Mexico. *Bulletin of Marine Science* 6:85–178.

Wallace, G.T., Jr., Mahoney, O.M., Dulmage, F., Storti, F., and Dudek, N. (1981) First-order removal of particulate aluminum in oceanic surface water. *Nature* 293:729–731.

Warming, E., (1891). De psammophile Formationer I Danmark. *Videnskabelige Meddelelser fra den Natuurhistorisch Forening i Kobenhavn* (Copenhagen), pp. 153–202.

Wiken, E.B., Broersma, K., LavKulich, L.M., Farstad, L. (1976). Biosynthetic alternation in British Columbia soil by ants (*Formica fusca Linné*). *Journal of Soil Science Society of America* 40:422–426.

Wilde, S.A., Youngberg, C.T., and Hovind, J.H. (1950). Changes in composition of ground water, soil fertility, and forest growth produced by the construction and removal of beaver dams. *Journal of Wildlife Management* 14:123–128.

Williams, S.T., and Gray, T.R.G. (1974). Decomposition of litter on the soil surface. In *Biology of Plant Litter Decomposition*, Vol. 2, C.H. Dickinson and G.J.F. Pugh, Eds. New York: Academic Press, pp. 611–632.

Womersley, H.B.S., and Bailey, A. (1969). The marine algae of Solomon Islands and their place in biotic reefs. *Philosophical Transactions of the Royal Society of London* 255:433–442.

Wright, J.P., Jones, C.G., and Flecker, A.S. (2002). An ecosystem engineer, the beaver, increases species richness at the landscape scale. *Oecologia* 132:96–101.

Yair, A., and Rutin, J. (1981). Some aspects of the regional variation in the amount of available sediments produced by isopods and porcupines, northern Negev, Israel. *Earth Surface Processes and Landforms* 6:221–234.

Yavitt, J.B., Angell, L.L., Fakey, T.J., Cirma, C.P., and Driscoll, C.T. (1992). Methane fluxes, concentrations and production in two Adirondack beaver impoundments. *Limnology and Oceanography* 37:1057–1066.

3

A NEW SPIRIT AND CONCEPT FOR ECOSYSTEM ENGINEERING?

William G. Wilson

3.1 ◦ INTRODUCTION

Controversy remains over the concept of ecosystem engineer (Jones et al. 1994, Wright and Jones 2006), although some signs indicate general resolution and acceptance (Stinchcombe and Schmitt 2006). The arguments for and against ecosystem engineering sometimes pertain to issues as trivial as whether or not *engineer* connotes motive (Power 1997a, Jones et al. 1997), with reservation of the term *engineer* to some entity having purpose (Power 1997b). One might assert that this level of semantic argument leaves ecology looking a bit pedantic, given that, say, physics discusses the flavors of quarks, including types called *charm* and *strange*. Even so, why should purpose be ascribed only to humans in their activities, and not, say, to beavers building dams?

More important arguments are that the ecosystem engineering (EE) concept has been so broadly defined as to include all species at all times in all situations (Reichmann and Seabloom 2002a, 2002b). If so, then the term becomes rather useless as an ecological concept, though some take this situation as a positive sign of its ubiquity (Wright and Jones 2006). After many discussions, however, the distinction between EE and abiotic interactions remains unclear.

I have three objectives with this chapter. First, I briefly present historical precedents of the term *abiotic interactions* and early discussions of the feedback between organisms and their environment. In no way is this a definitive review, but hopefully it's representative of the intellectual development. Interestingly, I found a particularly poignant remark by Cooper (1926) regarding ecology's buzzwords: "It almost seems that the moment one formulates a concept and provides it with a name and a terminology the spirit of it flies away and only the dead body remains." The "Cooper principle" is holding in this case: The spirit of ecosystem engineering has been around for at least 80 years as "environment modification," but its recent naming as *ecosystem engineering* and the assertions of its importance and ubiquity might be revealing its skeletal remains. I hope this historical perspective helps bring back the spirit.

My second objective is exploring the concept in pictures and math, which, I think, helps clarify the many words written on the topic. Only a few theoretical treatments exist, which have all been good, and it is this type of development that I explore here. Focusing on logistic growth-like mathematical models, I uncover what I believe is the most interesting situation deserving the label *ecosystem engineering* as unique from *environment modification*. Briefly, it is the situation where an organism modifies its environment to promote its own growth.

Third, I try to relate previous definitions of ecosystem engineering to other concepts, primarily abiotic interactions, keystone species, and niche construction. I avoid old ground, primarily for the insufficient reason that the semantic arguments are difficult for me to follow given the mathematical wiring of my brain, not to mention that Wright and Jones (2006) reviewed the verbal arguments quite extensively.

Summarizing my conclusions, I show that the concept of feedback between organisms and their environments, including knock-on effects to other organisms, has a long history dating back at least to the 1920s, and would perhaps best be dubbed *environment modification* after Solomon (1949). I argue that environment modification is a more general idea that encompasses recent definitions of ecosystem engineering. However, I'll use the two terms somewhat synonymously until the very end. Using a series of linearized models, I demonstrate a connection between ecosystem engineering and the *keystone species* concept (e.g., Power et al. 1996). Essentially, there's the organism, the activities it performs, and the consequences of its activities. Keystone species are those with large consequences, whereas advocates of ecosystem engineering focus on the activities irrespective of the consequence's magnitude. Finally, exploring the consequences of feedbacks between organisms and their environment reveals a situation in which natural selection leads to runaway growth and extreme environmental modifications.

3.2 ▬ A SHORT HISTORICAL PERSPECTIVE

An excellent comment by a reviewer to an earlier draft motivated an exploration of the history of the phrase, *abiotic interaction,* which, to me, implies the interaction between a species and an abiotic component of its environment. This pursuit also uncovered earlier mentions of ideas that might be called the spirit of ecosystem engineering.

ABIOTIC INTERACTIONS

The earliest use of the term "interaction" that I found occurred when Tyndall (1874) wrote (p. 47):

> There are two obvious factors to be here taken into account the creature and the medium in which it lives, or, as it is often expressed, the organism and its environment. Mr. Spencer's fundamental principle is that between these two factors there is incessant interaction. The organism is played upon by the environment, and is modified to meet the requirements of the environment. Life he defines to be "a continuous adjustment of internal relations to external relations."

Similarly, Ryder (1879) writes (p. 17): ". . . the static or dynamic environment and the organism are considered to be in a relation of retroactivity—in a state of interaction." One might also argue that the sentiment that the environment is modified by the organism is also expressed, dependent on the meaning of the term *retroactivity,* but I won't push this sentiment being present at this early date.

Twenty-some years later the attitude of environmental determinism seems to predominate. In his study of a glacial lake and its flora, Reed (1902) separately describes hydrodynamic factors, edaphic (soil) factors, and atmospheric factors, before describing (p. 137)

> Biotic factors—Fully as important as any of the fore going factors are the conditions imposed upon plants by the presence of other plants or animals; conditions to which they must adapt themselves if they survive.

It is not clear that he really considered biotic factors as "fully as important." Cowles (1899a) gives a similar litany of environmental factors, relegating to "Other Factors" the following statements (p. 110): "Animals do not appear to exert any dominating influences on the dune floras" and "The influence of plants . . . is relatively inconspicuous on the dunes," though he mentions that plants contribute organic matter that is quickly removed. However, he states later (Cowles 1899b, pp. 385–386),

> A great deal of physiographic work has been done in sand dune areas in total disregard of the plant life, although the results obtained from this study show that the vegetation profoundly modifies the topography.

Clements (1905) is more direct in pushing environmental determinism. In a section titled, "Cause and effect: habitat and plant," he writes (p. 17),

> . . . the habitat and the plant . . . is precisely the relation that exists between cause and effect, and its fundamental importance lies in the fact that all questions concerning the plant lead back to it ultimately. Other relations are important, but no other is paramount, or able to serve as the basis of ecology.

Later, Clements describes (p. 86):

> Biotic factors are animals and plants. With respect to influence they are usually remote, rarely direct. Nevertheless, they often play a decisive part in the vegetation.

Thus begins some conciliatory prose (p. 87):

> As a dead cover, vegetation is a factor of the habitat proper, but it has relatively little importance. . . . Its chief effects are in modifying soil temperature, . . . holding snow and rain, . . . adds humus to the soil. . . .

He further concedes, ". . . living vegetation reacts upon the habitat in a much more vital fashion, exerting a powerful effect upon every physical factor of the habitat." These quotes by Reed, Cowles, and Clements indicate a clear distinction between environmental and biotic factors, but environmental factors seem considered of such great importance that lumping them under "not biotic" would perhaps be too overarching and provide undue emphasis on biotic factors.

The first relevant mention of *abiotic* that I found was in a review by Faull (1919) of a book on tree diseases: "The first 4 chapters deal with such maladies of biotic and abiotic origin as are common to many kinds of trees. . . ." However, this reference is unclear without detailed examination of the book in question because, as I learned from this review, short-wavelength radiation was known as "abiotic rays" around this time. For example, Smith (1926) writes (p. 444): ". . . abiotic rays are known to be superficial in action, being unable to penetrate the human

epidermis . . . [but are] lethal to bacteria and other living organisms."
Biotic rays hurt people.

Clear use of the term as "not biotic" is seen when Trapnell (1933)
writes,

> The organisms themselves and the set of modified conditions they produce
> in their environment are here termed *internal factors.* Those concerned
> are the abiotic factors of plant-modified climate and of plant- and animal-
> modified soil, and the causal organisms, treated as phytobiotic and zoobi-
> otic factors.

Note that Trapnell (1933) mentions the modification of the environment
by organisms. Severtzoff (1934) uses the phrase ". . . increased resistance
of the individual to abiotic factors resulting in the lengthening of the
period between two plagues caused by abiotic factors." As a final example,
Smith (1935) writes (p. 874): ". . . the role played by biotic factors can not
well be considered separately from the abiotic factors." Smith (1935) also
writes (p. 897): ". . . that abiotic factors are of extreme importance in
relation to *fluctuations* in numbers." However, he states earlier that (p.
894),

> *density-dependent* factors are mainly biotic in nature, while *density-inde-*
> *pendent* factors are mainly physical or abiotic, and principally climatic.

The term *abiotic* doesn't seem to resonate with Clements and Shelford
(1939), who describe (p. 68)

> . . . two distinct types of interaction. The first of these is reaction, the effect
> of organisms upon the habitat; the second, coaction, or the influence of
> organisms upon each other. Such a distinction becomes of paramount
> importance when the biotic community is made the basis of treatment.

Solomon (1949) mentions the ongoing discussion (p. 24), "as to whether
biotic factors are more important than physical factors, and vice versa,"
and states that

> In general, the performance of a population depends both on its inherent
> qualities and on the nature of the environment, and it would be pointless
> to claim that either is more important than the other.

It is likely that most ecologists today agree with Solomon's sentiment that
the old abiotic versus biotic regulation dispute is a dull one. Whether it
is the environment or species interactions that regulate a population depends on the species, the community, and the environmental
context.

Twenty years later, the situation leads us to the merging of the terms to form "abiotic interactions." McGinnis et al. (1969) state in their first sentence (p. 697), "Humid tropical forest ecosystems are among the most complex terrestrial systems on earth in terms of their biotic and abiotic interactions." Likewise, Hamilton (1969) writes that (p. 588) ". . . culture experiments . . . found enhanced growth for various algae as a result of abiotic interactions between various compounds in an enrichment medium," though this phrase may refer to purely chemical interactions. Finally, Power et al. (1988) use the term when describing their general goal (p. 457) ". . . to summarize and discuss . . . the complexity of biotic–abiotic interactions in lotic ecosystems." Thus we have the progression from environmental determinism to the division of organismal interactions into those with biotic factors and those with abiotic factors.

ENVIRONMENT MODIFICATION AND ECOSYSTEM ENGINEERING

Excluding the possible allusion through "retroactivity" by Ryder (1879) as quoted earlier, Tansley (1920) clearly recognized the mutual interaction between organisms and the environment when he wrote (p. 120)

> Nevertheless it is clear, even to the most superficial observer, that the complex of interactions between plants and their environment does lead to a certain degree of order in the arrangement and characters of the resulting vegetations.

And speaking of succession, he says (p. 132) it

> . . . is carried out either by the mere simultaneous or successive migration of species into the habitat . . . or also by the reaction of the successive plant populations upon the habitat continuously modifying the latter. . . .

Similarly, Forbes (1922) states (p. 90), ". . . ecology is the science of the relation of organisms to their environment, including, of course, the *interactions* between the environment and the organism. . . ." So did Cooper (1926) when he wrote (p. 398)

> *The organisms* produce vegetational change through their effects upon the environment, and upon each other through the environment; and by the production of new forms, varieties, species, through evolution. The *environment* brings about changes in the vegetation because it, too, is undergoing constant change, partly inherent in itself but in part caused by the action of the vegetation upon it. Through selection it is also of great importance in the process of evolution.

Later, Tansley (1935) describes a biome (p. 301) as a

> *system*, of which plants and animals are components, . . . determined by climate and soil and in its turn reacts, sometimes and to some extent on climate, always on soil.

He later defines (p. 306) "the biome considered together with all the effective inorganic factors of its environment is the *ecosystem*." In a more modern way, Solomon (1949) writes (p. 17)

> *Modification of the environment by the population: Unfavorable* modification represents a reduction in one or more aspects of environmental capacity, the extent of the reduction increasing concurrently with density. The result is an increase in the intensity of the relevant density-dependent actions. *Favorable* modification represents an increase in environmental capacity concurrent with density. Such action is opposed to that of the other density-dependent factors.

It is clear that an appreciation of interactions and feedbacks between organisms and their environment existed long ago.

Given this historical context, we can consider the definition of ecosystem engineers by Jones et al. (1997, p. 1947):

> Physical ecosystem engineers are organisms that directly or indirectly control the availability of resources to other organisms by causing physical state changes in biotic or abiotic materials. Physical ecosystem engineering by organisms is the physical modification, maintenance, or creation of habitats. The ecological effects of engineering on other species occur because the physical state changes directly or indirectly control resources used by these other species.

Jones et al. (1997) also state that (p. 1949) "the direct uptake and utilization of an abiotic resource (light, water, nutrients) by an organism is not engineering." An example of an engineer is the provision of physical structure to another organism, for example, the habitat provided by a tree branch to an epiphyte makes the tree an ecosystem engineer. I will address this point in more detail in following text, but it is also stated that "many keystone species are engineers (e.g., beavers), but others (e.g., sea otters) are not."

Does ecosystem engineering differ from the "environment modification" that was developed progressively by Cooper (1926), Trapnell (1933), Tansley (1935), and Solomon (1949), paraphrased simply as the feedback between the biotic and abiotic components of an ecosystem? Cooper (1926) explicitly mentions effects on other organisms, and Solomon (1949), in the breadth of his paper, describes control through density-dependent processes. The primary distinction seemingly arises in the

Jones et al. (1997) definition with the specific modifier "causing physical state changes" and the subsequent control of resources due to the physical state changes, though this phrase sometimes takes on less importance in later definitions (Wright and Jones 2006).

From the Jones et al. (1997) definition I have difficulty distinguishing what is and is not ecosystem engineering. All interactions lead to a physical state change in one way or another. All organisms, by their very existence, impart physical state changes that affect resource availability to other organisms. Can an organism control the availability of resources to other organisms *without* causing physical state changes? *Pisaster* has been called an engineer because it generates open space that subdominant organisms can occupy, but it seems that a sea otter, excluded earlier, presumably also makes open space when it consumes mussels (but perhaps not sea urchins).

All of these difficulties would be solved by removing all references to "physical state change" from the Jones et al. (1997) definition, leaving what appears to be the more general historical concept of "environment modification" by Cooper (1926) and Solomon (1949) describing favorable or unfavorable environment modifications having evolutionary consequences. This historical precedent is a more general one because it includes all state change possibilities, not just physical states. Certainly there could be useful discussions about the many ways in which one organism or another modifies the many aspects of the environment, including those that modify, maintain, and create structural modifications, and the importance to the ecosystem of the subsequent effects in so doing. I like the spirit of environment modification.

3.3 ● A CONNECTION WITH *KEYSTONE SPECIES*?

How can the interactions between the biotic and abiotic components be represented? Two extremes are represented separately and together in Figure 3.1, while only considering within-species biotic interactions. The first panel, Figure 3.1A, represents a situation where the per capita birth and death rates, $B(E,n)$ and $D(E,n)$, depend on both the environment E and the species' population density, n, which can be represented mathematically as

$$
\begin{aligned}
\frac{1}{n}\frac{dn}{dt} &= B(E,n) - D(E,n) \\
&\approx (b_0(E) - b_1(E)n) - (d_0(E) + d_1(E)n) \\
&= K(E) - \alpha(E)n,
\end{aligned}
\tag{1}
$$

where in the second line, for the clarity of further development, I have assumed birth and death rates that have linear dependencies on species density (with environment-dependent coefficients) and then combined them into density-independent, $K(E) = b_0(E) - d_0(E)$, and density-dependent, $\alpha(E) = b_1(E) + d_1(E)$, terms (e.g., Wilson et al. 2003). Some ecologists take umbrage at such a shocking disregard for the true, nonlinear complexity of true ecological systems, yet pursue two-factor experiments to understand the true, nonlinear complexity of true ecological systems. A linearized mathematical model can sometimes highlight and enhance the understanding of more complicated systems, just like experiments on simplified situations.

Given the preceding model, its equilibrium population density is $n^* = K(E)/\alpha(E)$, an equilibrium dependent upon an extrinsically set environment E, mediated, in part, through species interactions $\alpha(E)$. The caricature of Figure 3.1A has the species population dynamics dependent on the environment, but the environment, as drawn, is unaffected by the presence or absence of the species denoted by n. With all respects, Clements would be quite pleased. The only purpose of this trivial model is to expose what was an implicit environmental dependence, making it conceptually explicit, which is something that everyone who might write down the logistic growth equation knows implicitly.

The second example, Figure 3.1B, demonstrates a species whose existence drives the production of the environmental component described by E, but is completely unaffected by this environmental component. This situation is represented mathematically as

$$\frac{1}{n}\frac{dn}{dt} = K_0 - \alpha_0 n \tag{2a}$$

$$\frac{dE}{dt} = f(n, E), \tag{2b}$$

where K_0 is the combination of density-independent birth and death rates, α_0 is the combined density-dependent per capita birth and death rates, and $f(n,E)$ details the environment and population density-dependent production and destruction of environmental factor E. At equilibrium, $n^* = K_0/\alpha_0$, which by assumption of this extreme case is unaffected by the environment, but the state of the environment is set by the implicit condition $f(n^*,E^*) = 0$, which involves the species' density. An important aspect of this example is that, unlike the situation in Figure 3.1A, the environment dependence of the species' per capita growth rate is removed. All that this model does is represent the state-

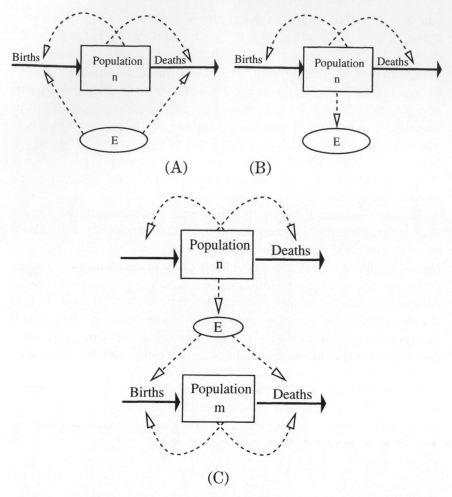

FIGURE 3.1 Essential components of biotic and abiotic interactions. The birth and death rates (solid arrows) for a population *n* are affected by interactions (dashed arrows), both biotic (emanating from the population) and abiotic (emanating from the environment E). (A) The environment affects the demographic rates, but is not affected by the population. (B) The population influences the environment, but the environment does not affect the population's demographic rates. (C) A population *n* affects the environment E, which, in turn, influences a population *m*.

ment that all living organisms affect some aspect of their environment. In both of the previous two cases there is no feedback between the species and the environmental feature under consideration.

A third example, Figure 3.1C, connects the preceding two situations, with one species denoted by density *n* affecting the environment, which in turn affects the dynamics of a second species denoted by density *m*,

$$\frac{1}{n}\frac{dn}{dt} = K_0 - \alpha_0 n \tag{3a}$$

$$\frac{dE}{dt} = f(n, E) \tag{3b}$$

$$\frac{1}{m}\frac{dm}{dt} = K_m(E) - \alpha_m(E)m. \tag{3c}$$

At equilibrium, the density of the first species is unaffected by the environmental feature, $n^* = K_0/\alpha_0$, in turn the environmental state is set implicitly by $f(n^*, E^*) = 0$, and, finally, the second species has an equilibrium density $m^* = K_m(E)/\alpha_m(E)$, dependent upon the environment set by the first species n. This model, I believe, precisely captures Cooper's (1926) ecological sentiments. Note that it doesn't concern species m whether the environment E is determined biotically by species n or determined by some arbitrary abiotic mechanism; it simply responds to the environment with no direct biotic interactions with n. All that this model does is represent that some species affect environmental variables that affect other species.

This third situation seemingly encompasses both ecosystem engineering and "keystoneness." Species n is the organism that does something, the something that it does is maintain E, and the consequence of the something is seen in species m. These are the engineer, engineering, and ecosystem effects, respectively. For example, if n represents tree density, then E might represent the nook density, and m might represent epiphyte density. Species n is then an ecosystem engineer providing spatial habitat E for another species m. Regarding the keystone species concept, Power et al. (1996) define the community importance (CI) of a species with density n as its relative effect on some arbitrary community trait (p. 609),

$$CI = \frac{1}{(\text{trait})}\frac{d(\text{trait})}{dn}, \tag{4}$$

and to qualify as a *keystone* species, its community importance has to be much greater than one, representing a high relative impact. Although they did not specifically state that a community trait can be a single species' density, they did allow it to be the density of a functional group of species. Then, if the trait is taken to be the density of species m (alternatively, I could consider m as the density of a functional group of species), it depends on E, which depends on n, or by the chain rule of calculus we see the purported keystone species do its work through the environment,

$$CI = \frac{1}{(\text{trait})} \frac{d(\text{trait})}{dE} \frac{dE}{dn}. \tag{5}$$

Again, species n is the organism that does something, the something that it does is mediated through E, and the consequence is seen in species m. These are the keystone species, consumption or interaction measure, and relative trait change, respectively.

Thus, using this third example from Figure 3.1C we can assign words from both ecosystem engineering and the keystone species concept to its elements, so exactly what is the distinction between an ecosystem engineer and a keystone species? There is agreement that the organism in question is species n, both admit abiotic and biotic factors for E, and both have impacts measured by m. There are apparent differences: E must be physical state changes for ecosystem engineering, but can be any abiotic or biotic interactions for keystone species (Power et al. 1996). The difference for m is that for a keystone species the impact must be relatively great, but the impact level, as long as there is one, isn't important for ecosystem engineering. One might say that a keystone species depends on its effect for its classification, and the ecosystem engineer depends on its process for its classification. The qualifying processes for keystone are broader than for EE, but the level of effect to qualify as EE ranges down to the ecologically insignificant.

If I have interpreted everything correctly, what is and is not an ecosystem engineer hinges critically on the phrase "physical state" and the change thereof. It seems that if an organism has any interaction that causes a physical state change in the ecosystem that affects another organism, then it is an ecosystem engineer. However, since the "direct uptake and utilization of an abiotic resource" is not included, something like nutrient concentration is not a physical state of the ecosystem. But nutrient concentration is a physical state of the environment, and then the definition of EE no longer hinges on the definition of "physical state," but rather on how the physical state is changed. Parsing processes into engineering and nonengineering then becomes seemingly problematic, and it is at this point in my understanding that the spirit of the concept flies away, and Cooper's principle rings true. Instead, I like the phrase "environment modification" as what goes on with E, irrespective of its impact on m, and irrespective of the modification process details.

3.4 ⟡ A UNIQUE FEATURE FOR ECOSYSTEM ENGINEERING?

In this section I propose a specific ecological situation that might well be reserved for the term *ecosystem engineering*. There is a feedback aspect of the species–environment interaction, as expressed by Solomon (1949), that hasn't garnered much apparent attention by the main proponents of ecosystem engineering, and is completely distinct from the keystone species concept. However, Gurney and Lawton (1996), Wright et al. (2004), and Cuddington and Hastings (2004) address this feedback feature quite well, and it is this feature, or some general representation of it, that I propose be reserved for ecosystem engineering. The feedback idea has also been addressed quite well by Cuddington et al. (unpublished), and my aim here is to present an abbreviated and specific pedagogical version of that latter model.

The caricature of the model for ecosystem engineering sits somewhere between the two concepts pictured in Figure 3.1A and 3.1B, combining species interactions and the environment and feedback between the two, and resulting in the interaction picture of Figure 3.2. In particular, the species "produces" the environmental variable at a per capita rate $\gamma > 0$, and the environment, in turn, affects either or both of the birth and death rates of the species. The representation used here combines

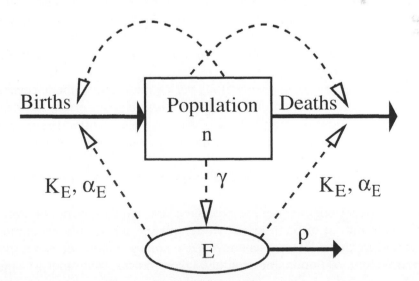

FIGURE 3.2 The presence of an environment affected by a species, which has demographic rates that are themselves affected by the environment, sets up a situation that, with a positive-feedback loop, can lead to extreme environmental modification and runaway species growth.

density-independent birth and death factors into a single term, and, likewise, the density-dependent factors; hence one set of interaction parameters, K_E and α_E, affect both birth and death rates. A mathematical representation takes the logistic growth equation and adds the effects of the coupled environment, giving

$$\frac{1}{n}\frac{dn}{dt} = K(E) - \alpha(E)n$$
$$= (K_0 + K_E E) - (\alpha_0 + \alpha_E E)n \tag{6}$$

$$\frac{dE}{dt} = f(n, E)$$
$$= -\rho(E - E_0) + \gamma n. \tag{7}$$

In the preceding equations I have taken the liberty of simplifying the species' interactions by making the density-dependent and density-independent coefficients have a linearized environmental dependence, similar to the linearization with density of the logistic growth model and to which umbrage may be taken. I have also assumed that the environment decays to a set point of $E = E_0 \geq 0$ in the absence of the engineering species, represented by $n = 0$. The environmental trait, E, is produced proportional to the species density, but production by n is independent of the level of E. Note that $\gamma = 0$ recovers Equation 1 describing Figure 3.1A, and $K_E = \alpha_E = 0$ recovers Equation 2a describing Figure 3.1B. For the sake of reducing the number of parameters, in the remainder of the analysis I will assume that the environmental set point is $E_0 = 0$ in the absence of species n.

Once again, my goal is conceptual clarity, and I follow the approach of Cuddington et al. (unpublished), but for only a single exemplary case with an evolutionary extension. In particular, I assume that the species has positive growth when rare in the absence of environmental feedback, or $K_0 > 0$.

Suppose now, for ease of analysis, that the environmental variable is one that adjusts rapidly compared with the population dynamics of the purported engineering species, such that we can always consider the environment to be in equilibrium with the species n's population density. In other words, whenever n changes a little or a lot, the environment responds immediately, taking on the value that satisfies $dE/dt = 0$. This is called a "quasiequilibrium" assumption. In the case described by Equation 7, the environment takes on the quasiequilibrium value, $E^* = \gamma n/\rho$. This value can then be placed into the equation for the population dynamics, under the assumption of instantaneous feedback,

$$\frac{1}{n}\frac{dn}{dt} = (K_0 + K_E E^*) - (\alpha_0 + \alpha_E E^*)n$$

$$= \left(K_0 + K_E \frac{\gamma n}{\rho}\right) - \left(\alpha_0 + \alpha_E \frac{\gamma n}{\rho}\right)n$$

$$= (K_0) + \left(\frac{K_E \gamma}{\rho} - \alpha_0\right)n - \left(\frac{\alpha_E \gamma}{\rho}\right)n^2. \tag{8}$$

The per capita growth rate as a function of density n is depicted in Figure 3.3 for two useful limiting situations. When the environment has no effect on density-dependent regulation, or $\alpha_E = 0$, the isolated effects of the environment on density-independent growth, shown in Figure 3.3A, has three levels of effect. When $K_E < 0$, the equilibrium population density is suppressed, and when $K_E > 0$ it is enhanced, but when $K_E > \rho\alpha_0/\gamma$, the engineering effects of density-independent growth overwhelm the unengineered density-dependent regulation, resulting in runaway environmental modification and positive per capita growth for all population densities.

The second limiting case examines effects of the environment on density-dependent regulation given fixed $K_E < \rho\alpha_0/\gamma$, shown in Figure 3.3B. The result is uneventful when $\alpha_E > 0$, and the environment increases density-dependent regulation, depressing the equilibrium population density of species n. However, an interesting scenario occurs for the case $\alpha_E < 0$, when interactions with the environment reduce the density-dependent regulation in the system. This case results in two changes. First, the stable equilibrium population density associated with the $\alpha_E = 0$ situation is enhanced. Second, a new upper equilibrium is introduced, but this equilibrium is unstable. If the population density were, somehow, to find itself above this upper equilibrium, runaway environmental modification and positive population growth would take place.

Thus, the scenario outlined by Solomon (1949) leads to some interesting ecological situations, particularly what he would call "favorable environmental modification." It is only an aside that, clearly, no ecologist, not even a theoretical one, argues for the realism of runaway growth. All that runaway growth means in a model is that, given the ingredients put into it, population regulation has been lost. Fortunately, there are other ecological ingredients that can regulate populations, but such regulating

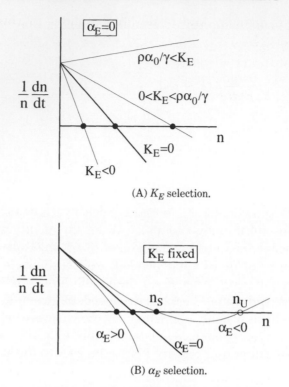

(A) K_E selection.

(B) α_E selection.

FIGURE 3.3 Net per capita growth rates under different scenarios for environment-dependent contributions to density-dependent (α_E) and density-independent (K_E) rates. It is assumed that $\gamma > 0$ and, thus, the environment takes on positive values. (A) If the environment only affects density-independent rates, mutants with larger values of K_E always have positive per capita growth rates when the wild-type, with lower values, are at equilibrium. If K_E exceeds $\rho\alpha_0/\gamma$, a runaway growth situation occurs. (B) Fixing K_E for clarity, it is seen that mutants with smaller values of α_E have positive per capita growth rates near the wild-type equilibrium. Interestingly, an unstable equilibrium occurs for higher density when $\alpha_E < 0$, meaning that a runaway growth situation can occur if the density ever exceeds this equilibrium.

ingredients are irrelevant to the question at hand. However, the conditions for runaway growth indeed indicate an interesting outcome worthy of understanding.

3.5 ● A SELECTIVE ARGUMENT FOR ECOSYSTEM ENGINEERING?

Suppose we have a wild-type species that is unaffected by the environment through either its density-independent or density-dependent interactions. This situation is represented mathematically in Equation 6 by specifying $K_E = \alpha_E = 0$. In words, this wild-type species affects the environment, but is unaffected by it, precisely like Figure 3.1B. The results that follow hold correct even if the wild-type's interaction terms are affected by the environment. At equilibrium we have

$$n^* = \frac{K_0}{\alpha_0} \tag{9}$$

$$E^* = \frac{\gamma K_0}{\rho \alpha_0}. \tag{10}$$

Now consider a mutant with a low population density, $n' \ll n$, which has a response to the environment at the level set by the wild-type species. It also interacts, in a population density-dependent way, with the wild-type species. Assuming that there are no mutations in the environment-free growth rate parameters, the mutant has a per capita growth rate given by

$$\begin{aligned}
\frac{1}{n'}\frac{dn'}{dt} &= (K_0 + K_E' E^*) - (\alpha_0 + \alpha_E' E^*)n^* \\
&= \left(K_0 + K_E' \frac{\gamma K_0}{\rho \alpha_0}\right) - \left(\alpha_0 + \alpha_E' \frac{\gamma K_0}{\rho \alpha_0}\right)\frac{K_0}{\alpha_0} \\
&= \frac{\gamma K_0}{\rho \alpha_0}\left[K_E' - \frac{K_0}{\alpha_0}\alpha_E'\right].
\end{aligned} \tag{11}$$

There are four interesting features coming out of this analysis. The first two features are that the strength of selection, by which I mean the factor in front of the brackets in Equation 11, $\gamma K_0 / \rho \alpha_0$, is greater when the wild-type species strongly affects the environment (the parameter γ is large), and is weaker when the environmental return time is fast (the parameter ρ is large). Hence, species that have little impact on a quickly recovering environmental variable have weak selection on their environmental-dependent interactions, but species that greatly affect an environmental variable that relaxes slowly to the unengineered state can have strong selection. The second two features are tantamount to the traditional r–K selection concepts (Pianka 1970, Reznick et al. 2002): Mutants that have a higher density-independent growth rate, $K_E' > 0$, or lower density-dependent interactions, $\alpha_E' < 0$, can increase their numbers when rare. In this model, by design, those changes come about through abiotic interactions with the "engineered" environmental variable E, and by design,

$E > 0$. Thus, selection on K_E and α_E move the species' traits into the parameter region having an enhanced equilibrium population and runaway growth, as discussed in the previous section. In other words, given the feedback outlined here, selection enhances the environmental modification effect, setting up the opportunity for species with extreme ecosystem engineering. Thus, we have linked Solomon's (1949) conception of biotic–abiotic feedback with Cooper's (1926) evolutionary ecology ideas.

3.6 ◗ DISCUSSION

Given the results of the conceptual and mathematical exploration just presented, I have a proposed definition for an ecosystem engineer that narrows its definition, but builds on Solomon's (1949) definition of an environment modifying organism:

> Ecosystem engineer: An ecosystem engineer modifies some aspect of the environment that feeds back to itself by enhancing its density-independent per capita growth rate or reducing its density-dependent regulation to have an over-all positive benefit to its per capita growth rate when compared to the unmodified environment.

Under this definition, it is insufficient for an engineer just to have an abiotic interaction that enhances an environmental variable, as depicted in Figure 3.1A. This situation would just be environment modification. It is insufficient for an engineer to have only a growth rate that depends on an aspect of the environment, as depicted in Figure 3.1B. This situation would just be an abiotic interaction affecting species growth. The combination of these two processes, as in Figure 3.1C, is also insufficient. This situation, if the impact is large, would just be a keystone situation. Indeed, this proposed definition makes a distinction between a negative engineering species and a positive engineering species: The effects on the environment by a negative engineering species degrade the environment for its own growth, but the effects of a positive engineering species promote its own growth. There is also natural selection, which further enhances positive ecosystem engineering traits. I'm torn between my preceding definition and reserving only the runaway growth and extreme environmental modification scenario for *ecosystem engineering*. Note that the preceding definition is independent of the impacts felt by other species in the community, from which, doubtless, some will benefit and some will suffer for any given environmental modification. This independence really makes the adjective *ecosystem* irrelevant, save for the involvement of an environmental component of the ecosystem (*sensu* Tansley 1935).

Given the definition just put forward, I am somewhat hard-pressed to distinguish between the concept of an ecosystem engineer as I have defined it here and the concept of "niche construction" (Odling-Smee et al. 1996, Laland et al. 1999, Wright and Jones 2006). Crain and Bertness (2006) state that the difference is essentially one of ecology versus evolution. The resolution may very well come down to what Cooper (1926) notes, "The environmental factors associated with the soil and the activities of the organisms themselves may bring about vegetation changes of any degree of speed." In essence, engineering happens quickly, niche construction happens slowly, but that is a rather weak difference. We can see one aspect of the speed idea in the preceding evolutionary analysis. Selection on the intensity of environmental modification, or more specifically on mutants having different values for γ in the preceding equations, may require environmental "localization" in that, even while rare, the mutant experiences the environmental effects of its own doing—rather than that of the wild-type—and these local effects feed back to its own per capita growth rate. The niche construction literature solves this problem as a process taking place over several generations (Odling-Smee et al. 1996, Laland et al. 1999), and the previous generations serve as the wild-type. Demonstrating this localization result for the within-generation situation is left as an exercise for the reader.

This proposed definition makes the purpose behind engineering somewhat validated, though my semantic arsenal is not rich, and I won't push this point too strongly. The engineering species benefits from environmental modification through the feedback by the engineering that is performed. Certainly this is not "purpose" as bestowed by a supernatural power, but natural selection can operate on this loop and enhance the engineering effects, thereby making the purpose of the engineering the enhancement of the species' per capita growth, which could be purpose as bestowed through a natural power.

Though previous definitions of ecosystem engineering emphasize its independence of subsequent impacts (Wright and Jones 2006), appeal is often made to situations in which the ecosystem impact is so complete that the entire community changes from one ecotype to another, the prototypical example being the beaver. I would also argue that this knock-on effect to other species is irrelevant to my proposed definition. If the environment modification does not feed back to the engineering species, then the species is not an ecosystem engineer; rather it's just a bull in a china shop causing environmental change. This species may or may not constitute a keystone species because of its potential downstream effects, but as soon as the species itself benefits from the environmental change it performs, the situation changes drastically from a natural selection perspective. Does *Pisaster* benefit from the space that it relieves of mussels? It doesn't, and from my perspective, that makes it a keystone species given its community-level impacts, but not an engineer because of the lacking feedback. Indeed, ecosystem engineers would now include hard-to-imagine species that have absolutely no impact on other species, but perform extreme habitat modifications to attain their own fitness benefits.

ACKNOWLEDGMENTS

Discussions with the National Center for Ecological Analysis and Synthesis (NCEAS) Ecosystem Engineering working group have been definitive and lively. Comments on an earlier draft by Justin Wright, Kim Cuddington, and an anonymous reviewer were extremely helpful and motivating. An excellent comment by Justin Wright prompted the entire historical perspective section. Support was provided by NCEAS, a center funded by NSF (DEB-0072909), the University of California, and the Santa Barbara campus.

REFERENCES

Clements, F.E. (1905). *Research Methods in Ecology*. Lincoln, NE: The University Publishing Co.

Clements, F.E., and Shelford, V.E. (1939). *Bio-ecology*. New York: John Wiley & Sons, Inc.

Cooper, W.S. (1926). The fundamentals of vegetational change. *Ecology* 7:391–413.

Cowles, H.C. (1899a). The ecological relations of the vegetation on the sand dunes of Lake Michigan. Part I.—Geographical relations of the dune floras. *The Botanical Gazette* 27:95–117.

Cowles, H.C. (1899b). The ecological relations of the vegetation on the sand dunes of Lake Michigan. Concluded. *The Botanical Gazette* 27:361–391.

Crain, C.M., and Bertness, M.D. (2006). Ecosystem engineering across environmental gradients: Implications for conservation and management. *Bioscience* 56:211–218.

Cuddington, K., and Hastings, A. (2004). Invasive engineers. *Ecological Modelling* 178:335–347.

Cuddington, K., Wilson, W.G., and Hastings, A. Population dynamics of ecosystem engineers. Unpublished manuscript.

Faull, J.H. (1919). Manual of tree diseases. *Botanical Gazette* 67:369.

Forbes, S.A. (1922). The humanizing of ecology. *Ecology* 3:89–92.

Gurney, W.S.C., and Lawton, J.H. (1996). The population dynamics of ecosystem engineers. *Oikos* 76:273–283.

Hamilton, D.H. (1969). Nutrient limitation of summer phytoplankton growth in Cayuga Lake. *Limnology and Oceanography* 14:579–590.

Jones, C.G., Lawton, J.H., and Shachak, M. (1994). Organisms as ecosystem engineers. *Oikos* 69:373–386.

Jones, C.G., Lawton, J.H., and Shachak, M. (1997). Positive and negative effects of organisms as physical ecosystem engineers. *Ecology* 78:1946–1957.

Laland, K.N., Odling-Smee, F.J., and Feldman, M.W. (1999). The evolutionary consequences of niche construction: A theoretical investigation using two-locus theory. *Journal of Evolutionary Biology* 9:293–316.

McGinnis, J.T., Golley, F.B., Clements, R.G., Child, G.I., and Duever, M.J. (1969). Elemental and Hydrologic Budgets of the Panamanian Tropical Moist Forest. *Bioscience* 19:697–700.

Odling-Smee, F.J., Laland, K.N., and Feldman, M.W. (1996). Niche construction. *The American Naturalist* 147:641–648.

Pianka, E.R. (1970). On *r*- and *K*-selection. *The American Naturalist* 104:592–597.

Power, M.E. (1997a). Estimating impacts of a dominant detritivore in a neotropical stream. *Trends in Ecology and Evolution* 12:47–49.

Power, M.E. (1997b). Ecosystem engineering by organisms: Why semantics matters. Reply from M. Power. *Trends in Ecology and Evolutionary Biology* 12:275–276.

Power, M.E., Stout, R.J., Cushing, C.E., Harper, P.P., Hauer, F.R., Matthews, W.J., Moyle, P.B., Statzner, B., and Wais DeBadgen, I.R. (1988). Biotic and abiotic controls in river and stream communities. *Journal of the North American Benthological Society* 7:456–479.

Power, M.E., Tilman, D., Estes, J.A., Menge, B.A., Bond, W.J., Mills, L.S., Daily, G., Castilla, J.C., Lubchenco, J., and Paine, R.T. (1996). Challenges in the quest for keystones. *Bioscience* 46:609–620.

Reed, H.S. (1902). A survey of the Huron River Valley. I. The ecology of a glacial lake. *Botanical Gazette* 34:125–139.

Reichman, O.J., and Seabloom, E.W. (2002a). Ecosystem engineering: A trivialized concept? Response from Reichman and Seabloom. *Trends in Ecology and Evolutionary Biology* 17:308.

Reichman, O.J., and Seabloom, E.W. (2002b). The role of pocket gophers as subterranean ecosystem engineers. *Trends in Ecology and Evolutionary Biology* 17:44–49.

Reznick, D., Bryant, M.J., and Bashey, F. (2002). *r*- and *K*-selection revisited: The role of population regulation in life-history evolution. *Ecology* 83:1509–1520.

Ryder, J.A. (1879). The gemmule vs. the plastidule as the ultimate physical unit of living matter. *The American Naturalist* 13:12–20.

Smith, F.F. (1926). Some cytological and physiological studies of mosaic diseases and leaf variegations. *Annals of the Missouri Botanical Garden* 13:425–484.

Smith, H.S. (1935). The role of biotic factors in the determination of population densities. *Journal of Economic Entomology* 28:873–898.

Solomon, M.E. (1949). The natural control of animal populations. *Journal of Animal Ecology* 18:1–35.

Stinchcombe, J.R., and Schmitt, J. (2006). Ecosystem engineers as selective agents: The effects of leaf litter on emergence time and early growth in *Impatiens capensis*. *Ecology Letters* 9:258–270.

Tansley, A.G. (1920). The classification of vegetation and the concept of development. *Journal of Ecology* 8:118–149.

Tansley, A.G. (1935). The use and abuse of vegetational concepts and terms. *Ecology* 16:284–307.

Trapnell, C.G. (1933). Vegetation types in Godthaab Fjord: In relation to those in other parts of West Greenland, and with special reference to Isersiutilik. *Journal of Ecology* 21:294–334.

Tyndall, J. (1874). Address Delivered Before the British Association Assembled at Belfast, with Additions. Longmans, Green, and Co., London: [darwin-online.org.uk].

Wilson, W.G., Lundberg, P., Vázquez, D.P., Shurin, J.B., Smith, M.D., Langford, W., Gross, K.L., Mittelbach, G.G. (2003). Biodiversity and species interactions: Extending Lotka–Volterra Theory to predict community properties. *Ecology Letters* 6:944–952.

Wright, J.P., Gurney, W.S.C., and Jones, C.G. (2004). Patch dynamics in a landscape modified by ecosystem engineers. *Oikos* 105:336–348.

Wright, J.P., and Jones, C.G. (2006). The concept of organisms as ecosystem engineers. Ten years on: Progress, limitations and challenges. *Bioscience* 56:203–209.

4

ECOSYSTEM ENGINEERING: UTILITY, CONTENTION, AND PROGRESS

Kim Cuddington

In the previous chapters, an overview of the different facets of the ecosystem engineer concept is provided through historical, conceptual, and mathematical analysis. By surveying the literature, Buchman et al. (Ch 2) provide a review of early empirical studies that may be included in the general area of ecosystem engineering. Wilson (Ch 3) surveys the historical development of the study of "abiotic interactions" in a more general sense, while Jones and Guttiérez (Ch 1) touch on recent controversy and development of the definition of ecosystem engineering. After situating the concept and related issues in the literature, all three sets of authors, to a greater or lesser extent, comment on the utility of the engineering concept. Finally, Jones and Guttiérez, and Wilson go on to provide analyses that they feel will clarify the concept even further.

The historical surveys reveal that the conceptual and empirical study of species-specific modification of the abiotic environment is not new, although the coining of a catchall phrase for this process and its consequences may be (i.e., the introduction of the term *ecosystem engineer* is relatively recent, Jones et al. 1994, 1997). Empirical studies of processes that can be encompassed in various definitions of ecosystem engineering date back to at least the mid 1800s (Buchman et al., Ch 2), while the importance of this general class of interactions has been recognized for at least 80 years (Wilson, Ch 3). It is not even true that biotic–abiotic interactions have been generally thought of as unimportant until recently. Indeed, Buchman et al. point out that key ecological concepts such as succession, facilitation, foundation species, and ecosystems contain implicit references to ecosystem engineering.

Why then introduce a new term to describe this general class of ecological interactions? Both Jones and Guttiérez, and Buchman et al. suggest that although these bidirectional interactions with the abiotic environment have long been recognized, in general, ecology has focused on trophic interactions. Buchman et al. claim that, by allowing us to draw analogies between disparate organisms with seemingly diverse relationships, the ecosystem engineer concept will enable us to generalize about species that affect the abiotic environment. Indeed, historically ecologists have used similar groupings to gain new understanding. The functional analogies between species that are important in determining soil processes are particularly well developed (Buchman et al., Ch 2). By moving beyond an idiosyncratic view of these biotic–abiotic interactions, ecologists may be able to generate a new conceptual framework for understanding population, community, and ecosystem dynamics and function (Buchman et al., Ch 2; Jones and Guttiérez, Ch 1).

This potential benefit of the concept does much to deflate objections that suggest ecosystem engineering is so broadly defined that it is useless (i.e., since all species affect the abiotic environment merely by existing). I am inclined to agree with Jones and Guttiérez that ubiquity does not inhibit the usefulness of a given concept. For example, energy flow is a ubiquitous feature of biotic communities, and yet, its ubiquity does not make it a less useful concept for describing how ecosystems function. Regardless of its commonness, engineering also seems to offer a useful point of view for understanding ecological relationships. Clearly we all are aware that a food web or a nutrient flow diagram does not capture all the roles that individual species play in a community and ecosystem. The concept provides the significant advantage of focusing our attention on these interactions that have been overlooked by trophic and energy flow approaches (Buchman et al., Ch 2).

Potential usefulness aside, there still are questions about the precise nature of ecosystem engineering. Jones and Guttiérez, and Wilson offer two different definitions. These two contributions seem to get at the heart of the controversy over ecosystem engineering. That is, do we define ecosystem engineering as an outcome or a process? Some authors have suggested that the term *ecosystem engineering* be restricted to those species that have large-magnitude effects (Reichman and Seabloom 2002). Some difficult issues arise from this type of outcome-based definition. If one defines ecosystem engineers as those species that have important, or large-magnitude, effects on community or ecosystem dynamics, how are either *important* or *large-magnitude* to be defined? For example, Jouquet et al. (2006) point out that some earthworms have a strong local influence while others have a more diffuse, but larger spatial influence on soils. Without a specific question in mind, it is difficult to classify the actions of only one of these groups as engineering using an outcome-based definition. Clearly, the definitions of such thresholds may turn on issues of spatial and temporal scale, and indeed, it recently has been suggested that progress in the application of the ecosystem engineer concept now depends on an explicit grappling with the issues of scale (Hastings et al. 2007).

Another related issue is the role of the environmental response in determining the impact of engineering. Ecosystem engineering refers to a species-specific, biotic–abiotic interaction. As such, the abiotic component of the interaction has dynamics of its own that can determine whether a species impact on population, community, or ecosystem dynamics is large or small. That is, the outcome of a particular biotic–abiotic interaction is context dependent. For example, in some environments burrowing fauna may not change erosion patterns significantly, while in other systems such activities may cause catastrophic effects (e.g., puffin burrowing on the island of Grassholm [UK], Furness 1991). Such context dependence suggests that outcome-based definitions of ecosystem engineering may miss the point. Similarly, it is clear that predation and competition could alter the ultimate impact of a species that has a particular abiotic influence. In this framework, an activity classified as ecosystem engineering in one environment may not be so defined in another environment. This context dependency is not merely a semantic point: The use of an outcome-based definition of ecosystem engineering may imply that the particular mechanism of environment modification is relatively unimportant, and further, that we could account for such effects in a trophic framework. However, the context dependency of these relationships suggests that, unless the environmental dynamics are known *a priori*, phenomenological descriptions of ecosystem engineering that roll the effects of engineering into explanatory frameworks based on trophic processes may be subject to large prediction error.

On the other hand, process-based definitions are not without difficulties. Wilson (Ch 3) notes that all interactions lead to a physical state change in the environment one way or another, and it is not clear why some of these changes qualify as engineering and others do not. In some cases it seems that loose definitions may lead to a long cascade of engineers. If one concludes that *Pisaster* is an engineer because it produces empty space by eating mussels, then mussels also should be considered engineers because they fill space. Indeed, it could become difficult to identify any rocky intertidal organism that is not an engineer. These difficulties seem to culminate in the problems of determining whether chemical changes to the environment qualify as ecosystem engineering. The quantity of nutrients available in a system, such as nitrogen, would seem to be a physical state of a system. If reductions due to plant assimilation are not ecosystem engineering, it seems likely that we also should not include increases in nitrogen due to fixation. However, the introduction of nitrogen-fixing plants can have profound effects on community structure (Vitousek 1986). Ecologists are still grappling with the relationship between chemical state changes and ecosystem engineering.

Jones and Guttiérez (Ch 1) clearly state that in their view, an outcome-based definition of ecosystem engineering is inadequate and weakens the value of the concept. In part, consequence-based definitions lead to the conflation of terms such as *ecosystem engineer* and *keystone species*. These authors define physical ecosystem engineering by referring specifically to the process involved: the modification of the abiotic environment through structural change. This emphasis on structural change allows them to distinguish this process from other processes that have similar effects (e.g., nitrogen increases in sediments that result from invertebrate excretion vs. increases that result from improved oxygen circulation in channels created by burrow construction). Jones and Guttiérez firmly reject suggestions that the term *ecosystem engineer* be restricted to those species that have large impacts, or positive impacts.

Wilson (Ch 3) notes that consequence-based definitions do indeed conflate keystone and ecosystem engineer species. However, he suggests that this confusion cannot be resolved with a process-based definition. Moreover, he claims that such definitions will artificially limit our studies of the ways in which species may alter the physical environment, sucking the life out of an inherently interesting topic. He suggests instead that we focus on environmental modification *per se*, irrespective of its impact or the particular process of modification. Finally, rather than using the term *ecosystem engineering* to refer to the process of environmental modification in general, Wilson claims that the most appropriate and parsimonious response is to limit the use of *ecosystem engineer* to those species that have a positive impact on their own population growth. That is, Wilson would restrict the use of *ecosystem engineer* to those species that Jones et al. (1997) describe as extended phenotype engineers.

The two contributions therefore take quite different approaches while simultaneously acknowledging the utility of the ecosystem engineer concept, and the importance of related issues such as a more explicit consideration of environmental modification in ecological studies. While Jones and Guttiérez claim that there is no fundamental difference between an engineering species affecting its own population dynamics and an engineering species affecting those of other species, Wilson suggests that *only* those species which affect their own population growth should be classified as engineers. Similarly, while Jones and Guttiérez explicitly include negative feedback in the engineering concept, Wilson specifically excludes it. The same debate about the inclusion of negative interactions has occurred in the niche-construction literature (e.g., Dawkins 2004, Brodie 2005).

Aside from the differences in definition, Jones and Guttiérez, and Wilson also illustrate two different approaches to resolving questions about the definition of ecosystem engineering. Jones and Guttiérez use a careful exegetical approach to defining the concept, while Wilson highlights the potential benefits of a mathematical approach. As this area of research matures, I suspect that such modeling approaches will be more prominent, as mathematical biologists become more aware of the ecosystem engineering concept and its potential utility.

Some may conclude that the conceptual disagreements expressed in these contributions are evidence that the ecosystem engineering concept has little value and little to contribute to ecological studies. I suggest just the opposite. Disagreement is almost always a clear sign that something of scientific import is being discussed, and such arguments are the main means to understanding. The contributions here crystallize points of disagreement and agreement, lending new clarity to an ongoing discussion.

REFERENCES

Brodie, E. (2005). Caution: Nice construction ahead. *Evolution* 59:249–251.

Dawkins, R. (2004). Extended phenotype—but not too extended. A reply to Laland, Turn and Jablonka. *Biology and Philosophy* 19:377–396.

Furness, R. (1991). The occurrence of burrow-nesting among birds and its influence on soil fertility and stability. *Symposium of the Zoological Society of London* 63:53–67.

Hastings, A., Byers, J.E., Crooks, J., Cuddington, K., Jones, C., Lambrinos, J., Talley, T., and Wilson, W. (2007). Ecosystem engineering in space and time. *Ecology Letters* 10:153–164.

Jones, C.G., Lawton, J.H., and Shachak, M. (1994). Organisms as ecosystem engineers. *Oikos* 69:373–386.

——. 1997. Positive and negative effects of organisms as ecosystem engineers. *Ecology* 78:1946–1957.

Jouquet, P., Dauber, J., Lagerlöf, J., Lavelle, P., and Lepage, M. (2006). Soil invertebrates as ecosystem engineers: Intended and accidental effects on soil and feedback loops. *Applied Soil Ecology* 32:153–164.

Reichman, O., and Seabloom, E.W. (2002). Ecosystem engineering: A trivialized concept? *Trends in Ecology and Evolution* 17:308.

Vitousek, P.M. (1986). Biological invasions and ecosystem properties. In *Biological Invasions of North America and Hawaii*, H.A. Mooney and J. Drake, Eds. New York: Springer, pp. 163–176.

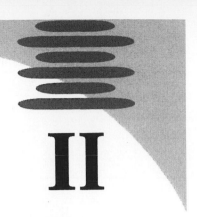

II

EXAMPLES AND APPLICATIONS

In this section, the chapters we have assembled detail the actions and effects of several prominent ecosystem engineers. We suggest that, in addition to their general interest, these thorough examples of ecosystem engineers aid greatly in understanding and thinking tangibly about the topics covered in the other portions of this book that deal with general concepts, mathematical representations, and conservation applications. The examples we have included purposefully span a wide spectrum of species and habitats, including aboveground and belowground, aquatic and terrestrial, extant and paleontological. Collectively, they emphasize the diversity and ubiquity of ecosystem engineering and the disparate systems to which the concept can be readily applied.

5

EARTHWORMS AS KEY ACTORS IN SELF-ORGANIZED SOIL SYSTEMS

Patrick Lavelle, Sebastien Barot, Manuel Blouin, Thibaud Decaëns, Juan José Jimenez, and Pascal Jouquet

5.1 ● INTRODUCTION

Earthworms are undoubtedly the most spectacular animal ecosystem engineers in all soils where neither prolonged drought nor toxic conditions occur. They are ancient organisms that have inhabited soils for very long periods of evolutionary time. Although the absence of fossils does not allow precise dating, we know that they were among the first aquatic organisms to colonize terrestrial environments ca. 200 million years ago. They can be found from several meters deep in soils to 20–30 m up in suspended soils of tropical tree canopies. Although they rarely have been acknowledged by societies for the major role they play in soils, scientists long have recognized their importance as "intestines of the soil" (Aristotle, 384–322 B.C.) or key actors of the "formation of vegetable mould" (Darwin 1881).

Three different functional groups comprise over 10,000 species of earthworms, from tropical forests and humid savannahs to boreal forests (Bouché 1977, Lavelle 1983). While small, brightly pigmented epigeic species specialize in the natural composting of organic debris deposited at the soil surface, large anecics with antero dorsal dark pigmentation inhabit semipermanent burrows in which they shelter most of the time.

Anecics leave their burrows during wet nights to collect litter that they accumulate close to the burrow entrance as "middens," or drag inside prior to ingesting it admixed with some mineral soil (Subler and Kirsch 1998, Bohlen et al. 2002). Endogeics, the third large functional group of earthworms, are unpigmented and seldom leave the soil. With geophagous feeding habits, they literally eat their way through the soil. Some of them, the "compacting" endogeics, transform unaggregated soil particles into solid macroaggregated structures, while others have the opposite "decompacting" effect (Blanchart et al. 1999).

Earthworms are clearly physical ecosystem engineers as defined by Jones et al. (1994): "organisms that directly or indirectly modulate the availability of resources to other species, by causing physical state changes in biotic or abiotic materials." They seem to be mostly extended phenotype engineers that build structures that likely maintain suitable conditions for their growth (Jouquet et al. 2006). The "extended phenotype engineers" are organisms that create structures or effects that directly influence the fitness of individuals. On the contrary, engineers creating biogenic structures that have no direct positive effect on themselves are accidental engineers (Jones et al. 1997). Some endogeics may, however, be considered as accidental engineers because the building of structures that they operate does not appear to have immediate positive feedback effects on their growth. It seems for example to be the case for species that ingest small soil aggregates that they further egest as large aggregates which they cannot re-ingest. Once having exhausted the stock of small aggregates, they are expected to suffer local extinction and/or move to nearby patches where soil is less macroaggregated (Rossi 2003).

Earthworm effect on soils may be summarized in a few striking statistics. Several hundreds of tonnes of soil annually transit through their guts in suitable environments (for example, 800 to 1300 $10^6 g ha^{-1} yr^{-1}$ in moist savannahs of the Ivory Coast, Lavelle 1978). Surface casts deposited range from 1–3 to 20–50 tonnes $ha^{-1} yr^{-1}$ (Lavelle and Spain 2006). They have the potential to ingest and bury all the leaf litter annually deposited in some forests. Populations of *Lumbricus terrestris*, for example, can consume the entire annual leaf fall; that is, $300 g m^{-2}$ in only 3 months in mixed forests of England (Satchell, 1967), or 94% of litter fallen in woodlands in Michigan (U.S.) in only 4 weeks (Knollenberg et al. 1985). Some estimates even indicate ingestion rates higher than the annual litter fall ($1071 g m^{-2}$ in evergreen oak forests in Japan [Sugi and Tanaka 1978, in Edwards and Bohlen 1996)]. These large mechanical impacts have profound effects on the soil environment and the organisms that live in it. Earthworms can also develop allelochemical interactions with plants (Rice 1984) by producing specific energetic or hormone-like chemicals and influencing their dispersal or germination patterns (Martin et al. 1987, Decaëns et al. 2003).

This chapter describes the drilosphere, which is defined as an earthworm population, all the biogenic structures that they build in the soil, and the communities of smaller organisms that inhabit these structures (Lavelle and Spain, 2006). We show that drilospheres have all the characteristics of self-organizing systems according to definitions given by Perry (1995) and Lavelle et al. (2006) and discuss the theoretical and practical meaning of this organization. We then address the interactions with micro-organisms and other inhabitants of the drilosphere that allow earthworms to derive energy from decomposing organic matter. Then we describe the physical domains created in soils by the accumulation and spatial array of biogenic structures (earthworm casts, galleries, voids, and middens), addressing the effects of this system on the soil environment and the possible positive feedbacks provided in return to earthworms and inhabitants of the drilosphere. Finally, we discuss how effective management of the drilospheres relates to sustained provision of soil ecosystem services, such as water infiltration and storage, C-sequestration, and nutrient cycling.

5.2 ● ADAPTATION OF EARTHWORMS AND OTHER ORGANISMS TO SOIL CONSTRAINTS: THE POWER OF MUTUALISM

Soils are highly constraining environments (Lavelle and Spain 2006). Movement is restricted since only 50–60% of the total volume is comprised of pores, at best. Pores have greatly diverse sizes, typically in the range of a few millimeters to microns or less. Their shapes are also greatly diverse, and connection is rarely achieved in a way that would allow easy and free movements for relatively large organisms. This porosity is filled with either air or water, in largely variable proportions according to climatic conditions.

Small microflora and microfauna (<0.2 mm on average) live in the water-filled soil pores. They mostly comprise bacteria, fungi, protists, and nematodes. Their capacity for movement is limited, and they have developed highly effective mechanisms for resisting dessication. Mesofaunal invertebrates (0.2 to 2 mm on average) live in the air-filled portion of soil pores and litter layers. Earthworms and a few other groups of larger soil invertebrate ecosystem engineers have the ability to dig the soil. This allows them to move freely by digging burrows and galleries while creating the voids they use for sheltering, feeding, and reproduction (Lavelle and Spain 2006).

The other major constraint in soil environments is the generally poor quality of food resources. Leaf and root litter and products of their successive stages of decomposition are the main food resources. They are often low in nutrients, creating important stoichiometric limitations on their use (Swift et al. 1979, Sterner and Elser 2002). Organic complexation of nutrients and simple dilution of organic matter in the mineral soil matrix are further constraints that require highly specialized breakdown systems for all soil organisms (Lavelle and Spain 2006). Bacteria and fungi are the only organisms that can process any organic material present in soils. Microorganisms usually operate via chain processing, with different generalist or specialist groups progressively breaking down even the most complex organic molecules while releasing metabolites that are then available to all the other soil organisms. Key components of the microbial community are the white rot fungi, a group of Basidiomycete that has the rare but essential capacity for breaking down polyphenol protein complexes that can immobilize over 80% of the nitrogen contained in decomposing leaves and roots (Toutain 1987).

Invertebrates seem to have rather limited proper digestive capabilities. A few studies demonstrate that part of the digestive enzymes present in their guts have been actually produced by microorganisms. This is the case for earthworms that have developed mutualist digestion systems in association with free-soil bacteria, as hypothesized by Lavelle et al. 1995 (also see Zhang et al. 1993, Lattaud et al. 1999, and Garvin et al. 2000). When geophagous earthworms ingest soil, they add an equivalent volume of water in their anterior gut plus 5 to 40% of the soil dry weight as intestinal mucus, a highly energetic product of the anterior gut wall (Martin et al. 1987, Barois et al. 1999). This mixture is energetically blended in the gizzard, which frees bacteria from soil micropores where they would be in dormant stages and gets them to full activity and enzymatic capacities within a remarkably short period of time. When the soil gets into the medium part of the gut, mucus that has not been metabolized by bacteria is removed, and bacteria start to digest soil organic matter for their own benefit and that of the earthworm (Lavelle et al. 1995). Experiments have actually shown a great increase in microbial activity in the posterior gut of earthworms (Barois and Lavelle 1986). Some of the enzymes found in the gut content are not produced by gut tissues, which supports the hypothesis of a microbial origin (Lattaud et al. 1999).

This digestion system allows earthworms to make use of very poor soil resources. An extreme case is represented by the endogeic African earthworm *Millsonia ghanensis* that feeds on soil from the deep (20–40 cm) horizons of sandy soils in savannahs of Central Ivory Coast. This soil only contains 0.6% organic matter on average, and it is known that organic matter at such depth in soil is significantly humified and therefore little digestible. The most common species in this savannah, *Reginaldia omodeoi*[1] Czusdi is a rather large animal, 15 to 20 cm in length at the adult stage, that may daily ingest up to ten times its weight in soil at maturity, and up to 30 times for recently born juveniles. Overall, worms of this species ingest 500 to 800 Mg soil ha^{-1} yr^{-1}, with a maximum during the rainy seasons and mostly in the upper 10 cm of soil. Only a few megagrams are deposited at the soil surface, the rest being deposited in the galleries that the worms just opened as they moved forward in search of the small organic rich aggregates that they ingest. The energy cost of this behavior, however, is enormous. For example, 96% of the energy assimilated by *R. omodeoi* Czusdi is spent in respiratory activities required by the daily ingestion of 10–25 times their own weight of soil and its further transformation into compact casts (Lavelle 1978).

Comparison of feeding regimes of earthworms across a latitudinal gradient from Western Africa to Northern Europe showed that as soon as temperature decreases, earthworms tend to feed on increasingly richer substrates (Lavelle 1983), presumably as the mutualist interaction with microflora is less efficient at lower temperatures (Lavelle et al. 1995), which forces them to use better quality material. Barois (1987) actually showed that the tropical geophagous earthworm *Pontoscolex corethrurus* was not able to grow when kept at 15 °C. The lack of growth was explained by a much smaller increase in microbial activity in the posterior gut than at 27 °C. This limited increase in microbial activity would not allow the worms to get enough assimilates from the poor soil they usually fed on. While endogeic geophages, which live on poor soil organic matter, dominate communities in the wet tropics, anecics that feed on a mixture of soil and litter are the dominant group in grasslands of France or England, and epigeics and the small polyhumic endogeics, which feed on leaf litter and soil organic accumulations, respectively, comprise most of the communities in Scandinavia and Iceland.

Interactions of earthworms with microorganisms allow them to derive significant amounts of energy from poor soil organic resources. In the Lamto savannah, for example, geophagous earthworms assimilate every year the equivalent of 1.2 Mg organic matter (24.8.10^6 J). They spend 96% of it in respiratory activities, and most of this energy is actually invested in the bioturbation of over 1000 Mg soil ha^{-1} yr^{-1} in their drilospheres. The

[1] Formerly known as *Millsonia anomala* Omodeoi.

result is the formation of a macroaggregated structure in the upper 20 cm of soil (Lavelle 1978).

5.3 ✦ THE DRILOSPHERE AS A SELF-ORGANIZING SYSTEM

Earthworms and other major soil ecosystem engineers create physical domains in soils that have all the characteristics of self-organized systems as defined by Perry (1995): Based on strong and rather specific interactions within physical boundaries, these systems change the constraints of their environment with positive feedbacks on their own living conditions (Lavelle et al. 2006; see Figure 5.1).

Soil constraints indeed have pushed soil organisms to develop intense interactions along evolutionary time, mostly of a mutualistic type (Lavelle and Spain 2006). These interactions operate within the boundaries of the rhizosphere of roots, drilosphere of earthworms, and termitosphere of termites, as well as a few other such domains (Lavelle 2002) that have

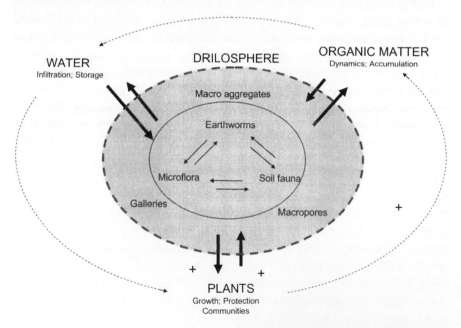

FIGURE 5.1 A general model of the drilosphere system. Within the boundaries (large dotted line) of their functional domain, the drilosphere, earthworms accumulate macro aggregates, galleries, and other pores that constitute the habitat for specific communities of microorganisms and invertebrates. Earthworms interact with these organisms. These interactions affect the external environment, especially plant growth and the composition of their communities, hydraulic conditions, and organic matter inputs and storage. Note that + signs indicate positive feedbacks.

more or less recognizable limits. These systems in turn have feedback effects on external constraints. Roots and earthworms, for example, significantly affect soil structure with known impacts on water availability and their own ability to further penetrate this environment.

These systems, finally, are in a metastable state of equilibrium: The engineer makes the environment on which it and other species depend, and this situation lasts as long as engineers produce structures that replace the ones that have been degraded or destroyed. The large amount of energy channeled into the drilosphere actually is invested in the building of solid aggregates and creation of voids that allow soils to provide ecosystem services at a high rate.

We shall first focus on the description of these individual structures before considering their assemblage in soils and their emerging effects on soil properties.

BIOGENIC STRUCTURES

Earthworm casts may have diverse shapes and sizes. A first classification separates granular casts formed from an accumulation of small, fragile, and fine-textured pellets from globular casts comprised of coalescent round or flattened units (Lee 1985). While soil texture has a great influence over the final shape and structure of casts, some anatomic features of the posterior part of the earthworm gut also influence the process. Some earthworms produce a continuous flow of small independent pellets that rarely stick together to form a globular mass. Others expel at discrete intervals rather large amounts of wet and plastic digested soil material that tends to form units of up to 1 cm depending on the species. These units when wet easily stick to others, forming sometimes large and solid structures after they have been dried at least once (Shipitalo and Protz 1988, Blanchart et al. 1993). When deposited and regularly accumulated at the soil surface, globular casts may form spectacular tower-shaped structures up to 10–15 cm high and several hundred g dry mass (Figure 5.2).

Granular casts are fragile structures easily whipped off by rain when deposited at the soil surface. Globular casts may persist for very long periods of time, especially when they have been deposited in the soil and stabilized by one or two drying–rewetting cycles (Shipitalo and Protz 1988, Marinissen and Dexter 1990). In the African savannah of Lamto (Ivory Coast), Blanchart et al. (1997) showed that the large casts that comprise the macroaggregated structure of these soils in the upper 15 cm can still be found almost intact 32 months after removal of earthworms by a 48 hr artificial flooding. Dry globular casts deposited at the

Figure 5.2 Tower-shaped earthworm cast in fallows in Vietnam. This structure is formed after weeks of daily deposition of casts at the top edge of the structure. Note dense colonization by fine roots. Photo P. Jouquet. (See color plate.)

soil surface can persist for periods of a few days to several weeks or months, depending on their own mineral constitution, the degree of protection by plant cover, and the intensity of rain and other climatic events (alternance of dry and moist periods; freezing/thawing) (Decaëns 2000, Le Bayon and Binet 1999). They may also be crushed by large mammals or broken by invertebrates that use them as shelter and/or food (Decaëns 2000).

When fresh, casts are the seat of intense microbial activities, and ammonium and other nutrients are found at relatively high concentrations (Lavelle et al. 1992, Blair et al. 1995). In fresh casts of the pantropical endogeic species *Pontoscolex corethrurus*, for example, ammonium concentrations in fresh casts vary from 67 to $1052\,\mu g\,g^{-1}$ soil depending on clay mineralogy and N (nitrogen) content of the soil they had ingested. This represents on average 4 to 10% of the ingested organic N (Lavelle and Spain 2006). Assimilable P (phosphorus) concentration is also multiplied by a factor of 2 to 8 in the same casts as compared to a nonin-gested control soil (Lopez-Hernandez et al. 1993, Chapuis-Lardy et al. 1996). Once dried, casts become a harsh environment for microorgan-isms. Porosity is often extremely reduced. Casts of *R. omodeoi*, for example, have a bulk density of 2.3 as compared to 1.4 on average in soil. A superficial 15 µm pellicle rich in clay minerals and polysaccharides seems to isolate the cast environment from the outside and limit water and air penetration (Blanchart et al. 1993). Laboratory incubations have shown that organic matter mineralization was reduced to almost zero in these structures after 30 days, while a control nonaggregated soil con-tinued to lose C (carbon) (Martin 1991). The quality of organic matter contained in earthworm casts is significantly different from the one in the nondigested control soil. Spectral signatures (Near Infrared Reflect-ance Spectrometry) allow separating them from aggregates produced by other biological or physical processes (Velasquez et al. 2007).

In soils favorable to earthworm activities, subterranean casts tend to accumulate as stable macroaggregates forming >40% of the total soil volume (Blanchart et al. 1999). Persistence and dynamics in time of these biogenic structures are still poorly documented. Highly unstable fresh globular casts can be easily dispersed or included into larger structures made by the addition of a number of similar structures.

The continuous deposition of casts at the soil surface is a response of earthworms to the general trend of soil to compact and a contribution to soil-forming processes. The proportion of the ingested soil that they deposit at the soil surface may vary from less than 5% to over 80% depending on species and soil conditions. Surface cast deposition is therefore a very poor indicator of earthworm activity; in the Lamto savannah, overall soil ingestion by endogeic earthworms estimated by a simulation model and surface cast depositions actually had opposite temporal patterns (Lavelle and Spain 2006). Surface cast deposition was maximum at the onset of the rainy season, while actual maximum soil ingestion by populations occurred several months later when surface deposition was very low.

Both categories of surface casts participate in the soil-creeping process, a general mechanism that transfers small-sized organic and mineral soil particles from the most elevated parts of the landscapes to low-lying areas where they accumulate (Nooren et al. 1995). Surface cast deposi-tion also contributes to the progressive burial of gravels and stones by covering each year the soil surface with a continuous layer of 0.25–0.50 mm (Darwin 1881) to 1–2 mm (Lavelle 1978).

Gallery networks and burrows made by anecic and a few endogeic species have been studied independently from aggregate assemblages.

In a 12-year-old pasture in France, Bastardie et al. (2005) made a thorough quantitative description of earthworm burrow systems by applying X-ray tomography to 12 soil cores 25 cm in diameter and 60 cm in depth. Earthworm mean density was 101+-3 S.D. individuals m^{-2} distributed among 8 species. Three were anecic, four endogeic, and one was epigeic. Total burrow length ranged from 687 to 1212 mm^{-3}. Volume represented 13.3 to 24.41m^{-3}, which is less than 2.5% of soil volume. Total area of internal burrow walls represented 1069 to 7237 $cm^2 m^{-2}$. Only 9–43% of the volume was connected to the soil surface, and large seasonal variations did occur.

Burrow systems seem to have species-specific shapes and organizations. The diameter of galleries, their branching, orientation, and the continuity of the burrow system significantly vary among species (Kretzschmar 1990, Lamparski et al. 1987, Lightart et al. 1993, Bastardie et al. 2005).

Earthworm burrowing activities are highly sensitive to soil compaction (Kretzschmar 1991) and such soil pollutants as heavy metals (Nahmani et al. 2005) or pesticides (imidacloprid also known as "gaucho"; Capowiez et al. 2005). Galleries may act as preferential ways of circulation for gases and water. Their walls are regularly recoated with cutaneous mucus and sometimes with cast deposits each time the worm passes through. Cast deposition occurs more frequently in deep soil strata than closer to the surface; as a result, continuity between gallery and porosity of the rest of the soil is much better achieved in upper soil horizons than in the deeper soil.

THE TOPOLOGY AND DYNAMICS OF DRILOSPHERIC ASSEMBLAGES

Recent studies have demonstrated a significant relationship between soil macroaggregation, especially the abundance and size of biogenic aggregates, and the presence of earthworms and other soil ecosystem engineers (Blanchart et al. 1999, Bossuyt et al. 2006, Velasquez et al. in press). In the Brazilian Amazonian region of Pará, pastures derived from a primary forest cut 6 years ago were planted to four different plant species and all possible combinations of them, in a complete randomly designed experiment replicated in three blocks. There were two shrub species, the local weed *Solanum nigrum* and the legume *Leucaena leucocephala*, and two herbaceous species, the legume *Arachis pintoi* and the African grass *Brachiaria bryzantha*, the same grass that had been planted 6 years ago when the pasture had been created. Soil macroinvertebrate communities significantly responded to the change that

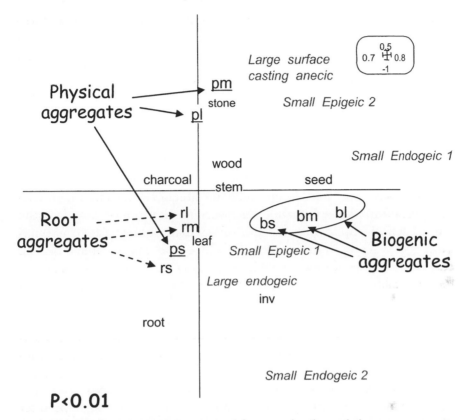

Figure 5.3 Significant co-inertia among soil fauna and soil morphology parameters in Amazonian pasture soils submitted to all possible combinations of four different plants. Projection on factorial plane 1-2 of fauna and soil morphology variables shows close location, in the right half of the figure, of earthworm species and biogenic aggregates, suggesting that they are formed by these earthworms. Key: pm, pl, ps: medium, large, and small physical aggregates; bs, bm, bl: biogenic aggregates of small, medium, and large size; rs, rm, rb: root aggregates of different sizes; inv: invertebrates found in small soil blocks.

occurred in vegetation. Soil macroaggregation also changed, and co-inertia analysis showed a significant relationship of this soil attribute with macrofauna communities (p < 0.01). Earthworms, especially endogeic species, were responsible for a great part of this aggregation as shown by location of their projections close to respective projections of biogenic aggregates in the factorial plan (Lavelle et al. unpublished data, Figure 5.3).

Drilosphere boundaries have never yet been directly described. They can be seen when examining thin sections of soil showing a discrete array of aggregates of different sizes and shapes; they also are felt when manually separating soil blocks into different classes of aggregates that further exhibit distinct spectral signatures (Velasquez et al. 2007). It will be particularly interesting to observe the frontiers of rhizospheres and drilospheres, two systems that otherwise develop very intense interactions.

Topology and dynamics of earthworm structure assemblages inside drilosphere boundaries are largely ignored. Laboratory and field experiments indicate only a few basic features that determine the spatial distribution of earthworms and the biogenic structures that they produce at spatial scales of a few centimeters to decameters.

First, endogeic earthworms do not seem to re-ingest casts of their own species unless they have been totally disintegrated (Lavelle 1978). This observation made on endogeic earthworms that produce globular casts has profound implications on the spatial distribution of their populations (Rossi 2003). In African moist savannahs at Lamto (Côte d'Ivoire), earthworms that produce globular casts (as the "compacting" species) have opposite patterns of horizontal distribution to species that produce granular casts ("decompacting" species). Patches with dominant decompacting populations actually had significantly lower bulk density (hence higher porosity) and a larger density of fine roots than patches predominantly occupied by "compacting" species. Statistical tests (partial Mantel test) showed that the nature of earthworm communities was responsible for these differences, not the opposite. The hypothesis that patches of opposite functional groups should move in time when transformation of soil has been completed has not been tested so far in the field. A modeling exercise predicts a shift in population distribution after 2–3 years of activity (Barot, in press).

Second, anecic earthworms seem to have rather sedentary and territorial ways of life (Edwards and Bohlen 1996). This allows soils that host dense populations to have rather regularly distributed vertical (and sometimes horizontal) drainage networks. A very interesting case was observed in rainforests of Madagascar that led to formulate a hypothesis on the role of anecic earthworms on soil conservation in these environments (Lavelle et al. unpublished data). The observed forest grows on highly unstable oxisols. Below a A_0 5 cm thick holorganic horizon, a 30 cm thick A1 horizon tops a 60 cm deep clayey B horizon. This B horizon has a special prismatic vertical structure that tends to disaggregate in case of physical disruptions like the one created by cutting a slope to create a road (Figure 5.4). Any excessive water infiltration in this soil layer is likely to generate horizontal disruptions leading to massive erosion events. Such events are prevented by absorption and drainage of the water in the upper 30 cm of soil, maintained by biological activities. The surface humic horizon, an accumulation of invertebrate fecal pellets

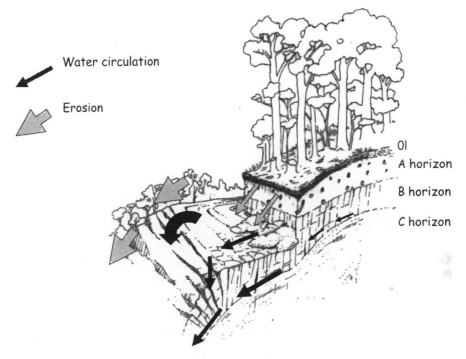

Water circulation

Erosion

Ol
A horizon
B horizon
C horizon

Figure 5.4 Hypothesized role of giant anecic earthworms in soil conservation of tropical rainforest in Madagascar (Ambohilero forest). Rainfall is first absorbed by a 10 cm thick surface organic layer and then enters soil of the A horizon, where a subhorizontal network of earthworm galleries stores and channels water toward low-lying areas. This prevents water from penetrating too much in unstable low-lying horizons. Cutting a road across the slope exposes the B horizon (with a prismatic fragile structure) and C horizon made of highly dispersable alterites, eliminates the natural drainage system and accelerates massive soil erosion and the occurrence of major landslides ("lavakas"). Drawing by R.L. Andriamarisoa.

(mainly *Diptera* larvae, in that case), acts like a sponge able to absorb the equivalent of approximately 100 mm rainfall. Below this spongelike structure, giant anecic earthworms create a dense network of regularly distributed horizontal galleries that seem to act as a pipe network allowing water to store and convey by a horizontal transfer to low-lying areas and natural effluents (Figure 5.5). If confirmed, this hypothesis would explain how the destruction of these self-organized systems may lead to such spectacular landslides at the landscape scale.

Another consequence of the relative sedentarity of anecic and some endogeic earthworms is the accumulation over time of surface casts at the same place, which end up making rather large, sometimes tower-like, structures at the soil surface (Edwards and Bohlen 1996). In predominantly herbaceous fallows in Vietnam, individual cast accumulations

Figure 5.5 Section of the A horizon in soil of the Ambohilero Forest (Madagascar) showing anecic earthworm horizontal galleries (arrows and smaller photo) forming a pipe network allowing subhorizontal drainage. (Photos by P. Lavelle).

may amount to several hundred g dry weight, and total mass deposited at the soil surface may be approximately 10 kg dry mass m^2 (Jouquet unpublished data). Anecics also may collect litter deposited around the mouth of their burrow, creating "middens" colonized by a dense and rather specific fauna and microflora (Hamilton and Sillman 1989, Subler and Kirsch 1998, Bohlen et al. 2002). This community is thought to achieve a preliminary decomposition of litter before earthworm ingestion. This process achieves an "external rumen" type of digestion as defined by Swift et al. (1979).

All these observations still are rather isolated and need to be supported by more fieldwork and modeling exercises and extended to a larger number of species and situations to be considered as general features. They also need to be interpreted in terms of the interactions that earthworms and other organisms develop inside the drilosphere.

BIOLOGICAL INTERACTIONS INSIDE THE DRILOSPHERE

With a few exceptions, studies show increased microbial activities in gut contents, fresh casts, and burrow walls produced by earthworm activities (Parle 1963, Barois and Lavelle 1986, Scheu 1987, Daniel and Anderson 1992, Fischer et al. 1995, Karsten and Drake 1997, Lattaud et al. 1997, Winding et al. 1997, Zhang 2000, Tiunov et al. 2001, Kersante et al. 2006). This activity—largely associated with earthworm digestion processes—is mostly that of soil-dwelling microorganisms. They used to be in resting stages in the soil and took advantage of optimal conditions created by the earthworm in its foregut to resume their activity. This process has been described as the "Sleeping Beauty" paradigm, which states that most microbial activity in soils occurs in specific microsites created by the activities of macroorganisms. Microbial communities in soil are mostly in resting stages, waiting for these "Prince Charmings" to reactivate them (Lavelle et al. 1995). Earthworm guts, gallery walls, and fresh casts are the drilospheric microsites where such activations do occur. There is growing evidence that only part of microorganisms are stimulated in this process, and more research is required to know how specific this interaction is and whether activated microbial communities differ among earthworm species and among the different soil functional domains (Lavelle et al. 2005).

Drilosphere structures are the habitat of very diverse communities of invertebrates of all sizes, while fine roots often concentrate in this specific environment. Decaëns et al. (1999a) showed a rather fast colonization of casts of the anecic neotropical earthworm *Martiodrilus carimaguensis* by fine roots and a diverse community of invertebrates of the macro- and mesofauna. Drilospheres are also highly favorable habitats for Collembola and Acari (Marinissen and Bok 1988, Loranger et al. 1998). In a pasture of Martinique (French West Indies), patches densely colonized by the earthworm *Polypheretima elongata* had 28 instead of 23 species of Collembola; population density was 13,000 m^{-2} instead of 9000 outside the patches, and the Shannon index of diversity for their communities was 3.53 instead of 2.74 outside the patches.

FEEDBACK EFFECTS OF THE DRILOSPHERE ON SOIL CONDITIONS AND OTHER ORGANISMS

Feedback effects of biological interaction on environment constraints are expected to occur in drilosphere as a result of self-organization (Perry 1995). The accumulation of earthworm biogenic structures in soils has significant effects on soil physical properties that may, or may not, have positive feedback effects on earthworms through changes in moisture regime in soils, a fundamental determinant of earthworm activities (Edwards and Bohlen 1996).

Many experiments in laboratory and observations in field conditions have indicated such effects (Edwards and Bohlen 1996, Chauvel et al. 1999, Decaëns et al. 1999b, Hallaire et al. 2000). Clay mineralogy seems to be one clue to the occurrence of significant influence of earthworms with more pronounced and lasting effects in soils with kaolinitic 1:1 type clays than with smectitic 2:1 clay materials (Blanchart et al. 2004).

Pontoscolex corethrurus, a very active endogeic invasive earthworm that produces globular casts, has been claimed to be responsible for soil compaction in sweet potato cultures (Rose and Wood 1980), maize crops (Hallaire et al. 2000), and recently installed Amazonian pastures following conversion of primary forest to pasture (Chauvel et al. 1999). *P. corethrurus* is actually a clear example of invasive engineer as defined by Cuddington and Hastings (2004), as it is able to survive in conditions that many native species cannot withstand and modify the habitat in ways that make other species' return more difficult (Lavelle et al. 1987, Lapied and Lavelle 2003).

Unlike compacting species, decompacting filiform endogeics significantly decrease soil bulk density when kept alone in experimental soils (Blanchart et al. 1999). Although more data clearly are needed to reach a conclusion, it seems that single earthworm species are not able to maintain alone suitable physical conditions in most cases. They probably need to interact with other earthworm and other invertebrate engineer species, or natural physical processes, in order to achieve this feature. These findings support the view of Jouquet et al. (2006) that endogeic earthworms may be accidental rather than extended phenotype ecosystem engineers. However, regulations obviously occur at the scale of communities, and positive feedback of soil structure maintained by a community may further affect each of the species in the community. Earthworms seem to re-ingest casts of other species, thus converting certain types of structures (e.g., casts of anecics rich in organic residues, or compact casts of large endogeic species) into other types (epigeic or other litter invertebrates, Scheu and Wolters 1991; loose granular casts of epigeics, Mariani et al. 2001), and thus exerting regulatory effects on the proportions of each type and preventing the accumulation in excess of a single category of casts.

Compacting species seem to exert negative feedbacks on their own survival by reducing the porous space and eating out the small aggregates that are their feeding resource; however, decompacting species develop at the same time opposite effects in adjacent patches. The resulting effect of the two functional groups likely has positive feedbacks on both groups at the scale of the ecosystem.

The experiment conducted in natural field conditions in Brazilian Amazonia (Figure 5.3) by Lavelle et al. (unpublished data) and observations of Velasquez et al. (2006) showed that earthworms actually may be responsible for a significant part of aggregation in the upper 10–20 cm of many soils. Soils that have improved biogenic aggregation are less compact and likely present improved hydraulic properties in the upper few cm below surface. This result, however, largely depends on the diversity and composition of earthworm and other soil engineer communities and the nature of their respective biogenic structures.

Feedback effects on direct or indirect competitors comprise the decrease in litter dwelling arthropods when anecic populations increase and adverse effects on communities of plant parasitic nematodes (Yeates 1981, Lavelle et al. 2004). As regards microbial communities, drilospheres tend to be colonized by bacteria rather than fungi (Hendrix et al. 1986). Mutualist digestion systems developed in earthworm guts seem to involve only bacteria, and there is slight evidence that earthworm cutaneous mucus sprayed over litter accumulated in "middens" or in burrow walls might have some fungistatic effects (Tiunov et al. 2001). Positive effects on plant growth also are likely to increase the amount of food available to earthworms (Brown et al. 1999, Scheu 2003).

ALLELOCHEMICAL AND BIOLOGICAL EFFECTS ON PLANT HEALTH AND COMMUNITIES

Interactions among earthworms and plants are intense and involve a rather diverse range of mechanisms. Plants' growth and resistance to parasites are improved in the presence of earthworms. Their communities also may be affected by the selective effect of earthworms on the germination of the soil seed bank. Several hundreds of laboratory and field experiments have shown significant increases in plant production in over 70% of cases (Brown et al. 1999, Scheu 2003). The sense and intensity of this effect vary with plant and earthworm species. Shoot and grain productions are generally significantly enhanced, while root production remains unaffected or decreases (Brown et al. 1999). Effects generally are greater in poor than in fertile soils. This supports the hypothesis that earthworm effects are constant and proportional to their overall activity; their contribution is less visible when plant production is not limited by soil constraints. Five mechanisms likely explain earthworm effects:

the release of nutrients in fresh casts and their uptake by fine roots;
favorable effects on soil physical properties;
enhanced activities of mutualist microorganisms, mycorhizae, and N-fixing bacteria;
direct protection from belowground parasites; indirect protection from aboveground parasites;
hormone-like effects on plant growth.

Allelochemical effects are clearly involved in the last mechanism and likely operating to some extent in mechanisms 3 and 4.

Recently, Blouin et al. (2006) have shown that the enhancement of rice growth in the presence of *Reginaldia omodeoi* was due neither to an enhanced nutrient availability, nor to any change in soil physical properties. Plants received different amounts of mineral-N fertilizer, from 0 to $1600 \mu mol\,l^{-1}$. In the presence of earthworms, a rather constant increase was observed, whatever the mineral-N concentration. Since the experiment did not allow parasites or specific root mutualists to act, and because no limitation in water availability or other nutrients was present, they concluded that a "hormone-like effect" probably was responsible for the observed effects (Figure 5.6). This effect, first mentioned by Tomati et al. (1988), has been found in *Eisenia fetida* lombricompost extracts (Atiyeh et al. 2002, Arancon et al. 2003).

Earthworm effect therefore is more than a simple indirect effect of their physical engineering activities on plants. This was shown again in an experiment where rice plants (*Oryza* sativa) had been infested with a cyst-forming nematode with or without earthworms (*R. omodeoi*) in the soil (Blouin et al. 2005). Earthworm activities changed the expression of stress-responsive genes in the leaves of rice plants and allowed them to become tolerant instead of drying out, as was observed when earthworms were absent (Blouin et al. 2005). This systemic response of plants to earthworm activities recently has been confirmed with *Arabidopsis thaliana* interacting with the *Lumbricidae Aporrectodea caliginosa* (U. Jana, A. Reppelin, Y. Zuily-Fodil unpublished data). It is an indication that highly sophisticated communication and interactions among earthworms and plants have been selected by evolution. The exact nature of the interaction—the signal molecules likely involved and their origin (produced by the earthworm or by specific microbes activated by the earthworm)—is not known.

Another example of a systemic response of plants to earthworm activities is observed in tea plants restored with the FBO (Fertilisation Bio-Organique) patented method in South China (Senapati et al. 1999, P. Lavelle, J. Dai, E. Velasquez, and N. Ruiz-Camacho unpublished data).

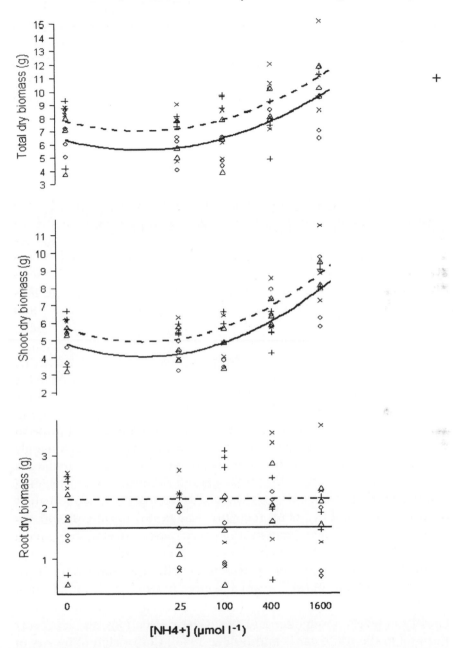

Figure 5.6 Response of rice (*Oriza sativa*) to increasing inputs of mineral nitrogen, in the presence (dotted line) and absence (solid line) of earthworms. The effect of earthworms on plant growth (distance between the two curves) is constant whatever the nutrient status of soil, which allows rejecting the hypothesis of enhanced mineralization to explain the gain in plant growth observed in their presence.

Following the inoculation of earthworms and stimulation of their activities by organic amendments in soil, tea quality evaluated by systematic tasting assessment was significantly improved.

At the larger scale of a pasture plot, several studies have shown that earthworms have significant effects on the germination of seed banks (Decaëns et al. 2003, Milcu et al. 2006). Other examples show how earthworms and other soil organisms may influence the composition of plant communities and their natural successions through different effects (Bernier and Ponge 1994, De Deyn et al. 2003).

5.4 ➤ HARNESSING THE DRILOSPHERE TO RESTORE ECOSYSTEM FUNCTIONS IN DEGRADED SOILS

Drilospheres are one of several self-organized systems operated by ecosystem engineers that drive soil function. They have significant effects on soil-based supporting and regulating ecosystem services, especially plant production, carbon sequestration, and water infiltration and storage (Lavelle et al. 2006). Although drilospheric effects on organic matter dynamics are complex, and may have opposite directions depending on scales and specific organic fractions, there is some evidence that earthworm activities have positive effects on carbon sequestration in the long term (Lavelle and Spain 2006). Much research still is required, however, to answer this critical issue.

Drilospheric effects on plant growth and health and on soil physical properties are much better understood. Earthworms are legitimately considered important actors in the maintenance of adequate hydraulic properties in the upper 20 cm of soils, where infiltration and most detoxification processes operate. For these reasons, earthworms and their drilospheres long have been recognized as useful resources, and the potential for their management in agroecosystems is vast and diverse (Lavelle et al. 1999, Jimenez and Thomas 2001).

Soil degradation is most often associated with a depletion in biodiversity and abundance of earthworm and other invertebrate communities. Earthworms are greatly sensitive to land use intensification at plot and landscape levels. Ploughing and pesticide applications are especially harmful to them (Edwards and Bohlen 1996). Conversion of forests to pastures and cropped land and habitat fragmentation may eliminate a large proportion of native species where they still exist, although with variable and still poorly understood patterns (Fragoso et al. 1997, Lavelle and Lapied 2003, Hendrix et al. 2006). Natives are partly replaced by communities of exotic species, less than 50 species that form similar assemblages worldwide in comparable environment conditions. We thus are losing at a very fast rate the extraordinary diversity of native communities generated by exceptional rates of endemism in highly sedentary organisms. In Amazonia, for example, the average ratio of local species richness to regional richness has been estimated at approximately 1%, compared to 20–30% on average for ants and termites and 80% for Sphingidae moths (Lavelle and Lapied 2004).

Communities of peregrine species or locally adapted species in turn tend to disappear when soil management makes the environment too difficult for them to survive. They suffer most from a lack of organic resources on which to feed, frequent destruction of their populations and habitats by ploughing, poisoning by pesticides, and water stress in soils when reduced plant and litter covers are reduced. In these soils, most ecosystem services associated with drilospheric activities tend to decline, even plant production that may require increasing amounts of chemical and other inputs to achieve the same crop yield level.

Management of ecosystem engineers is an important option in ecosystem restoration (Byers et al. 2006). Reconstitution of drilospheres,[2] particularly, is an action to consider when re-creating or restoring soils, and several options already have been proposed to achieve this purpose (Senapati et al. 1999). The FBO (Fertilisation Bio Organique) patented method[3] used in tree plantations creates hot spots of high fertility where organic residues of different qualities are buried in a specific order in the soil and inoculated with appropriate earthworms. A recent application of this method in India and China has allowed biodiversity of invertebrate communities and soil aggregation to significantly improve, while the organic tea thus produced had a significantly improved gustative quality (Pradeep Panigrahi unpublished data; Patrick Lavelle, Jun Dai, Nuria Ruiz-Camacho, Elena Velasquez unpublished data).

In general, drilosphere establishment first requires a restoration of organic inputs that provide adequate and sufficient feeding resources for the earthworms (Lavelle et al. 2001). Maintenance of permanent plant covers and organic amendments are practices that allow achieving, at least partly, this objective, provided the quality and location of organic materials are adequate for the earthworm species present. Such feeding resources eventually will allow local relict populations or inoculated earthworms to develop their digestive interactions with local microorganisms and increase their population density.

The quality of biological interactions within the new drilosphere thus created must be considered. In some cases, earthworm inoculation does

[2] This process should not, however, be confounded with vermicomposting, which is the transformation of raw organic matter into a high-value compost by the Lumbricidae earthworm *Eisenia fetida*. This process is done outside the soil, and these worms are epigeics that cannot dig the soil.

[3] PCT/FR 97/01363 for Sri Lanka; W0 98/03447.

fail, probably due to the inability of earthworms to adapt a microbial community that is too different from that of their original soil. This has been observed, for example, by Gilot-Villenave (1994), who showed that adult worms from the species *R. omodeoi* taken from a savannah site would not survive if transplanted in soil of an adjacent gallery forest; young worms issued from cocoons produced in savannah would survive, however, if they did hatch in the forest soil and interact from the beginning with the local soil microflora.

Once the interactions of organisms inside the drilosphere are re-established, the system will start to function and expand as biogenic structures are created. Interactions with other functional domains (drilospheres of other earthworm species, rhizospheres) restore the essential mechanisms that allow soils to provide the large range of functions used by human populations as ecosystem services.

Interactions of earthworms with plant roots are another important process to consider at that stage. It is likely that the topology of root systems that greatly differs among plant species has much to do with plant response to earthworm activities. Plants that have dense systems with a large proportion of very fine roots—e.g., the well-known tropical American plant *Bixa orelana* L. used by American Indians for their traditional face paintings—best respond to earthworm inoculations; conversely, plants with short systems or rather thick roots, such as the palm tree *Bactris gasipaes* Kunth, have limited responses (Brown et al. 1999).

5.5 ✧ CONCLUSION

The recognition that earthworms are key operators of self-organized systems (SOS) in soils has important theoretical and practical implications. As regards soil ecology theory, we found a clear correspondence between the main characteristics of drilospheres and those of SOS as precisely defined by Perry (1995). This means that other characteristics of SOS that are more difficult to observe or assess also may be applicable to drilospheres and should be explored.

The shape and localization in soils of system spatial boundaries (for example, the limits among drilospheres and rhizospheres), the existence of discrete time boundaries at which different SOS interact (for example, critical stages during successional or invasive processes where earthworms trigger massive nutrient releases from organic reserves; Bernier and Ponge 1994, McLean and Parkinson 1997), the exact nature of the hierarchical organization of SOS in soils, and the place of drilospheres in them all are research topics that should be addressed in the future in order to better understand soil ecological function.

There is also a great need to depict and understand the nature of interactions among different SOS in soil. Interactions of soil invertebrates with plants and the vast domain of belowground–aboveground interactions are another research field that still is in its infancy and requires increased research efforts (Hooper et al. 2000, Blouin et al. 2006).

These new research questions should help address still-ignored mechanisms and patterns that affect soil function. We might thus find truly adapted concepts and theories for soils where current "aboveground" general theories often prove to be poorly applicable. Ecosystem engineering and self-organization are clearly levers that allow organisms to thrive in soils, and plants to grow better while having strong interactions with all soil organisms. Models based on purely trophic vertical (i.e., along food webs) or horizontal (i.e., among organisms with comparable ecological niches) interactions are unlikely to explain much of the soil function, except in soils where ecosystem engineers have been eliminated or never existed (Lavelle 2002).

On the other hand, the research field of plant–soil invertebrate interactions and roles played by microorganisms in them seems highly promising. As indicated by the SOS theory, these interactions that are critical in sustaining soil functions should take place at discrete scales of space and time that we need to discover while chemical and other mechanisms involved will be described.

The view of drilospheres as ecological "modules" that could be added when lacking, or replaced and/or repaired if damaged, has great practical consequences. To start, drilosphere restoration requires adequate environmental conditions and significant energy inputs (Senapati et al. 1999). In practical terms, there is a need to know basic soil and climate conditions that are required by species considered for reintroduction or enhancement (Barois et al. 1999, Fragoso et al. 1999). There is also a need to understand the way introduced individuals will interact with local microflora in order to establish efficient mutualist relationships (Gilot-Villenave 1994). The energy budget of the operation is also fundamental to determine how much organic matter should be brought, and in which form, to sustain drilospheric activities to a level that produces significant improvements in target ecosystem services (Lavelle et al. 2001).

Management of drilospheres, although highly promising, still requires important research and technological developments. Plant physiologists and soil fertility and soil ecology experts must coordinate their efforts to optimize the production of ecosystem services by a sustainable management of drilospheres and other soil self-organized systems. The best naturally selected or genetically modified plants will never achieve their potential for production in a degraded soil, nor will poorly productive traditional cultivars produce more in the most ecologically active soil. Conservation of soil biodiversity, water infiltration and storage, and C-sequestration also will have to be efficient in any of these systems to meet the rapidly growing demand for soil ecosystem services. In this necessary effort to optimize the provision of all soil ecosystem services at higher rates, all approaches need to be used in a comprehensive way.

REFERENCES

Arancon, N.Q., Lee, S., Edwards, C.A., and Atiyeh, R. (2003). Effects of humic acids derived from cattle, food and paper-waste vermicomposts on growth of greenhouse plants. *Pedobiologia* 47:741–744.

Atiyeh, R.M., Lee, S., Edwards, C.A., Arancon, N.Q., and Metzger, J.D. (2002). The influence of humic acids derived from earthworm-processed organic wastes on plant growth. *Bioresource Technology* 84:7–14.

Barois, I. (1987). Interactions entre les vers de terre (oligochaeta) tropicaux géophages et la microflore pour l'exploitation de la matière organique du sol, PhD thesis, Université de Paris VI.

Barois, I., and Lavelle, P. (1986). Changes in respiration rate and some physicochemical properties of a tropical soil during transit through *Pontoscolex corethrurus* (Glossoscolecidæ, Oligochæta). *Soil Biology Biochemistry* 18(5):539–541.

Barois, I., Lavelle, P., Brossard, M.J., Tondoh, M., Martinez, A., Rossi, J.P., Senapati, B.K., Angeles, A., Fragoso, C., Jimenez, J.J., Decaëns, T., Lattaud, C., Kanyonyo, K.K., Blanchart, E., Chapuis, L., Brown, G.G., and Moreno, A. (1999). Ecology of earthworm species with large environmental tolerance and/or extended distributions. In *Earthworm Management in Tropical Agroecosystems*, P. Lavelle, L. Brussaard, and P. Hendrix, Eds. Wallingford, UK: CAB International, pp. 57–86.

Barot, S., Rossi, J.P., and Lavelle, P. (in press). Self-organization in a simple consumer resource system, the example of earthworms. *Soil Biology and Biochemistry*.

Bastardie, F., Capowiez, Y., and Cluzeau, D. (2005). 3D characterisation of earthworm burrow systems in natural soil cores collected from a 12-year-old pasture. *Applied Soil Ecology* 30:34–46.

Bernier, N., and Ponge, J.F. (1994). Humus form dynamics during the sylvogenetic cycle in a mountain spruce forest. *Soil Biol. Biochemistry* 26:183–220.

Blair, J.M., Parmelee, R.W., and Lavelle, P. (1995). Influences of earthworms on biogeochemistry in North American ecosystems. In *Earthworm Ecology in Forest, Rangeland and Crop Ecosystems in North America*, P.H. Hendrix, Ed. Chelsea, MI: Lewis Publishing, pp. 127–158.

Blanchart, E., Albrecht, A., Alegre, J., Duboisset, A., Pashanasi, B., Lavelle, P., and Brussaard, L. (1999). Effects of earthworms on soil structure and physical properties. In *Earthworm Management in Tropical Agroecosystems*, P. Lavelle, L. Brussaard, and P. Hendrix, Eds. Wallingford, UK: CAB International, pp. 139–162.

Blanchart, E., Albrecht, A., Brown, G.G., Decaëns, T., Duboisset, A., Lavelle, P., Mariani, L., and Roose, E. (2004). Effects of tropical endogeic earthworms on soil erosion. *Agriculture, Ecosystems and Environment* 104:303–315.

Blanchart, E., Bruand, A., and Lavelle, P. (1993). The physical structure of casts of Millsonia anomala (Oligochaeta:Megascolecidae) in Shrub savanna soils (Côte d'Ivoire). *Geoderma* 56:19–132.

Blanchart, E., Lavelle, P., Braudeau, E., Le Bissonais, Y., and Valentin, C. (1997). Regulation of soil structure by geophagous earthworm activities in humid savannas of Côte d'Ivoire. *Soil Biol. Biochem.* 29:431–439.

Blouin, M., Barot, S., and Lavelle, P. (2006). Earthworms (*Millsonia anomala*, Megascolecidae) do not increase rice growth through enhanced nitrogen mineralization. *Soil Biol. Biochemistry* 38:2063–2068.

Blouin, M., Zuily-Fodil, Y., Pham-Thi, A.T., Laffray, D., Reversat, G., Pando, A., Tondoh, J., and Lavelle, P. (2005). Belowground organism activities affect plant aboveground phenotype, inducing plant tolerance to parasites. *Ecol Lett.* 8:202–208.

Bohlen, P.J., Edwards, C.A., Zhang, Q., Parmelee, R.W., and Allen, M. (2002). Indirect effects of earthworms on microbial assimilation of labile carbon. *Applied Soil Ecology* 20:255–261.

Bossuyt, H., Six, J., and Hendrix, P.F. (2006). Interactive effects of functionally different earthworm species on aggregation and incorporation and decomposition of newly added residue carbon. *Geoderma.* 130:14–25.

Bouché, M.B. (1977). Stratégies lombriciennes. *Ecology Bulletin* 25:122–132.

Brown, G., Pashanasi, B., Gilot-Villenave, C., Patron, J.C., Senapati, B.K., Giri, S., Barois, I., Lavelle, P., Blanchart, E., Blakemore, R.J., Spain, A.V., and Boyer, J. (1999). Effects of earthworms on plant growth in the Tropics. In *Earthworm Management in Tropical Agroecosytems*, P. Lavelle, L. Brussaard, and P. Hendrix, Eds. Wallingford, UK: CAB International, pp. 87–148.

Byers, T.E., Cuddington, K., Jones, C.G., Talley, T.S., Hastings, A., Lambrinos, J.G., Crooks, J.A., and Wilson, W.G. (2006). Using ecosystem engineers to restore ecological systems. *Trends Ecol. Evolution* 21:494–500.

Capowiez, Y., Rault, M., Costagliola, G., and Mazzia, C. (2005). Lethal and sublethal effects of imidacloprid on two earthworm species (*Aporrectodea nocturna* and *Allolobophora icterica*). *Biology and Fertility of Soils* 41:135–143.

Chapuis-Lardy, L., Brossard, M., Lavelle, P., and Laurent, J.Y. (1996). Digestion of a vertisol by an endogeic earthworm (*Polypheretima elongata*, Megascolecidae) increases soil phosphate extractibility. *European Journal of Soil Biology* 32(2):107–111.

Chauvel, A., Grimaldi, M., Barros, E., Blanchart, E., Desjardins, T., Sarrazin, M., and Lavelle, P. (1999). Pasture degradation by an Amazonian earthworm. *Nature* 389: 32–33.

Cuddington, K., and Hastings, A. (2004). Invasive engineers. *Ecol. Model.* 178:331–447.

Daniel, O., and Anderson, J.M. (1992). Microbial biomass and activity in contrasting soil materials after passage through the gut of the earthworm *Lumbricus rubellus* Hoffmeister. *Soil Biology Biochemistry* 24:465–470.

Darwin, C. (1881). *The Formation of Vegetable Mould Through the Action of Worms with Observations on Their Habits.* London: Murray.

Decaëns, T. (2000). Degradation dynamics of surface earthworm casts in grasslands of the eastern plains of Colombia. *Biol. Fertil. Soils* 32:149–156.

Decaëns, T., Mariani, L., and Lavelle, P. (1999a). Soil surface macrofaunal communities associated with earthworm casts in grasslands of the Eastern Plains of Colombia. *Applied Soil Ecology* 13:87–100.

Decaëns, T., Jiménez, J., and Lavelle, P. (1999b). Effect of exclusion of the anecic earthworm *Martiodrilus carimaguensis*—Jiménez and Moreno on soil properties and plant growth in grasslands of the eastern plains of Colombia. *Pedobiologia* 43:1–7.

Decaëns, T., Mariani, L., Betancourt, N., and Jiménez, J.J. (2003). Seed dispersion by surface casting activities of earthworms in Colombian grasslands. *Acta Oecologica* 24:175–185.

De Deyn, G.B., Raaijmakers, C.E., Zoomer, H.R., Berg, M.P., de Ruiter, P.C., Verhoef, H.A., Bezemer, T.M., and van der Putten, W.H. (2003). Soil invertebrate fauna enhances grassland succession and diversity. *Nature* 422:711–713.

Edwards, C.A., and Bohlen, P.J. (1996). *Biology and Ecology of Earthworms*, 3rd ed. London: Chapman and Hall.

Fischer, K., Hahn, D., Amann, R.I., Daniel, O., and Zeyer, J. (1995). In situ analysis of the bacterial community in the gut of the earthworm *Lumbricus terrestris* L by whole-cell hybridization. *Canadian Journal of Microbiology* 41:666–673.

Fragoso, C., Brown, G.G., Patron, J.C., Blanchart, E., Lavelle, P., Pashanasi, B., Senapati, B., and Kumar, T. (1997). Agricultural intensification, soil biodiversity and agroecosystem function in the tropics: The role of earthworms. *App. Soil Ecol.* 6:7–37.

Fragoso, C., Kanyonyo ka Kajondo, J., Blanchart, E., Moreno, A., Senapati, B., Rodriguez, C. (1999). A survey of tropical earthworms: Taxonomy, biogeography and environmental plasticity. In *Earthworm Management in Tropical Agroecosystems*, P. Lavelle, L. Brussaard, and P. Hendrix, Eds. Wallingford, UK: CAB International, pp. 1–27.

Garvin, M.H., Lattaud, C., Trigo, D., and Lavelle, P. (2000). Activity of glycolytic enzymes in the gut of *Hormogaster elisae* (Oligochaeta, Hormogastridae). *Soil Biology and Biochemistry* 32:929–934.

Gilot-Villenave, C. (1994). Determination of the origin of the different growing abilities of two populations of *Millsonia anomala* (Omodeo and Vaillaud), a tropical geophagous earthworm. *European Journal of Soil Biology* 39:125–131.

Hallaire, V., Curmi, P., Duboisset, A., Lavelle, P., Pashanasi, B. (2000). Soil structure changes induced by the tropical earthworm *Pontoscolex corethrurus* and organic inputs in a Peruvian ultisol. *European Journal of Soil Biology* 36:35–44.

Hamilton, W.E., and Sillman, D.Y. (1989). Influence of earthworm middens on the distribution of soil microarthropods. *Biol. Fertil. Soils* 8:279–284.

Hendrix, P.F., Baker, G.H., Callaham, M.A., Damoff, G.A., Fragoso, C., Gonzalez, G., James, S.W., Lachnicht, S.L., Winsome, T., and Zou, X. (2006). Invasion of exotic earthworms into ecosystems inhabited by native earthworms. *Biological Invasions* 8(6):1287–1300.

Hendrix, P.F., Parmelee, R.W., Crossley, Jr., D.A., Coleman, D.C., Odum, E.P., and Groffman, P.M. (1986). Detritus food webs in conventional and non-tillage agroecosystems. *BioSci.* 36:374–380.

Hooper, D.U., Bignell, D.E., Brown, V.K., Brussaard, L., Dangerfield, J.M., Wall, D.H., Wardle, D.A., Coleman, D.C., Giller, K.E., Lavelle, P., Van der Putten, W.H., De Ruiter,

P.C., Rusek, J., Silver, W.L., Tiedje, J.M., and Wolters, V. (2000). Interactions between aboveground and belowground biodiversity in terrestrial ecosystems: Patterns, mechanisms, and feedbacks. *BioScience* 50:1049–1061.

Jimenez, J.J., and Thomas, R. (Eds.). (2001). *The Nature's Plow: Soil Macroinvertebrate Communities in the Neotropical Savannas of Colombia.* Cali, Colombia: CIAT.

Jones, C.G., Lawton, J.H., and Shachak, M. (1994). Organisms as ecosystem engineers. *Oikos* 69:373–386.

Jones, C.G., Lawton, J.H., and Shachak, M. (1997). Positive and negative effects of organisms as physical ecosystem engineers. *Ecology* 78:1946–1957.

Jouquet, P., Lagerlof, D.J., Lavelle, P., and Lepage, M. (2006). Soil invertebrates as ecosystem engineers: Intended and accidental effects on soil and feedback loops. *Applied Soil Ecology* 32:153–164.

Karsten, G.R., and Drake, H.L. (1997). Denitrifying bacteria in the earthworm gastrointestinal tract and in vivo emission of nitrous oxide (N_2O) by earthworms. *Applied Environmental Microbiology* 63:1878–1882.

Kersante, A., Martin-Laurent, F., Soulas, G., and Binet, F. (2006). Interactions of earthworms with Atrazine-degrading bacteria in an agricultural soil. *FEMS Microbiology Ecology* 57:192–205.

Knollenberg, W.G., Merritt, R.W., and Lawson, D.L. (1985). Consumption of leaf litter by *Lumbricus terrestris* (Oligochaeta) on a Michigan woodland floodplain. *American Midland Naturalist* 113:1–6.

Kretzschmar, A. (1990). Experimental burrow system: Pathway patterns and building behavior for the earthworm *Aporrectodea longa. Review of Ecol. Biol. Sol* 27:299–306.

Kretzschmar, A. (1991). Burrowing ability of the earthworm *Aporrectodea longa* limited by soil compaction and water potential. *Biol. Fert. Soils* 11:48–51.

Lamparski, F., Kobel-Lamparski, A., and Kaffenberger, R. (1987). The burrows of *Lumbricus badensis* and *Lumbricus polyphemus*. In *On Earthworms*, A.M. Bonvicini Pagliai and P. Omodeo, Eds. Modena: Mucchi Editore, pp. 131–140.

Lapied, E., and Lavelle, P. (2003). The peregrine earthworm *Pontoscolex corethrurus* in the east coast of Costa Rica. *Pedobiologia* 47:471–474.

Lattaud, C., Mora, P., Garvin, M., Locati, S., and Rouland, C. (1999). Enzymatic digestive capabilities in geophagous earthworms—origin and activities of cellulolytic enzymes. *Pedobiologia* 43:842–850.

Lattaud, C., Zhang, B.G., Locati, S., Rouland, C., and Lavelle, P. (1997). Activities of the digestive enzymes in the gut and in tissue culture of a tropical geophagous earthworm, *Polypheretima elongata* (Megascolecidae). *Soil Biology Biochemistry* 29:335–339.

Lavelle, P. (1978). *Les Vers de Terre de la Savane de Lamto (Côte d'Ivoire): Peuplements, Populations et Fonctions dans L'écosystème.* Thèse d'Etat. Université de Paris VI. Publication du Laboratoire de Zoologie de l'ENS.

Lavelle, P. (1983). The structure of earthworm communities. In *Earthworm Ecology: From Darwin to Vermiculture*, J.E. Satchell, Ed. London: Chapman & Hall, pp. 449–466.

Lavelle, P. (2002). Functional domains in soils. *Ecol. Res.* 17:441–450.

Lavelle, P., Aronson, J., Lowry, P., and Tongway, D. (2005). Unpublished data shown in Figure 5.4 in this text.

Lavelle, P., Barois, I., Fragoso, C., Cruz, C., Hernandez, A., Pineda, A., and Rangel, P. (1987). Adaptative strategies of *Pontoscolex corethrurus* (Glossoscolecidæ, Oligochæta), a peregrine geophagous earthworm of the humid tropics. *Biol. Fert. Soils* 5:188–194.

Lavelle, P., Barros, E., Blanchart, E., Brown, G., Desjardins, T., Mariani, L., and Rossi, J.P. (2001). Soil organic matter management in the tropics: Why feeding the soil macrofauna? *Nutr. Cycl. Agro* 61:53–61.

Lavelle, P., Blouin, M., Boyer, J., Cadet, P., Laffray, D., Pham-Thi, A.T., Reversat, G., Settle, W.H., and Zuily-Fodil, Y. (2004). Plant parasite control and soil fauna diversity. *C.R. Biologies* 327:629–638.

Lavelle, P., Brussaard, L., and Hendrix. P. (Eds.). (1999). *Earthworm Management in Tropical Agroecosystems.* Wallingford, UK: CAB-International.

Lavelle, P., Decaëns, T., Aubert, M., Barot, S., Blouin, M., Bureau, F., Margerie, F., Mora, P., and Rossi, J.P. (2006). Soil invertebrates and ecosystem services. *European Journal of Soil Biology* 42:S3–S15.

Lavelle, P., and Lapied, E. (2003). Endangered earthworms of Amazonia. *Pedobiologia* 47:419–427.

Lavelle, P., Lattaud, C., Trigo, D., and Barois, I. (1995). Mutualism and biodiversity in soils. In *The Significance and Regulation of Soil Biodiversity*, H.P. Collins, G.P. Robertson, and M.J. Klug, Eds. Dordrecht, The Netherlands: Kluwer Academic Publishers, pp. 23–33.

Lavelle, P., Melendez, G., Pashanasi, B., and Schaefer, R. (1992). Nitrogen mineralization and reorganization in casts of the geophagous tropical earthworm *Pontoscolex corethrurus* (Glossoscolecidae). *Biol. Fertil. Soils* 14:49–53.

Lavelle, P., Rouland, C., Binet, F., Diouf, M., and Kersanté, A. (2005). Regulation of microbial activities by roots and soil invertebrates. In *Microorganisms in Soils: Roles, Genesis and Functions*, Soil Biology series 3, F. Buscot and A. Varma, Eds. Berlin: Springer Verlag, pp. 291–305.

Lavelle, P., and Spain, A.V. (2006). *Soil Ecology*, 2nd ed. Amsterdam: Kluwer Scientific Publications.

Lavelle, P., Velasquez, E., Rendeiro, C., and Martins, M. Unpublished data shown in Figure 5.3 in this text.

Le Bayon, R.C., and Binet, F. (1999). Rainfall effects on erosion of earthworm casts and phosphorus transfers by water runoff. *Biol. Fert. Soil.* 30:7–13.

Lee, K.E. (1985). *Earthworms: Their Ecology and Relationships with Soils and Land Use.* Sydney: Academic Press.

Ligthart, T.N., Peek, G.J.W.C., and Taber, E.J. (1993). A method for the three-dimensional mapping of earthworm burrow systems. *Geoderma* 57:129–141.

Lopez-Hernandez, D., Fardeau, J.C., and Lavelle, P. (1993). Phosphorus transformations in two P-sorption contrasting tropical soils during transit through *Pontoscolex corethrurus* (Glossoscolecidae, Oligochaeta). *Soil Biology Biochemistry* 25:789–792.

Loranger, G., Ponge, J.F., Blanchart, E., and Lavelle, P. (1998). Impact of earthworms on the diversity of microarthropods in a vertisol (Martinique). *Biol. Fert. Soils* 27:21–26.

McLean, M.A., and Parkinson, D. (1997). Soil impacts of the epigeic earthworm *Dendrobaena octaedra* on organic matter and microbial activity in lodge pole pine forest. *Canadian Journal of Forest Restoration* 27:1907–1913.

Mariani, L., Bernier, N., and Jimenez, J. (2001). Régime alimentaire d'un ver de terre anécique des savanes colombiennes. Une remise en question des types écologiques. *C. R. Biologie*. 324:733–742.

Marinissen, J.C.Y., and Bok, J. (1988). Earthworm-amended soil structure: Its influence on Collembola populations in grassland. *Pedobiologia* 32:243–252.

Marinissen, J.C.Y., and Dexter, A.R. (1990). Mechanisms of stabilization of earthworm casts and artificial casts. *Biol. Fert. Soil* 9:163–167.

Martin, A. (1991). Short- and long-term effects of the endogeic earthworm *Millsonia anomala* (Omodeo) (Megascolecidae, Oligochaeta) of tropical savannas, on soil organic matter. *Biol. Fertil Soils* 11:234–238.

Martin, A., Cortez, J., Barois, I., and Lavelle, P. (1987). Les mucus intestinaux de Ver de Terre, moteur de leurs interactions avec la microflore. *Rev. Ecol. Biol. Sol* 24:549–558.

Milcu, A., Schumacher, J., Scheu, S. (2006). Earthworms (*Lumbricus terrestris*) affect plant seedling recruitment and microhabitat heterogeneity. *Functional Ecology* 20:261–268.

Nahmani, J., Capowiez, Y., and Lavelle, P. (2005). Effects of metal pollution on soil macroinvertebrate burrow systems. *Biol. Fertil. Soils* 42:31–39.

Nooren, C.A.M., Breemen, N., van Stoorvogel, J.J., and Jongmans, A.G. (1995). The role of earthworms in the formation of sandy surface soils in a tropical forest in Ivory Coast. *Geoderma* 65:135–148.

Parle, J.N. (1963). Micro-organisms in the intestines of earthworms. *Journal of Gen. Microbiology* 31:1–11.

Perry, D.A. (1995). Self-organizing systems across scales. *Trends Res. Ecol. Evol.* 10:241–245.

Rice, E.L. (1984). *Allelopathy*, 2nd ed. Orlando, FL: Academic Press.

Rose, C.J., and Wood, A.W. (1980). Some environmental factors affecting earthworm populations and sweet potato production in the Tari Basin, Papua New Guinea highlands. *Papua New Guinea Agricultural Journal* 31:1–13.

Rossi, J.P. (2003). The spatiotemporal pattern of a tropical earthworm species assemblage and its relationship with soil structure. *Pedobiologia* 47:497–503.

Satchell, J.E. (1967). Lumbricidæ. In *Soil Biology*, A. Burges and F. Raw, Eds. London: Academic Press, pp. 259–322.

Scheu, S. (1987). Microbial activity and nutrient dynamics in earthworm casts (Lumbricidae). *Biol. Fert. Soil* 5:230–234.

Scheu, S. (2003). Effects of earthworms on plant growth: Patterns and perspectives. *Pedobiologia* 47:846–856.

Scheu, S., and Wolters, V. (1991). Influence of fragmentation and bioturbation on the decomposition of C-14-labelled beech leaf litter. *Soil Biology Biochemistry* 23:1029–1034.

Senapati, B.K., Lavelle, P., Giri, S., Pashanasi, B., Alegre, J., Decaëns, T., Jimenez, J.J., Albrecht, A., Blanchart, E., Mahieux, M., Rousseaux, L., Thomas, R., Panigrahi, P., and Venkatachalam, M. (1999). Soil earthworm technologies for tropical agroecosystems. In *The Management of Earthworms in Tropical Agroecosytems*, P. Lavelle, L. Brussaard, and P. Hendrix, Eds. Wallingford, UK: CAB International, pp. 189–227.

Shipitalo, M.J., and Protz, R. (1988). Factors influencing the dispersibility of clay in worm casts. *Soil Science Society of America Journal* 52:764–769.

Sterner, R.W., and Elser, J.J. (2002). *Ecological stoichiometry: The Biology of Elements from Molecules to the Biosphere*. Princeton, NJ: Princeton University Press.

Subler, S., and Kirsch, A.S. (1998). Spring dynamics of soil carbon, nitrogen, and microbial activity in earthworm middens in a no-till cornfield. *Biol. Fertil. Soils* 26:243–249.

Swift, M.J., Heal, O.W., and Anderson, J.M. (1979). *Decomposition in Terrestrial Ecosystems*. Oxford: Blackwell Scientific.

Tiunov, A.V., Bonkowski, M., Alphei, J., and Scheu, S. (2001). Microflora, protozoa and nematoda in *Lumbricus terrestris* burrow walls: A laboratory experiment. *Pedobiologia* 45:46–60.

Tomati, U., Grappeli, A., and Galli, E. (1988). The hormone-like effect of earthworm casts on plant growth. *Biol. Fertil. Soil* 5:288–294.

Toutain, F. (1987). Activité biologique des sols, modalités et lithodépendance. *Biol. Fertil. Soils* 3:31–38.

Velasquez, E., Pelosi, C., Brunet, D., Grimaldi, M., Martins, M., Rendeiro, A.C., Barrios, E., and Lavelle, P. (2007). This ped is my ped: Visual separation and NIRS spectra allow determination of the origins of soil macro-aggregates. *Pedobiologia* 51: 75–87.

Winding, A., Ronn, R., and Hendriksen, N.B. (1997). Bacteria and protozoa in soil microhabitats as affected by earthworms. *Biol. Fert. Soil* 24:133–140.

Yeates, G.W. (1981). Soil nematode populations depressed in the presence of earthworms. *Pedobiologia* 22:191–198.

Zhang, B.G., Li, G.T., Shen, T.S., Wang, J.K., and Sun, Z. (2000). Changes in microbial biomass C, N, and P and enzyme activities in soil incubated with the earthworms *Metaphire guillelmi* or *Eisenia fetida*. *Soil Biol. Biochemistry* 32:2055–2062.

Zhang, B.G., Rouland, C., Lattaud, C., and Lavelle, P. (1993). Activity and origin of digestive enzymes in gut of the tropical earthworm *Pontoscolex corethrurus*. *European Journal of Soil Biology* 29:7–11.

6

MICROHABITAT MANIPULATION: ECOSYSTEM ENGINEERING BY SHELTER-BUILDING INSECTS

John T. Lill and Robert J. Marquis

6.1 ⬤ INTRODUCTION

Plant-feeding insects that construct shelters on their food plants provide ample opportunities for examining the impacts of allogenic ecosystem engineering on nature's most diverse group of organisms, the arthropods. Shelters serve as habitats for a variety of plant-dwelling arthropods that exploit a range of available resources within these constructs. The small size and somewhat ephemeral nature of these constructs at first may suggest their effects on animal community composition and ecosystem processes are trivial (see criteria in Jones et al. 1997). A growing body of observational and experimental studies, however, demonstrates that shelter-builders are "microhabitat manipulators" that construct a large number of structures that are sufficiently persistent to permit colonization by a wide variety of secondary inhabitants. For example more than 25 species of arthropods from nine different orders have been recorded inside leaf rolls constructed on cottonwood trees (Martinsen et al. 2000). Thus, shelter-builders can play a pivotal role in structuring arthropod communities. Unlike the constructs of larger animals, whose engineering behaviors are difficult to simulate, leaf shelters and their occupants are easily manipulated. As such, they provide an ideal system for testing some of the more general theories regarding the scaling of engineering impacts on species richness (Wright et al. 2002, Castilla et al. 2004), the integration of trophic and engineering impacts (Wilby et al. 2001, Wright and Jones 2006), and the influence of engineering on community-level interactions or "interaction networks" (Proulx et al. 2005).

6.2 ⬥ SHELTERS AND SHELTER-BUILDERS

A wide variety of arthropods, including spiders, caterpillars, sawflies, weaver ants, thrips, tree crickets, and beetles, use plant foliage to construct domiciles, partially or fully enclosed retreats within which they reside for all or a portion of their life (Wagner and Raffa 1993, Berenbaum 1999, Taylor and Jackson 1999, Anderson and McShea 2001, Fukui 2001, Marquis and Lill 2006). These constructs can take the form of leaf rolls, leaf webs (clusters of leaves connected by or enveloped in silk), leaf ties (sandwiches of overlapping leaves sewn together with silk), leaf folds (folding all or part of a leaf onto itself, held fast with silk), and leaf tents (e.g., the constructs of some skipper larvae, in which a flap of leaf is cut, folded over, and fastened to the main leaf surface with silken "guy wires"; Lind et al. 2001). In addition to these leaf shelters, many other groups of arthropods create habitats on or in plants by consuming plant tissues (i.e., gall formers, leaf miners, and various types of borers) (Marquis and Lill 2006). While we make occasional reference to the literature examining the engineering effects of these internal feeders, this chapter focuses primarily on the externally feeding insects that use silk to construct shelters on plant foliage.

The shelter-building habit is perhaps most widespread within the Lepidoptera (Scoble 1992), in which the larvae of species in at least 24 families use silk to construct shelters out of live foliage (Jones 1999). While widespread within the order, most engineering species are microlepidoptera (Powell 1980). Several species of gregariously feeding sawfly larvae (Hymenoptera: Symphyta) in the family Pamphiliidae (e.g., species of *Neurotoma, Cephalacia,* and *Acantholyda;* Wagner and Raffa 1993) also use silk to form shelters out of foliage in an analogous manner. We refer to these lepidopteran and sawfly species collectively as leaf *shelter-builders.* Representatives of several other lepidopteran families (e.g., Incurvariidae, Psychidae, and Tineidae) use silk, leaf material, and sometimes frass to make "portable" shelters; we group these species with larvae that construct cases (e.g., larvae in the Coleophoridae) and larvae that hide under silk mats or within frass tubes on the surface of plant foliage into a guild we call *concealed feeders.* The distinction between shelter-builders and concealed feeders is that species in the latter category typically build shelters sufficiently large to house themselves alone. As such, these constructs are likely to provide relatively little in the way of usable habitat for other organisms. In contrast, the constructs of shelter-builders are often (but not always) available for colonization by other arthropods.

To our knowledge, the feeding strategies of regional Lepidoptera faunas have been characterized in only two regions: Great Britain and Canada. In Great Britain, the microlepidoptera constitute approximately 60% of all species (Table 3 of Gaston et al. 1992). Among the microlepidoptera, 40% of the >1100 species construct leaf shelters; this amounts to approximately 25% of all the species of Lepidoptera endemic to Britain, not including the shelter-building macrolepidoptera. The second regional example comes from the data collected by the Canadian Forest Insect Survey (McGugan 1958; Prentice 1962, 1963, 1965). Among the more than 950 species characterized by the survey, shelter-builders comprise approximately 20%. Fifty percent of all microlepidoptera (dominated by the families Tortricidae, Gelechiidae, Oecophoridae, and Pyralidae) are shelter-builders. Among the Canadian macrolepidoptera, shelter-builders are most common in the families Hesperiidae (skippers) and Lasiocampidae (tent caterpillars), and also include a few species of Nymphalidae, Notodontidae, Noctuidae, and Arctiidae. At a more local scale, shelter-building caterpillars comprise approximately 25% of the leaf-chewing herbivore species on oaks (*Quercus* sp.) in Missouri (R.J. Marquis unpublished data) and 28% of the species on American beech (*Fagus grandifolia*) in Maryland (J.T. Lill unpublished data).

In evaluating the importance of shelter building in structuring arthropod communities, the abundance, size, and spatial distribution of shelter-builders and their constructs within plant canopies are likely to be more important than their species richness. Unfortunately, relatively few large-scale studies of herbivore communities have been conducted that include counts of microlepidoptera (but see Diniz and Morais 1997, Marquis et al. 2000, Forkner et al. 2006); the majority of community studies have focused on externally feeding macrolepidoptera, which are generally much easier to identify and count. To begin to address this issue, we present here a portion of our own data, compiled from multiple studies of temperate forest trees, each of which involved the same sampling method (visual sampling of fixed quantities of understory foliage), but which varied in the identity of the focal host plant. Densities of shelter-builders, expressed as a percentage of the leaf-chewing fauna recorded in a given census, were quite variable among host plant species and censuses (Figure 6.1). It is evident from these data, however, that these shelter-building insects, though small in size, are nonetheless

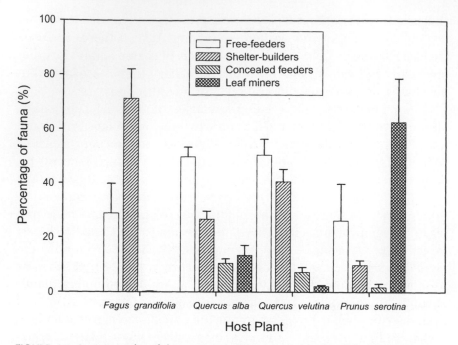

FIGURE 6.1 Summary plot of the mean (±1 SE) percentage of leaf-chewing herbivores recorded in each of four feeding guilds on foliage of four understory tree species. Bars are means of multiple visual censuses conducted over several years (N = 6 censuses, *F. grandifolia*; N = 12, *Q. alba*; N = 12, *Q. velutina*; N = 5, *P. serotina*). Data are compiled from multiple studies conducted by the authors in Missouri, Maryland, and Washington, D.C. Mean total herbivores (±1 SE) recorded per census = 1056 ± 410 (*F. grandifolia*); 2668 ± 364 (*Q. alba*); 662 ± 91 (*Q. velutina*); 740 ± 99 (*P. serotina*).

important (and sometimes dominant) components of the herbivore faunas of these tree species. Similar results have been reported for other tree species, including willows (*Salix* spp.) in Japan (Nakamura and Ohgushi 2004), cottonwoods (*Populus* sp.) in the U.S. (Furniss and Carolin 1977), and at least one species of tropical tree (*Xylopia aromatica*) in Brazil (Barosela 1999, Costa and Varanda 2002), all of which routinely support very high relative abundances of shelter-builders. In addition, many species of shelter-builders are important economic pests of a variety of timber, fruit tree, and crop plants and periodically undergo outbreaks, dramatically increasing the availability of shelter habitats (Table 6.1). If these patterns are representative of tree species in general, it suggests that the microhabitats constructed by these high-density insects are a widespread and dependable feature in forested environments, factors which should increase their "importance value" as ecosystem engineers (Jones et al. 1994).

TABLE 6.1 Common shelter-building crop and forest pests.

Common Name	Species	Family	Plants Attacked	Reference
Pine false webworm	*Acantholyda erythrocephala* (L.)	Phamphiliidae	Various pine species	Baker (1972)
Blueberry leaftier	*Acleris curvalana* (Kearfott)	Tortricidae	Blueberry	Dixon and Carl (2002)
Oak leaftier	*Acleris semipurpurana* (Kearfott)	Tortricidae	Oaks	Morin et al. (2004)
Strawberry leafroller	*Anclyis comptana* (Froelich)	Tortricidae	Strawberry	Landis et al. (2005)
Fruit tree leafroller	*Archips argyrospila* (Walker)	Tortricidae	Fruit trees, bald cypress	Moore et al. (1982), Dreistadt et al. (1994)
Uglynest caterpillar	*Archips cerasivorana* (Fitch)	Tortricidae	Cherry trees	Douglas and Cowles (2006)
Boxelder leafroller	*Archips negundana* (Dyar)	Tortricidae	Manitoba maple	Ives and Wong (1988)
Oak leafroller	*Archips semiferana* (Walker)	Tortricidae	Oaks	Baker (1972)
Hickory leafroller	*Argyrotaenia juglandana* (Fernald)	Tortricidae	Hickory trees	Baker (1972)
Redbanded leafroller	*Argyrotaenia velutinana* (Walker)	Tortricidae	Apple trees	Pfeiffer et al. (1993)
Jack pine budworm	*Choristoneura pinus* (Freeman)	Tortricidae	Jack pine	Conway et al. (1999)
Large aspen tortrix	*Choristoneura conflictana* (Walker)	Tortricidae	Aspen trees	Beckwith (1973)
Spruce budworm	*Choristoneura fumiferana* (Clemens)	Tortricidae	Spruce, larch, hemlock	Simard and Payette (2005)
Oblique-banded leafroller	*Choristoneura rosaceana* (Harris)	Tortricidae	Apple trees	Trimble and Appleby (2004)
Rice leaf-folder	*Cnaphalocrucis medinalis* (Guenee)	Pyralidae	Cultivated rice	Fraenkel and Fallil (1981)
Cabbage webworm	*Crocidolomia binotalis* (Zeller)	Pyralidae	India mustard	Singh and Rawat (1979)
Palmer-worm	*Dichomeris ligulella* (Hbn.)	Gelechiidae	Oaks	Baker (1972)
Aspen two-leaf tier	*Enargia decolor* (Walker)	Noctuidae	Trembling aspen	Jones et al. (1996)
Spruce needle-miner	*Endothenia albolineana* (Kearfott)	Tortricidae	Spruce	Ives and Wong (1988)
Light brown apple moth	*Epiphyas postvittana* (Walker)	Tortricidae	Pome fruit, grapes	Danthanarayana (1983)
Turf grass webworm	*Herpetogramma licarsisalis* (Walker)	Pyralidae	Grasses	Murdoch et al. (1990)

TABLE 6.1 *Continued*

Common Name	Species	Family	Plants Attacked	Reference
Mimosa webworm	*Homadaula anisocentra* (Meyr.)	Plutellidae	Mimosa and honey locust	Bastian and Hart (1991)
Fall webworm	*Hyphantria cunea* (Drury)	Arctiidae	Variety of tree species	Baker (1972)
Poplar tentmaker	*Ichthyura inclusa* (Hbn.)	Notodontidae	Poplar and willow	Baker (1972)
Alfalfa webworm	*Loxostege commixtalis* (Walker)	Pyralidae	Alfalfa	Capinera et al. (1981)
Eastern tent caterpillar	*Malacosoma americana* (Fab.)	Lasiocampidae	Various cherry species	Carmona and Barbosa (1983)
Forest tent caterpillar	*Malacosoma disstria* (Hbn.)	Lasiocampidae	Variety of tree species	Roland (1993)
Turf webworm	*Maruca tesulalis* (Geyer)	Pyralidae	Turf grasses	Vishakantaiah and Jagdeesh Babu (1981)
Plum spinning webworm	*Neurotoma inconspicua* (Norton)	Pamphiliidae	Cherry species	Baker (1972)
Bruce spanworm	*Operophtera bruceata* (Hulst)	Geometridae	Maples, oaks, cherries	Moskal and Franklin (2004)
Apple pandemis	*Pandemis pyrusana* (Kearfott)	Tortricidae	Fruit trees	Miliczky and Calkins (2002)
Variegated leafroller	*Platynota flavedana* (Clemens)	Tortricidae	Fruit trees, blackberry, strawberry	Pfeiffer et al. (1993)
Beet webworm	*Pyrausta sticticalis* (L.)	Pyralidae	Beets	Leonova and Slynko (2004)
Nantucket pine moth	*Rhyacionia frustrana* (Comst.)	Tortricidae	Pine trees	Baker (1972)
Maple webworm	*Tetralopha asperatella* (Clemens)	Pyralidae	Maples	Horsley et al. (2002)
Pine webworm	*Tetralopha robustella* (Zeller)	Pyralidae	Pine trees	Hertel and Benjamin (1979)
Celery leaftier	*Udea rubigalis* (Guenee)	Pyralidae	Celery, sugar beets, ornamentals	Lange (1987)

6.3 ✏ LEAF SHELTERS AS HABITATS FOR ARTHROPODS

To assess the impact of a particular ecosystem engineer, it is important to evaluate the quality of the habitat patches it creates relative to non-engineered patches; the greater the difference in quality (or in the types of resources provided) among patches, the greater the impacts are likely to be (Jones et al. 1997). For shelter-occupying arthropods, habitat quality is perhaps best assessed from the point of view of the shelter-builder itself; after all, at least some of the resources that it creates for itself are likely to mirror those utilized by other arthropods. Multiple hypotheses have been proposed to explain the adaptive nature of the shelter-building habit, all of which enjoy some experimental support (Damman 1993, Fukui 2001, Danks 2002). Chief among these are the amelioration of harsh abiotic conditions, the manipulation of plant tissue quality, and decreased levels of attack from natural enemies. None of these hypotheses are mutually exclusive, and the benefits and costs attributed to shelter construction and occupancy are likely to vary among environments.

AMELIORATION OF ABIOTIC CONDITIONS

The microenvironments created by shelter-builders differ markedly from ambient conditions encountered on nonengineered leaves or other plant surfaces with regard to four important variables: temperature, humidity, wind, and solar radiation (Willmer 1980, Joos et al. 1988, Hunter and Willmer 1989). Because shelters consist of transpiring foliage, the relative humidity inside of shelters is typically increased relative to ambient conditions. The drying effects of wind also are decreased or eliminated, increasing the ability of the shelter-builders (and other occupants) to maintain water balance (Willmer 1980, 1982; Hunter and Willmer 1989). Because desiccation is a significant threat for many arthropods (Carne 1969, Zalucki et al. 2002), its amelioration has probably been a selective factor supporting the independent evolution of shelter building in various arthropod lineages. However, insects also must maintain thermal balance, and shelter types differ in their abilities to moderate temperature fluctuations (Hensen 1958a). At least one species of shelter-builder (*Choristoneura conflictana*) has been observed to abandon its leaf rolls when internal temperatures exceed 36°C (Hensen 1958a). Other species, such as colonial tent caterpillars (*Malacosoma* spp.; Fitzgerald 1995), fall webworms (*Hyphantria cunea*), and leaf-webbing Pieridae (Fitzgerald and Underwood 2000), construct larger, more complex structures offering a range of thermal environments that can be behaviorally exploited by occupants to meet their particular thermal needs (Costa 1997). In addition, shelter-builders often may select the site and orientation of the shelter to increase the quality of the resulting internal microclimate (Alonso 1997).

Few studies of the micrometeorology of different shelter types have been conducted (Table 6.2), and almost all of these have come from regions with temperate climates, making generalization difficult. If protection from desiccation is a principle adaptive feature of shelter building, the incidence of shelter-builders in regional faunas should be an inverse function of average relative humidity (holding other factors constant). Similarly, the high relative humidity inside many types of leaf shelters would be expected to enhance the engineering effect for desiccation-prone arthropods in dry environments. Larsson et al. (1997) tested this hypothesis with larvae of the non-shelter-building beetle, *Galerucella lineola*, which responded to decreasing humidity by increasing their affinity for, and performance in, artificially constructed willow leaf rolls. While Fernandes and Price (1992) showed a link between decreased humidity and increased incidence of gall formation, studies of shelter-builders show higher incidences of shelter building in the tropics than those reported for temperate regions. For example, Diniz and Morais (1997) found that 65% of the externally feeding caterpillars in the cerrado of Brazil build shelters, while Janzen (1988) reported more than 60% of the caterpillars in a Costa Rican dry forest were either shelter-builders or leaf miners. While both of these tropical habitats have strong dry seasons, average relative humidity during the growing season (when most larvae are present) is high, suggesting other selective factors may play a more important role in determining feeding habits in these faunas.

While shelters built by different species can vary substantially in their capacities to modulate aspects of the abiotic environment (Table 6.2), studies suggest that the following physical variables are useful in predicting the microclimate inside of particular shelter types: shelter exposure to incident radiation (a function of position in the canopy and/or proximity to edges or light gaps), orientation (lateral vs. vertical), the surface area of exposed foliage relative to shelter volume, whether the structure is sealed or ventilated, and absorptive properties of the foliage (Hensen 1958a, Willmer 1982, Fukui 2001). Comparative studies examining the differential colonization and use of these varying microenvironments by secondary users should help to identify associations between particular groups of arthropods and the shelter types they prefer to inhabit. Combined with data on the abundances of primary shelter-builders, such studies could then be used to

TABLE 6.2 Abiotic features of various shelters.

Shelter Type	Species of Shelter–Builder/User	Physical Property Examined; Finding	Reference
Leaf ties	*Choristoneura conflictana* (Tortricidae) (neonates and early instars)	Temperature inside of leaf ties; no cooling effect observed even at high temperatures.	Hensen (1958a)
Leaf rolls	*Choristoneura conflictana* (later instars)	Temperature inside roll; enhanced evaporative cooling at higher temperatures due to chimney effects when vertically oriented only.	Hensen (1958a)
Leaf rolls	*Anacampsis* (=*Compsolechia*) *niveopulvella* (Cham.) (Gelechiidae)	Temperature inside roll; increases linearly with radiation, but small volume of roll prevents roll from reaching excessive temperatures. Larvae avoid making holes in shelters and feeding rate is increased inside humid environment of shelter.	Hensen (1958a, 1958b)
Leaf rolls	*Galerucella lineola* (Chrysomelidae)	Relative humidity (RH) inside of leaf rolls; beetle larvae had increased affinity for rolls with decreasing RH, higher growth rates inside leaf rolls than outside, and increased survival under high RH conditions.	Larsson et al. (1997)
Leaf rolls	*Pleuroptya ruralis* (Crambidae)	Relative humidity inside leaf rolls; leaf rolls exposed developing larvae to very high humidities (ca. 100%), allowing them to maintain constant hemolymph osmolality in the face of large diurnal fluctuations in RH (external to rolls). Larvae were highly desiccation prone when removed from rolls.	Willmer (1980)

TABLE 6.2 *Continued*

Shelter Type	Species of Shelter-Builder/User	Physical Property Examined; Finding	Reference
Leaf rolls	*Tortrix viridana* (Tortricidae)	Relative humidity inside of leaf rolls; RH was significantly greater inside than outside leaf rolls and the difference was greater on dry days. Larvae inside rolls maintained constant hemolymph osmolality whereas larvae removed from rolls rapidly desiccated. Rolls made from damaged leaves had decreased RH and required more time to construct than rolls on undamaged leaves.	Hunter and Willmer (1989)
Silken tents	*Malacosoma* spp.	Temperature and relative humidity inside of tents; high RH inside of tents protects larvae from desiccation. Thermodynamic properties of tent (minimal convective heat loss) allow larvae to increase body temperature at colder temperatures.	Fitzgerald (1995)
Webbed foliage (bolsa)	*Eucheira socialis* (Pieridae)	Temperature inside of bolsa; varying thermal environment of shelters enable larvae to maintain low body temperatures, which facilitates nocturnal foraging bouts.	Fitzgerald and Underwood (2000)
Leaf tents	*Yponomeuta mahalbella* (Yponomeutidae)	Temperature inside of tents; larvae constructing tents oriented to the south and southeast experienced the greatest increase in temperature relative to ambient and had increased growth, survival, and adult mass.	Alonso (1997)
Leaf curls	*Papilio glaucus* (Papilionidae)	Tempearture in leaf curl; larvae inhabiting leaf curls on exposed (sunny) leaves increased their rate of development.	Grossmueller and Lederhouse (1985)

predict which systems are likely to elicit the greatest engineering responses.

MANIPULATION OF TISSUE QUALITY

A number of studies indicate that arthropods occupying leaf shelters have enhanced access to high-quality food resources. The nutritional benefits of concealed feeding may result from reduced exposure to light, which can alter tissue quality (e.g., decreased leaf toughness and phenolic concentrations in rolled leaves; Sagers 1992, Oki and Varanda 2000, Fukui et al. 2002) or from the toxic effects (both pre- and post-ingestive) of foliage consumption in the presence of light (Berenbaum 1978, Sandberg and Berenbaum 1989, Champagne et al. 1996). The direct influence of foliage manipulation on structural defenses (e.g., lignins and leaf toughness) and other quantitative carbon-based defenses with low turnover or mobility (e.g., polyphenolics; Feeny 1970) is likely to be greatest for shelters constructed with young, expanding foliage, because the light environment in which these tissues develop can affect the accumulation of these compounds (Coley et al. 1985, Dudt and Shure 1994, Ruohomaki et al. 1996, but see Costa and Varanda 2002 for a counterexample). For example, leaf rolls constructed by oecophorid moth larvae on immature foliage of *Byrsonima intermedia* in Brazil had almost a threefold reduction in tannin concentrations relative to immature foliage exposed to full sun (Oki and Varanda 2000), and Japanese lilac (*Syringa reticulata*) leaf rolls constructed by the tortricid *Rhopobota naevana* had significantly lower concentrations of total phenolics than nonrolled leaves (Fukui et al. 2002).

While few studies have quantified foliage quality of shelter vs. nonshelter leaves directly, a number of studies have performed bioassays to assess the preference for, or performance on, shelter vs. nonshelter foliage. Three studies have found that shelter-building caterpillars preferentially construct shelters using leaves of lower nutritional quality than those available, suggesting that food quality effects are secondary to other ecological factors in these systems. The larvae of both *Achlya flavicornis* (Thyatiridae) and *Omphalocera munroei* (Pyralidae) preferentially select younger and older leaves of their host plants, respectively, for use in shelter construction and feeding; in the absence of enemies, these choices reduced larval development time (both species) and reduced pupal mass and survival (*A. flavicornis*) relative to larvae reared on available higher quality foliage (Damman 1987, Reavey 1991). Similarly, Hunter (1987) found that larvae of *Diurnea fagella* (Oecophoridae) preferentially built shelters on oak leaves that had been previously damaged, even though these leaves reduced survival in the absence of enemies. In all three of these cases, the authors argue that the choice of suboptimal foliage is offset by other advantages provided by the foliage (e.g., ease of construction, longevity of the construct, or reduced predation during or after construction). Several other studies suggest that the suitability of foliage as a building material for making constructs is of greater importance than nutritional quality *per se* (Ruehlmann et al. 1988, Mueller and Dearing 1994, Loeffler 1996a, Dorn et al. 2001). For example, Forkner et al. (2004) suggest that higher abundance of spring leafrollers found on high-tannin black oak canopy foliage compared with low-tannin white oak foliage may reflect a preference for the larger black oak leaves, which enable the caterpillars to form rolls with more whorls.

Because most shelter-builders feed upon the leaves in their shelter, herbivorous shelter-builders may alter the quality of the foliage to which they have access if the host plant is capable of localized induced responses (Karban and Baldwin 1997). While induced responses to feeding by shelter-builders have not been examined explicitly, many shelter-builders change shelters prior to exhausting the food resources present within a shelter (e.g., Ide 2004), possibly because of feeding-related declines in food quality (Edwards and Wratten 1983, Ruehlmann et al. 1988). However, shelter-building that occurs on lactiferous plants may increase the palatability of the affected foliage when trenching behaviors, such as severing of leaf veins, are required to facilitate shelter construction (Dussourd 1993).

Because other herbivores are common secondary occupants of leaf shelters (Cappuccino 1993; Fukui 2001; Lill and Marquis 2003, 2004), the foliage quality available in existing shelters may influence the arthropod response to these engineered constructs. Clearly, additional studies comparing the nutritional ecology of shelter and nonshelter foliage from a variety of different shelter types using leaves of different ages are necessary to assess the general importance of altered food quality for subsequent occupation and use of shelters by other herbivores.

PROTECTION FROM NATURAL ENEMIES

Perhaps the most commonly assumed adaptive feature of shelter-making is increased protection against natural enemies. Studies have demonstrated the protective function of shelters against particular guilds of natural enemies, including ants (Fowler and MacGarvin 1985, Heads and Lawton 1985, Vasconcelos 1991, Loeffler 1996b, Jones et al. 2002, Eubanks et al. 1997), birds (Atlegrim 1989, Atlegrim 1992, Sipura 1999, Low and Connor 2003), ladybird beetles (Messina et al. 1997), spiders

(Loeffler 1996b), predatory wasps (Damman 1987, Jones et al. 2002), and fish (Mueller and Dearing 1994, Dorn et al. 2001). Several studies have demonstrated that experimentally removing shelter-builders from their shelter increases mortality (Damman 1987, Cappuccino 1993), or that high rates of mortality occur during construction (Mueller and Dearing 1994). However, because many species spend little or no time outside of their shelters (once constructed), assessing mortality in unsheltered locations can be problematic. Such studies may require tethering larvae to foliage (e.g., Damman 1987), a factor that may cause larvae to behave abnormally (i.e., many shelter-building larvae will not feed until they have located an existing shelter or built their own). Many secondary occupants of leaf shelters, however, are facultative shelter-users, occur-ring naturally both inside and outside of shelters, facilitating experimen-tal tests of the importance of shelters in reducing mortality from enemies (e.g., Larsson et al. 1997, Messina et al. 1997).

Once occupied, shelters offer several potential antipredator benefits to residents. One of the most evident is the ability to conceal feeding from visually oriented predators that cue in on movement (Bernays 1997). Although not all shelter-dwellers feed on the tissue of their shel-ters, those that do can take advantage of this protection and feed rela-tively continuously, including diurnally, when many free-living herbivores avoid feeding or moving altogether (Wagner 2005). Such continuous feeding can result in more rapid development, minimizing the window of exposure for larvae to natural enemies (Clancy and Price 1987, Benrey and Denno 1997, Lill and Marquis 2001).

While most studies documenting the beneficial effects of shelters on enemies point to decreased detection as the primary means of escape, at least one study has demonstrated that frass incorporated into shelters contains toxic compounds derived from cyanogenesis, thus repelling predatory ants (Peterson 1986). Because many species of shelter-build-ers incorporate frass into their shelters (Weiss 2006), it is possible that repellant effects of this sort are more widespread, calling for increased investigation.

While a number of studies have highlighted the realized or potential antipredator benefits of shelters, other studies have demonstrated that shelter occupants suffer high rates of predation and parasitism. For example, a number of bird species actively forage on leaf rolls or dead curled leaves, using the constructs as visual cues to efficiently locate prey (Robinson and Holmes 1982, Remsen and Parker 1984, Murakami 1999), a behavior that has been shown to have both innate and learned (via reinforcement) components (Greenberg 1987). Similarly, social wasps (*Polistes* spp.) will readily locate and attack skipper larvae inside their shelters once they learn they contain prey (Weiss et al. 2004). Besides the visual cues presented by the construct itself, damage to the shelter leaves and the release of plant and frass volatiles are additional "token stimuli" used by a variety of natural enemies for prey location (Heinrich 1979, Heinrich and Collins 1983, Odell and Goodwin 1984, Steiner 1984, Loeffler 1996b, Jones et al. 2002, Weiss 2003). Some shelter-builders behaviorally reduce damage or frass-related cues by feeding and defecating outside of their shelters (e.g., many skippers and some Pantheidae, such as *Charadra deridens* and *Colocasia* spp.; Wagner 2005), presumably to reduce attack by predators and parasitoids (Mattiacci and Dicke 1995). These and other antipredator behaviors exhibited by shelter-builders (e.g., chemical defense secretions produced by tent-making caterpillars; Darling et al. 2001) imply that simply residing in leaf shelters does not ensure protection from enemies. In addition, a number of studies have found high rates of secondary occupancy of leaf shelters by generalist predators (e.g., Martinsen et al. 2000, Fournier et al. 2003), potentially facilitating encounters between shelter-occupants and natural enemies.

We are aware of no studies that have demonstrated reduced parasitism inside of shelters compared with outside. On the contrary, shelter-builders, like other concealed feeders and leaf miners, often have high rates of attack and support diverse parasitoid assemblages (Pasek and Kearby 1984, Hawkins and Sheehan 1994, Lill 1999). For example, Le Corff et al. (2000) found that parasitism of highly concealed spring leaf-rolling caterpillars on oaks was equal to that of free-feeding caterpillars in each of two years (between 30 and 40%), despite large differences in exposure. So while occupants of some shelters may gain a measure of protection against a variety of generalist predators, others may be obvious targets subject to high mortality from enemies.

6.4 ⬝ ENGINEERING EFFECTS ON ARTHROPOD COMMUNITIES

The construction of leaf shelters on plants has been shown to have strong effects on the abundance (Martinsen et al. 2000, Bailey and Whitham 2003, Fournier et al. 2003, Nakamura and Ohgushi 2003, Lill and Marquis 2004), species richness (Walz and Whitham 1997, Martinsen et al. 2000, Bailey and Whitham 2003, Lill and Marquis 2003), and community structure (Bailey and Whitham 2003; Lill and Marquis 2003, 2004) of plant-dwelling arthropods. While the number of studies examining community-level arthropod responses is relatively small, the resulting engineering effects have been strong and positive, suggesting that the resources provided by shelters are readily exploited. The organisms affected represent a variety of feeding guilds and trophic levels. As a consequence, the nontrophic engineering effects of shelter-building frequently elicit coupled trophic interactions, which may accentuate or dampen net engineering effects on communities, depending on the organisms involved (Wilby et al. 2001). In addition, a number of studies have examined the biotic and abiotic factors that regulate where and when engineering can occur, acting as "regulators" of these coupled engineering–trophic interaction webs.

ENGINEERING EFFECTS ON BIOTIC AND ABIOTIC RESOURCE FLOWS

Leaf shelters increase the habitat heterogeneity that exists on plants (Lawton 1983) and thus would be expected to have positive effects on species richness and abundance if the engineered habitats are used by a different set of species than use corresponding nonengineered habitats (Jones et al. 1997). Secondary occupants include other shelter-builders (Carroll and Kearby 1978; Carroll et al. 1979; Cappuccino 1993; Cappuccino and Martin 1994; Bailey and Whitham 2003; Lill 2004; Lill and Marquis 2003, 2004) and non-shelter-building arthropods (Morris 1972; Carroll and Kearby 1978; Carroll et al. 1979; Hajek and Dahlsten 1986; Martinsen et al. 2000; Bailey and Whitham 2003; Fournier et al. 2003; Nakamura and Ohgushi 2003; Lill and Marquis 2003, 2004; Lill 2004; Crutsinger and Sanders 2005) from a wide variety of taxa (reviewed in Fukui 2001, Marquis and Lill 2006). Secondary occupation of existing leaf shelters can occur while the original shelter-builder is still present or following the abandonment of the shelter by the engineer (Cappuccino 1993, Lill 2004).

The resources provided to secondary occupants of shelters are even more varied than those created for use by the primary occupant; these may include each of the habitat features detailed in preceding text, as well as prey items for predators and parasitoids, detritus (e.g., frass, exuvia, uneaten dead portions of leaves, and accompanying fungi) fed upon by scavengers, and honeydew produced by aphids or other sucking insects occupying (or creating) leaf shelters (Fukui 2001, Nakamura and Ohgushi 2003). Some secondary occupants take up permanent residence in leaf shelters, while others use them temporarily as resting sites, oviposition sites (e.g., many adult moths and beetles; Lill and Marquis 2004), or protected sites in which to molt (Fukui 2001). In systems in which shelters are occupied by a succession of shelter-builders (of the same or different species), the engineering effects are likely to be magnified because the structural integrity of the shelters can be extended well beyond the residency of the initial constructor(s). Such positive feedback loops between successive generations of shelter-builders may increase local densities of these species, potentially resulting in outbreaks (e.g., of some agricultural and forestry pests; Table 6.1).

Increased habitat heterogeneity created by shelter-builders can alter arthropod diversity in a variety of ways. Shelters can provide recruitment sites for species with specialized microhabitat requirements that otherwise may not be found on the plant. For example, in our study of the effects of leaf ties on the arthropod communities of white oak saplings, we found that several species of low-density shelter-builders occurred only on trees containing preexisting shelters (Lill and Marquis 2003). The habitats provided by shelter-builders also can increase the abundance and diversity of habitat generalists capable of occupying both engineered and nonengineered habitats. For example, spiders often use leaf constructs when building webs or selecting nesting sites. As a result, plants with high densities of leaf shelters may have higher densities of spiders than plants with low densities of shelters due to the increased structural complexity offered by these plants (Fournier et al. 2003). Specialist predators and parasitoids also would be predicted to increase, tracking the increased diversity of their prey or hosts. Non-shelter-using arthropods (shelter-avoiders) could be negatively impacted by shelter-constructors if the availability of non-shelter habitats becomes limited (Marquis and Lill 2006). Because there is no *a priori* reason to expect that the habitats produced by shelter-builders will support a greater or lesser diversity of arthropods than similar-sized patches of nonengineered foliage (Jones et al. 1997), any net positive effects of engineering on diversity measures should stem from the availability of multiple habitat types, each with its own set of associated arthropod species.

Recruitment of arthropods to engineered habitat patches frequently occurs in conjunction with reproduction, whereby arthropods of various types lay eggs or give birth (e.g., many parthenogenetic aphids) on foliage incorporated into a shelter (Cappuccino and Martin 1994, Lill and Marquis 2004) or lay eggs into host arthropods residing within shelters (Pasek and Kearby 1984, Lill 1999). Shelters used temporarily by more mobile arthropods (e.g., many adult beetles) may increase the "residence time" of these arthropods on the plant by providing concentrated, high-quality resources. The fidelity, survivorship, and residence time of arthropods secondarily occupying leaf shelters of different types require further study so that we can move beyond simply documenting shelter use to gaining insights into the ecological factors determining arthropod responses to habitat creation.

ENGINEERING EFFECTS ON INTERSPECIFIC INTERACTIONS

In addition to the engineering effects just described, the secondary occupation of leaf shelters by different species of arthropods can influence the outcomes of a variety of interspecific interactions among arthropods, including those between competitors, predators and prey, and mutualists. These interactions thus can have important effects on community outcomes.

Competitive interactions

Because shelters are frequently occupied by multiple arthropods (either concurrently or sequentially), there is considerable potential for both intraspecific and interspecific competition. Among herbivores sharing shelters, competition can be quite intense because the food resources available inside leaf shelters can be limited due to high densities of occupants. When food resources are exhausted, herbivores that have not completed development will be forced either to expand their shelter by pulling in additional foliage or to relocate and establish a new shelter on less-damaged foliage. However, studies have found that previously occupied (and damaged) leaf ties are as equally attractive to new herbivores as unoccupied ties, suggesting that food resources may not be the limiting factor in site selection (Lill 2004, Lill and Marquis 2004, Lill et al. 2007).

While the negative effects of resource competition within shelters remain to be examined, several studies have documented antagonistic behavioral interactions occurring among shelter occupants. For example, the leaf webs constructed by *Depressaria pastinacella* on parsnip are aggressively defended from being usurped by conspecifics displaced from their own webs (Berenbaum et al. 1993). Similarly, cherry leafroller caterpillars (*Caloptilia serotinella*) utilize vibrational signals (i.e., leaf scraping) in response to intruders in what are hypothesized to be territorial disputes (Fletcher et al. 2006). While the mechanism is unclear, one species of leaf-tying caterpillar (*Psilocorsis quercicella*) attained higher pupal mass when reared singly in a leaf shelter than when reared with two conspecifics when food was not limiting (Lill et al. 2007), suggesting that such negative behavioral interactions can influence fitness measures. The very existence of these aggressive behaviors underscores the prominent role competition likely has played in shaping the traits of shelter-dwellers.

Predator–prey interactions

As with primary occupants, secondary occupation of leaf shelters can reduce or increase the risk of predation. Secondary occupants can experience increased rates of predation due to cues left behind by the primary constructor, enhanced cues to natural enemies, or because shelters increase their contact with natural enemies using the shelters for other reasons (i.e., as nesting sites, or as shelters from the elements or their own predators). To our knowledge, only two studies (Larsson et al. 1997, Lill et al. 2007) have examined the consequences of secondary occupation of leaf shelters for attack by natural enemies, and these studies found either weak or inconsistent effects.

Mutualistic interactions

Shelter-building also can impact mutualistic interactions. For example, aphids and other sap-sucking insects (e.g., lace bugs) are some of the most common secondary occupants of leaf shelters in a variety of systems (Hajek and Dahlsten 1986, Nakamura and Ohgushi 2003, Crutsinger and Sanders 2005). Aphids can reach very high densities inside of leaf shelters, although it is unclear what benefits they receive from occupying shelters. The honeydew produced by large colonies of aphids can attract tending ants, which in turn can influence the abundance or distribution of other arthropods on the plant. The use of floral structures by some species of shelter-builders also raises the possibility that shelter-building can influence plant–pollinator mutualisms. Even small amounts of damage to flowers can decrease visitation by pollinators (Mothershead and Marquis 2000), so the potential exists for such interactions. In the case of flower feeding by the inflorescence-webbing caterpillar *Depressaria pastinacella* on *Pastinaca sativa*, part of the reduction in seed production is due to reduced pollinator service associated with avoidance of webbed inflorescences (Lohman et al. 1996).

Predicting engineering effects of shelter-builders

While comparative studies examining the engineering effects of different shelter types (or species of shelter-builders) have yet to be conducted, several features might be expected to be important and deserve consideration when planning experiments involving shelter-builders.

These include the persistence, size, number, and accessibility of shelters, as well as their physical properties. Persistent shelters, or shelter types that are routinely colonized by other shelter-builders (e.g., many leaf ties and rolls) should have greater effects on arthropod communities than shelters that disassemble following abandonment by the primary engineer (Jones et al. 1997). Larger structures may also have greater engineering impacts by permitting secondary occupation by a greater number of arthropods with a wider range of body sizes. For solitary shelter-builders, shelter size is at least partially constrained by the size of the foliage of their host plants. The number of leaf shelters constructed (and abandoned) by an individual arthropod over its life should be positively related to its engineering impacts. Leaf shelters also vary considerably in their accessibility to secondary occupants, with some shelters (e.g., many leaf folds and rolls) being "sealed shut" while others (e.g., many loose leaf webs) are quite open to potential colonists. Finally, shelters providing a high-quality microenvironment that ameliorates the most important climatic threats facing many arthropods in a particular environment are likely to have greater engineering effects than shelters with microenvironments that are little changed from ambient (non-sheltered) habitats or that provide microclimates used by only a small subset of potential colonists.

Scaling effects of shelter-builders

In addition to these local effects, shelter-builders may have larger regional effects by influencing population dynamics. For example, because many species of shelter-building caterpillars are bivoltine or multivoltine, shelter-provisioning by one generation can facilitate establishment of future generations of both conspecifics and heterospecifics (Cappuccino 1993) resulting in regional differences in caterpillar densities and diversities. Such feedback loops between engineers have been proposed by Jones et al. (1997) and modeled by Gurney and Lawton (1996). We suggest that the rapid generation times of these insects make them ideal candidates for empirical study; such studies will provide much-needed parameters useful in testing and refining these models.

While multiple studies have indicated that shelter-building tends to increase arthropod species richness, the spatial scales used in sampling differ among studies, leaving open the question of how the engineering effects on species richness change with the number of habitat "patches" sampled (comparing engineered and nonengineered habitat patches). Studies have found strong effects of shelters on arthropod richness at relatively small scales (saplings or branches of larger trees with a relatively small number of habitat patches; Martinsen et al. 2000, Lill and Marquis 2003), but at larger scales (i.e., landscapes) these engineering effects on richness might be expected to attenuate, because the proportion of obligate shelter-users is relatively small compared with non-shelter-using arthropod species in the regional fauna.

If shelter-building has a compounding effect on plant damage by attracting more herbivores, which has been shown in some systems (Marquis et al. 2002, Marquis and Lill unpublished data), and damage influences plant fitness, then shelter-builders have the potential to alter plant community structure. If plant damage is concentrated on a dominant competitor, engineers could indirectly increase plant community diversity (or decrease diversity if focused on inferior competitors). A variety of studies have documented the strong engineering effects of herbivores on plant communities (e.g., Wilby et al. 2001), but most of these have focused on soil disturbance by animals and its influence on recruitment. Few shelter-builders appear to routinely defoliate plants (outbreaking species in Table 6.1 are exceptions), but the successive use of shelters by other herbivores has the potential to compound plant damage over the season.

Controls on engineering

Because shelter-builders have cascading impacts on arthropod communities through both their engineering effects and their resulting trophic interactions, it is salient to consider what factors regulate their abundance on different plant species or in different habitats. Leaf age appears to be a limiting factor for some shelter-builders, due to changes in leaf toughness or foliage size, both of which can affect the ability of shelter-builders to build and maintain their constructs (Damman 1987, Mueller and Dearing 1994, Dorn et al. 2001). For example, many leaf-rollers are restricted to using young foliage, which may be available only at a certain time of year in seasonal environments with synchronously flushing plant species (Coley and Barone 1996). Similarly, the creation of other shelter types (e.g., leaf ties) may be possible only after leaf expansion is near completion and leaf overlap occurs (Marquis et al. 2002).

Marquis et al. (2002) found that plant architecture, specifically the number of touching leaves, determines the density of leaf ties formed on white oak (*Quercus alba*) saplings. Bailey and Whitham (2003) found that elk browsing of aspen (*Populus tremuloides*) decreased leaf fold galls formed by the sawfly *Phyllocolpa bozemanii*, an important aspen microhabitat used by a variety of arthropods. Hunter (1987) found that early-season damage by two caterpillar species (*Tortrix viridana* and

Operophtera brumata) increased the leaf-rolling ability of late-season *Diurnea fagella* caterpillars. Finally, Seyffarth et al. (1996) found that fire in the cerrado vegetation of Brazil increased the abundance of leaf-rollers (unidentified species) on the common host plant *Ouratea hexasperma*, as a result of refoliation following the burn. The growing recognition of the importance of such "interaction webs" requires that engineering be integrated into more traditional models involving trophic interactions alone in formulating predictive models of the network of factors influencing community and ecosystem dynamics (Proulx et al. 2005).

6.5 ⬤ PROSPECTUS

Looking forward, there are a number of open paths of inquiry into the ecology of shelter-builders. Studies that quantify the engineering impacts of different shelter types on arthropod abundance, diversity, and community structure are needed for a much wider variety of plant species, growth forms, and climatic regimes. Most studies conducted to date (and all of those examining community-level responses) have focused on leaf shelters constructed on temperate tree species. Studies documenting engineering effects for both herbaceous and woody plants growing in other parts of the world (especially in the tropics) are needed to test the generality of the results obtained thus far. Replicated experiments across taxa or sites are imminently feasible because most shelter types can be created by investigators. Within a particular habitat, it would be prudent to compare the effects of leaf shelters on plant species with well-developed vs. poorly developed shelter-building faunas; such a comparison would shed light on whether arthropod responses to these constructs are driven by evolved, canalized habitat selection behaviors or are more serendipitous and behaviorally plastic.

Because modification of the abiotic environments inside of shelters is thought to have played an important role in the evolution of the shelter-building habit, faunal studies examining the incidence of shelter building along an abiotic stress gradient, or that examine the phylogeography of shelter-builders, would help to test this assertion and perhaps illuminate which environmental variable(s) are the most commonly altered by the construction of the shelter. In addition, more detailed autecological studies of the physical and chemical properties of different shelter types would help circumscribe the niche space produced and occupied by shelter-builders and their associates (Odling-Smee et al. 2003). These include more detailed studies of temperature and humidity fluctuations inside of shelters, further examination of how and when shelter-building influences leaf quality, and how the microclimate of a shelter influences arthropod physiology and fitness.

The biotic interactions that occur within leaf shelters also are likely to be important drivers of subsequent engineering effects. For example, the importance of inter- and intra-specific competition for food resources by shelter-inhabiting herbivores can determine how frequently shelter-builders move and thus the quantity of sheltered habitat available for use by other animals. The relatively small size of most shelters and the large amount of damage they incur is compelling evidence that direct competition for limiting food resources is a common occurrence, but this requires confirmation through carefully designed field and/or laboratory experiments. Moreover, because leaf shelters are frequently occupied by more than one arthropod at a time, studies of species interactions occurring within these shelters are likely to provide much-needed insights into how engineering and trophic effects interact to influence the size and diversity of arthropod assemblages. For any particular shelter type, are the resident arthropods random collections of potential colonists, or do certain species tend to co-occur? Are shelter-dwelling predators important sources of mortality for cohabitants? Can residents behaviorally exclude potential colonists? Do arthropods sharing a shelter partition the space to reduce interactions? How frequently do shelters provide enemy-free space (Jeffries and Lawton 1984) and against what types of predators? To address many of these questions, "windows" into the dynamics of within-shelter interactions are needed and may require some creativity. For leaf ties, we have had some success with using artificial leaves of clear acetate that can be clipped to leaves to provide one surface to visualize interactions occurring with the ties.

The engineering impact of multiple shelter types constructed on the same plant is another area that warrants investigation. Many plants host a variety of different shelter types as well as other types of plant modifiers (e.g., concealed feeders and internal feeders: miners, borers, and gallers). All of these structures have the potential to increase habitat heterogeneity (Lawton 1983), but it is not known whether their engineering effects are additive or nonadditive. In addition, the relationships between shelter density and community responses are totally unknown (e.g., are they linear, saturating, or unimodal, and over what scales?; Marquis and Lill 2006). We argue that the ease of manipulation of these constructs relative to other types of ecosystem engineers holds great promise for addressing both system-specific questions and more general questions posed by theoreticians and those seeking to integrate ecosystem engineering more fully into ecological and evolutionary studies (Wright and Jones 2006).

ACKNOWLEDGMENTS

We thank John Landosky, John Flunker, Nick Barber, Banak Gamui, and Beto Dutra for comments on an earlier version, and acknowledge USDA grant 99-35302-8017 for financial support.

REFERENCES

Alonso, C. (1997). Choosing a place to grow. Importance of within-plant abiotic microenvironment for *Yponomeuta mahalabella*. *Entomologia Experimentalis et Applicata* 83:171–180.

Anderson, C., and McShea, D.W. (2001). Intermediate-level parts in insect societies: Adaptive structures that ants build away from the nest. *Insectes Sociaux* 48:291–301.

Atlegrim, O. (1989). Exclusion of birds from bilberry stands: Impact on insect larval density and damage to the bilberry. *Oecologia* 79:136–139.

——. (1992). Mechanisms regulating bird predation on a herbivorous larva guild in boreal coniferous forests. *Ecography* 15:19–24.

Bailey, J.K., and Whitham, T.G. (2003). Interactions among elk, aspen, galling sawflies, and insectivorous birds. *Oikos* 101:127–134.

Baker, W.L. (1972). *Eastern Forest Insects*. Washington, D.C.: USDA Forest Service.

Barosela, J.R. (1999). Herbivoria foliar em *Xylopia aromatica* (Lam.) Mart. de três fisionomias de Cerrado e sua relação com o teor de taninos, valor nutritivo e entomofauna asociada. Dissertasção de maestrado, Universidade Federal, Sao Carlos.

Bastian, R.A., and Hart, E.R. (1991). Temperature effects on developmental parameters of the mimosa webworm (Lepidoptera, Plutellidae). *Environmental Entomology* 20:1141–1148.

Beckwith, R.C. (1973). *The Large Aspen Tortrix*. Washington, D.C.: USDA Forest Service.

Benrey, B., and Denno, R.F. (1997). The slow-growth-high-mortality hypothesis: A test using the cabbage butterfly. *Ecology* 78, 987–999.

Berenbaum, M.R. (1978). Toxicity of a furanocoumarin to armyworms: A case of biosynthetic escape from insect herbivores. *Science* 201:532–534.

——. (1999). Shelter-making caterpillars: Rolling their own. *Wings* 22:7–10.

Berenbaum, M., Green, E., and Zangerl, A.R. (1993). Web costs and web defense in the parsnip webworm (Lepidoptera: Oecophoridae). *Environmental Entomology* 22:791–795.

Bernays, E.A. (1997). Feeding by lepidopteran larvae is dangerous. *Ecological Entomology* 22:121–123.

Capinera, J.L., Renaud, A.R., and Naranjo, S.E. (1981). Alfalfa webworm *Loxostege commixtalis* foliage consumption and host preference. *Southwestern Entomologist* 6:18–22.

Cappuccino, N. (1993). Mutual use of leaf shelters by lepidopteran larvae on birch. *Ecological Entomology* 18:287–292.

Cappuccino, N., and Martin, M.-A. (1994). Eliminating early-season leaf-tiers of paper birch reduced abundance of mid-summer species. *Ecological Entomology* 19:399–401.

Carmona, A.S., and Barbosa, P. (1983). Overwintering egg mass adaptations of the eastern tent caterpillar, *Malacosoma americanum* (Fab.) (Lepidoptera: Lasiocampidae). *Journal of the New York Entomological Society* 9:68–74.

Carne, P.B. (1969). On the population dynamics of the Eucalypt-defoliating sawfly *Perga affinis* Kirby (Hymenoptera). *Australian Journal of Zoology* 17:113–141.

Carroll, M.R., and Kearby, W.H. (1978). Microlepidopterous oak leaftiers (Lepidoptera: Gelechioidea) in central Missouri. *Journal of the Kansas Entomological Society* 51:457–471.

Carroll, M.R., Wooster, M.T., Kearby, W.H., and Allen, D.C. (1979). Biological observations on three oak leaftiers: *Psilocorsis quercicella*, *P. reflexella*, and *P. cryptolechiella* in Massachusetts and Missouri. *Annals of the Entomological Society of America* 72:441–447.

Castilla, J.C., Lagos, N.A., and Cerda, M. (2004). Marine ecosystem engineering by the alien ascidian *Pyura praeputialis* on a mid-intertidal rocky shore. *Marine Ecology Progress Series* 268:119–130.

Champagne, D.E., Arnason, J.T., Philogene, B.J.R., Morand, P., and Lam, J. (1996). Light-mediated allelochemical effects of naturally occurring polyacetylenes and thiophenes from Asteraceae on herbivorous insects. *Journal of Chemical Ecology* 12:835–858.

Clancy, K.M., and Price, P.W. (1987). Rapid herbivore growth enhances enemy attack: Sublethal plant defenses remain a paradox. *Ecology* 68:733–737.

Coley, P.D., and Barone, J.A. (1996). Herbivory and plant defenses in tropical forests. *Annual Review of Ecology and Systematics* 27:305–335.

Coley, P.D., Bryant, J.P., and Chapine, F.S., III. (1985). Resource availability and plant anti-herbivore defense. *Science* 230:895–899.

Conway, B.E., McCullough, D.G., and Leefers, L.A. (1999). Long-term effects of jack pine budworm outbreaks on the growth of jack pine trees in Michigan. *Canadian Journal of Forest Restoration* 29:1510–1517.

Costa, A.A., and Varanda, E.M. (2002). Building of leaf shelters by *Stenoma scitiorella* Walker (Lepidoptera: Elachistidae): Manipulation of host plant quality. *Neotropical Entomology* 31:537–540.

Costa, J.T. (1997). Caterpillars as social insects. *American Scientist* 85:150–159.

Crutsinger, G.M., and Sanders, N.J. (2005). Aphid-tending ants affect secondary users in leaf shelters and rates of herbivory on *Salix hookeriana* in a coastal dune habitat. *American Midland Naturalist* 154:296–304.

Damman, H. (1987). Leaf quality and enemy avoidance by the larvae of a pyralid moth. *Ecology* 68:88–97.

——. (1993). Patterns of interaction among herbivore species. In *Caterpillars: Ecological and Evolutionary Constraints on Foraging*, N.E. Stamp and T.M. Casey, Eds. New York: Chapman and Hall, pp. 132–169.

Danks, H.V. (2002). Modification of adverse conditions by insects. *Oikos* 99:10–24.

Danthanarayana, W. (1983). Population ecology of the light brown apple moth, *Epiphyas postvittana* (Lepidoptera: Tortricidae). *Journal of Animal Ecology* 52:1–33.

Darling, D.C., Schroeder, F.C., Meinwald, F., Eisner, M., and Eisner, T. (2001). Production of a cyanogenic secretion by a thyridid caterpillar (*Calindoea trifascialis*, Thyrididae, Lepidoptera). *Naturwissenschaften* 88:306–309.

Diniz, I.R., and Morais, H.C. (1997). Lepidopteran caterpillar fauna of cerrado host plants. *Biodiversity and Conservation* 6:817–836.

Dixon, P.L., and Carl, K. (2002). *Croesia curvalana* (Kearfott), blueberry leaftier (Lepidoptera: Tortricidae). In *Biological Control Programmes in Canada, 1981–2000*, P.T. Mason and J.T. Huber, Eds. New York: CABI, pp. 87–89.

Dorn, N.J., Cronin, G., and Lodge, D.M. (2001). Feeding preferences and performance of an aquatic lepidopteran on macrophytes: Plant hosts as food and habitat. *Oecologia* 128:406–415.

Douglas, S.M., and Cowles, R.S. (2006). *Plant Pest Handbook: A Guide to Insects, Diseases, and Other Disorders Affecting Plants.* The Connecticut Agricultural Experiment Station. Available at http://www.caes.state.ct.us/PlantPestHandbookFiles/pphIntroductory/pphfront.htm.

Dreistadt, S.H., Clark, J.K., and Flint, M.L. (1994). *Pests of Landscape Trees and Shrubs: An Integrated Pest Management Guide.* (Publication 2259.) Oakland, CA: University of California Division of Agriculture and Natural Resources.

Dudt, J.F., and Shure, D.J. (1994). The influence of light and nutrients on foliar phenolics and insect herbivory. *Ecology* 75:86–98.

Dussourd, D.E. (1993). Foraging with finesse: Caterpillar adaptations for circumventing plant defenses. In *Caterpillars: Ecological and Evolutionary Constraints on Foraging*, N.E. Stamp and T.M. Casey, Eds. New York: Chapman and Hall, pp. 92–131.

Edwards, P.J., and Wratten, S.D. (1983). Wound induced defenses in plants and their consequences for patterns of insect grazing. *Oecologia* 59:88–93.

Eubanks, M.D., Nesci, K.A., Petersen, M.K., Liu, Z., and Sanchez, H.B. (1997). The exploitation of an ant-defended host-plant by a shelter-building herbivore. *Oecologia* 109:454–460.

Feeny, P. (1970). Seasonal changes in oak leaf tannins and nutrients as a cause of spring feeding by winter moth caterpillars. *Ecology* 51:565–581.

Fernandes, G.W., and Price, P.W. (1992). The adaptive significance of insect gall distribution: Survivorship of species in xeric and mesic habitats. *Oecologia* 90:14–20.

Fitzgerald, T.D. (1995). *The Tent Caterpillars.* Ithaca: Cornell University Press.

Fitzgerald, R.D., and Underwood, D.L.A. (2000). Winter foraging patterns and voluntary hypothermia in the social caterpillar *Eucheira socialis*. *Ecological Entomology* 25:35–44.

Fletcher, L.E., Yack, J.E., Fitzgerald, R.D., and Hoy, R.R. (2006). Vibrational communication in the cherry leaf roller caterpillar *Caloptilia serotinella* (Gracillarioidea: Gracillariidae). *Journal of Insect Behavior* 19:1–18.

Forkner, R.E., Marquis, R.J., and Lill, J.T. (2004). Feeny revisited: Condensed tannins as anti-herbivore defences in leaf-chewing herbivore communities of *Quercus*. *Ecological Entomology* 29:174–187.

Forkner, R.E., Marquis, R.J., Lill, J.T., and Le Corff, J. (2006). Impacts of alternative timber harvest practices on leaf-chewing herbivores of oak. *Conservation Biology* 20:429–440.

Fournier, V., Rosenheim, J.A., Brodeur, J., Laney, L.O., and Johnson, M.W. (2003). Herbivorous mites as ecological engineers: Indirect effects on arthropods inhabiting papaya foliage. *Oecologia* 135:442–450.

Fowler, S.V., and MacGarvin, M. (1985). The impact of hairy wood ants, *Formica lugbris*, on the guild structure of herbivorous insects on birch, *Betula pubescens*. *Journal of Animal Ecology* 54:847–855.

Fraenkel, G., and Fallil, F. (1981). Spinning (stitching) behavior of the rice leaf-folder, *Cnaphalocrosis medinalis*. *Entomologia Experimentalis et Applicata* 29:138–146.

Fukui, A. (2001). Indirect interactions mediated by leaf shelters in animal-plant communities. *Population Ecology* 43:31–40.

Fukui, A., Murakami, M., Konno, K., Nakamura, M., and Ohgushi, T. (2002). A leaf-rolling caterpillar improves leaf quality. *Entomological Science* 5:263–266.

Furniss, R.L., and Carolin, V.M. (1977). Western forest insects. In *Miscellaneous Publication 1339*. Washington, D.C.: United States Forest Service.

Gaston, K.J., Reavey, D., and Valladares, G.R. (1992). Intimacy and fidelity: Internal and external feeding by the British microlepidoptera. *Ecological Entomology* 17:86–88.

Greenberg, R. (1987). Development of dead leaf foraging in a tropical migrant warbler. *Ecology* 68:130–141.

Gross, P. (1993). Insect behavioral and morphological defenses against parasitoids. *Annual Review of Entomology* 38:251–273.

Grossmueller, D.W., and Lederhouse, R.C. (1985). Oviposition site selection: An aid to rapid growth and development in the tiger swallowtail butterfly, *Papilio glaucus*. *Oecologia* 66:68–73.

Gurney, W.S.C., and Lawton, J.H. (1996). The population dynamics of ecosystem engineers. *Oikos* 76:273–283.

Hajek, A.E., and Dahlsten, D.L. (1986). Coexistence of three species of leaf-feeding aphids (Homoptera) on *Betula pendula*. *Oecologia* 68:380–386.

Hawkins, B.A., and Sheehan, W. (1994). *Parasitoid Community Ecology*. Oxford: Oxford University Press.

Heads, P.A., and Lawton, J.H. (1985). Bracken, ants, and extrafloral nectaries. III. How insect herbivores avoid ant predation. *Ecological Entomology* 10:29–42.

Heinrich, B. (1979). Foraging strategies of caterpillars: Leaf damage and possible predator avoidance strategies. *Oecologia* 42:325–337.

Heinrich, B., and Collins, S.L. (1983). Caterpillar leaf damage and the game of hide and seek with birds. *Ecology* 64:592–602.

Hensen, W.R. (1958a). The effects of radiation on the habitat temperatures of some poplar-inhabiting insects. *Canadian Journal of Zoology* 36:463–478.

——. (1958b). Some ecological implications of the leaf-rolling habit in *Compsolechia niveopulvella* Chamb. *Canadian Journal of Zoology* 36:809–819.

Hertel, G.D., and Benjamin, D.M. (1979). Biology of the pine webworm *Tetralopha robustella* in Florida USA slash pine *Pinus elliottii* var. *elliottii* plantations. *Annals of the Entomological Society of America* 72:816–819.

Horsley, S.B., Long, R.P., Bailey, S.W., Hallett, R.A., and Wargo, P.M. (2002). Health of eastern North American sugar maple forests and factors affecting decline. *Northern Journal of Applied Forestry* 19:34–44.

Hunter, M.D. (1987). Opposing effects of spring defoliation on late season oak caterpillars. *Ecological Entomology* 12:373–382.

Hunter, M.D., and Willmer, P.G. (1989). The potential for interspecific competition between two abundant defoliators on oak: Leaf damage and habitat quality. *Ecol. Entomology* 14:267–277.

Ide, J.-Y. (2004). Leaf trenching by Indian red admiral caterpillars for feeding and shelter construction. *Population Ecology* 46:275–280.

Ives, W.G.H., and Wong, H.R. (1988). *Trees and Shrub Insects of the Prairie Provinces.* Edmonton: Canadian Forestry Service.

Janzen, D.H. (1988). Ecological characterization of a Costa Rican dry forest caterpillar fauna. *Biotropica* 20:120–135.

Jeffries, M.J., and Lawton, J.H. (1984). Enemy-free space and the structure of ecological communities. *Biological Journal of the Linnean Society* 23:269–286.

Jones, C.G., Brodersen, H., Czerwinski, E.J., Evans, H.J. and Keizer, A.J. (1996). *Results of Forest Insect and Disease Surveys in the Northeast Region of Ontario, 1995.* Canadian Forest Service Information Report O-X I-VIII, pp. 1–27.

Jones, C.G., Lawton, J.H., and Shachak, M. (1994). Organisms as ecosystem engineers. *Oikos* 69:373–386.

——. (1997). Positive and negative effects of organisms as ecosystem engineers. *Ecology* 78:1946–1957.

Jones, M.T. (1999). Leaf shelter-building and frass ejection behavior in larvae of *Epargyreus clarus* (Lepidoptera: Hesperiidae), the silver-spotted skipper. Washington, D.C.: Georgetown University, M.S. thesis.

Jones, M.T., Castellanos, I., and Weiss, M.R. (2002). Do leaf shelters always protect caterpillars from invertebrate predators? *Ecological Entomology* 27:753–757.

Joos, B., Casey, T.M., Fitzgerald, T.D., and Buttemer, W.A. (1988). Roles of the tent in behavioral thermoregulation of eastern tent caterpillars. *Ecology* 69:2004–2011.

Karban, R., and Baldwin, I.T. (1997). *Induced Responses to Herbivory.* Chicago: University of Chicago Press.

Landis, D.A., Menalled, F.D., Costamagna, A.C., and Wilkinson, T.K. (2005). Manipulating plant resources to enhance beneficial arthropods in agricultural landscapes. *Weed Science* 53:902–908.

Lange, W.H. (1987). Insect pests of sugar beet. *Annual Review of Entomology* 32:341–360.

Larsson, S., Haggstrom, H.E., and Denno, R.F. (1997). Preference for protected feeding sites by larvae of the willow-feeding leaf beetle *Galerucella lineola. Ecological Entomology* 22:445–452.

Lawton, J.H. (1983). Plant architecture and the diversity of phytophagous insects. *Annual Review of Entomology* 28:23–39.

Le Corff, J., Marquis, R.J., and Whitfield, J.B. (2000). Temporal and spatial variation in a parasitoid community associated with the herbivores that feed on Missouri *Quercus. Environmental Entomology* 29:181–194.

Leonova, I.N., and Slynko, N.M. (2004). Life stage variations in insecticidal susceptibility and detoxification capacity of the beet webworm, *Pyrausta sticticalis* L. (Lep., Pyralidae). *Journal of Applied Entomology* 28:419–425.

Lill, J.T. (1999). Structure and dynamics of a parasitoid community attacking larvae of *Psilocorsis quercicella* (Lepidoptera: Oecophoridae). *Environmental Entomology* 28:1114–1123.

——. (2004). Seasonal dynamics of leaf-tying caterpillars on white oak. *Journal of the Lepidopterists' Society* 58:1–6.

Lill, J.T., and Marquis, R.J. (2001). The effects of leaf quality on herbivore performance and attack from natural enemies. *Oecologia* 126:418–428.

——. (2003). Ecosystem engineering by caterpillars increases insect herbivore diversity on white oak. *Ecology* 84:682–690.

——. (2004). Leaf ties as colonization sites for forest arthropods: An experimental study. *Ecological Entomology* 29:300–308.

Lill, J.T., Marquis, R.J., Walker, M., and Peterson, L. (2007). Ecological consequences of shelter-sharing by leaf-tying caterpillars. *Entomological Experimentalis et Applicata.* doi: 10.1111/j.1570-7458.2007.00546.x.

Lind, E.M., Jones, M.T., Long, J.D., and Weiss, M.R. (2001). Ontogenetic changes in leaf shelter construction by larvae of *Epargyreus clarus* (Hesperiidae), the Silver-spotted skipper. *Journal of the Lepidopterists' Society* 54:77–82.

Loeffler, C.C. (1996a). Adaptive trade-offs of leaf folding in *Dichomeris* caterpillars on goldenrods. *Ecological Entomology* 21:34–40.

——. (1996b). Caterpillar leaf-folding as a defense against predation and dislodgement: Staged encounters using *Dichomeris* (Gelechiidae) larvae on goldenrods. *Journal of the Lepidopterists' Society* 50:245–260.

Lohman, D.J., Zangerl, A.R., and Berenbaum, M.R. (1996). Impact of floral herbivory by parsnip webworm (Oecophoridae: *Depressaria pastinacella* Duponchel) on pollination and fitness of wild parsnip (Apiaceae: *Pastinaca sativa* L.). *American Midland Naturalist* 136:407–412.

Low, C., and Connor, E.F. (2003). Birds have no impact on folivorous insect guilds on a montane willow. *Oikos* 103:579–589.

Marquis, R.J., Forkner, R.E., Lill, J.T., and Le Corff, J. (2000). Impact of timber harvest on species accumulation curves for oak herbivore communities of the Missouri Ozarks. In *Proceedings of the Second Missouri Ozark Ecosystem Project Symposium: Post-Treatment Results of the Landscape Experiment*, S.R. Shifley and J.M. Kabrick, Eds. Saint Paul, MN: USDA Forest Service, North Central Research Station, pp. 183–195.

Marquis, R.J., and Lill, J.T. (2006). Effects of arthropods as physical ecosystem engineers on plant-based trophic interaction webs. In *Indirect Interaction Webs: Nontrophic Linkages Through Induced Plant Traits*, T. Ohgushi, T.P. Craig, and P.W. Price, Eds. Cambridge: Cambridge University Press, pp. 246–274.

Marquis, R.J., Lill, J.T., and Piccinni, A. (2002). Effect of plant architecture on colonization and damage by leaftying caterpillars of *Quercus alba*. *Oikos* 99:531–537.

Martinsen, G.D., Floate, K.D., Waltz, A.M., Wimp, G.M., and Whitham, T.G. (2000). Positive interactions between leafrollers and other arthropods enhance biodiversity on hybrid cottonwoods. *Oecologia* 123:82–89.

Mattiacci, L., and Dicke, M. (1995). The parasitoid *Cotesia glomerata* (Hymenoptera: Braconidae) discriminates between first and fifth larval instars of its host *Pieris brassicae* on the basis of contact cues from frass, silk, and herbivore-damaged leaf tissue. *Journal of Insect Behavior* 8:485–497.

McGugan, B.M. (1958). *Forest Lepidoptera of Canada, Papilionidae to Arctiidae*, Vol. 1. Ottawa: Department of Forestry of Canada.

Messina, F.J., Jones, T.A., and Nielson, D.C. (1997). Host-plant effects on the efficacy of two predators attacking Russian wheat aphids (Homoptera: Aphididae). *Environmental Entomology* 26:1298–1404.

Miliczky, E.R., and Calkins, C.O. (2002). Spiders (Araneae) as potential predators of leafroller larvae and egg masses (Lepidoptera: Tortricidae) in central Washington apple and pear orchards. *Pan-Pacific Entomologist* 78:140–150.

Moore, W.S., Koehler, C.S., and Frey, L.S. (1982). *Fruittree Leafroller on Ornamentals and Fruit Trees*. Leaflet 21053c. Oakland: University of California Division of Agriculture and Natural Resources.

Morin, R.S., Jr., Liebhold, A.M., and Gottschalk, K.W. (2004). Area-wide analysis of hardwood defoliator effects on tree conditions in the Alleghany Plateau. *Northern Journal of Applied Forestry* 21:31–39.

Morris, R.F. (1972). Predation by insects and spiders inhabiting colonial webs of *Hyphantria cunea*. *Canadian Entomologist* 104:1197–1207.

Moskal, L.M., and Franklin, S.E. (2004). Relationship between airborne multispectral image texture and aspen defoliation. *International Journal of Remote Sensing* 25: 2701–2711.

Mothershead, K., and Marquis, R.J. (2000). Fitness impacts of herbivory through indirect effects on plant-pollinator interactions in *Oenothera macrocarpa*. *Ecology* 81:30–40.

Mueller, U.G., and Dearing, D. (1994). Predation and avoidance of tough leaves by aquatic larvae of the moth *Parapoynx rugosalis* (Lepidoptera: Pyralidae). *Ecological Entomology* 19:155–158.

Murakami, M. (1999). Effect of avian predation on survival of leaf-rolling lepidopterous larvae. *Research Population Ecology* 41:135–138.

Murdoch, C.L., Tashiro, H., Tavares, J.W., and Mitchell, W.C. (1990). Economic damage and host preference of lepidopterous pests of major warm season turfgrasses of Hawaii, USA. *Proceedings of the Hawaiian Entomological Society* 30:63–70.

Nakamura, M., and Ohgushi, T. (2003). Positive and negative effects of leaf shelters on herbivorous insects: Linking multiple herbivore species on a willow. *Oecologia* 136:445–449.

———. (2004). Species composition and life histories of shelter-building caterpillars on *Salix miyabeana*. *Entomological Science* 7:99–104.

Odell, T.M., and Goodwin, P.A. (1984). Host selection by *Blepjaripa pratensis*, a tachinid parasite of gypsy moth. *Journal of Chemical Ecology* 10:311–320.

Odling-Smee, F.J., Laland, K.N., and Feldman, M.W. (2003). *Niche Construction: The Neglected Process in Evolution*. Princeton, NJ: Princeton University Press.

Oki, Y., and Varanda, E.M. (2000). Lepidoptera rollers and *Byrsonima intermedia* relationship: The role of secondary metabolites and nutritional status. In XXIth International Congress of Entomology Abstracts. Iguassu Falls, Brazil: Embrapa.

Pasek, J.E., and Kearby, W.H. (1984). Larval parasitism of *Psilocorsis* spp. (Lepidoptera: Oecophoridae), leaftiers of central Missouri oaks. *Journal of the Kansas Entomological Society* 57:84–91.

Peterson, S.C. (1986). Breakdown products of cyanogenesis: Repellency and toxicity to predatory ants. *Naturwissenschaften* 73:627–628.

Pfeiffer, D.G., Kaakeh, W., Killian, J.C., Lachance, M.W., and Kirsch, P. (1993). Mating disruption to control damage by leafrollers in Virginia apple orchards. *Entomologia Experimentalis et Applicata* 67:47–55.

Powell, J.A. (1980). Evolution of larval food preferences in microlepidoptera. *Annual Review of Entomology* 25:133–159.

Prentice, R.M. (1962). *Forest Lepidoptera of Canada, Nycteolidae, Notodontidae, Noctuidae, Liparidae*, Vol. 2. Ottawa: Department of Forestry of Canada.

——. (1963). *Forest Lepidoptera of Canada, Lasiocampidae, Drepanidae, Thyatiridae, Geometridae*, Vol. 3. Ottawa: Department of Forestry of Canada.

——. (1965). *Forest Lepidoptera of Canada, Microlepidoptera*, Vol. 4. Ottawa: Department of Forestry of Canada.

Proulx, S.R., Promislow, D.E.L., and Phillips, P.C. (2005). Network thinking in ecology and evolution. *Trends in Ecology and Evolution* 20:345–353.

Reavey, D. (1991). Do birch-feeding caterpillars make the right feeding choices? *Oecologia* 87:257–264.

Remsen, J.V., Jr., and Parker, T.A., III. (1984). Arboreal dead-leaf-searching birds of the Neotropics. *The Condor* 86:36–41.

Robinson, S.K., and Holmes, R.T. (1982). Foraging behavior of forest birds: The relationships among search tactics, diet, and habitat structure. *Ecology* 63:1918–1931.

Roland, J. (1993). Large-scale forest fragmentation increases the duration of tent caterpillar outbreak. *Oecologia* 93:25–30.

Ruehlmann, T.E., Matthews, R.W., and Matthews, J.R. (1988). Roles for structural and temporal shelter-changing by fern-feeding lepidopteran larvae. *Oecologia* 75:228–232.

Ruohomaki, K., Chapin, F.S., III, Haukioja, E., Neuvonen, S., and Suomela, J. (1996). Delayed inducible resistance in mountain birch in response to fertilization and shade. *Ecology* 77:2301–2311.

Sagers, C.L. (1992). Manipulation of host plant quality: Herbivores keep leaves in the dark. *Functional Ecology* 6:741–743.

Sandberg, S.L., and Berenbaum, M.R. (1989). Leaf-tying by tortricid larvae as an adaptation for feeding on phototoxic *Hypericum perforatum*. *Journal of Chemical Ecology* 15:875–885.

Scoble, M.J. (1992). *The Lepidoptera: Form, Function and Diversity*. Oxford: Oxford University Press.

Seyffarth, J.A.S., Calouro, A.M., and Price, P.W. (1996). Leaf rollers in *Ouratea hexasperma* (Ochnaceae): Fire effect and the plant vigor hypothesis. *Revista Brasiliera de Biologia* 56:135–137.

Simard, M., and Payette, S. (2005). Reduction of black spruce seed bank by spruce budworm infestation compromises postfire stand regeneration. *Canadian Journal of Forest Restoration* 35:1686–1696.

Singh, O.P., and Rawat, R.R. (1979). Assessment of losses to mustard *Brassica juncea* ssp *juncea* by the cabbage webworm *Crocidolomia binotalis* (Lepidoptera Pyralidae). *Indian Journal of Agricultural Science* 49:967–969.

Sipura, M. (1999). Tritrophic interactions: Willows, herbivorous insects and insectivorous birds. *Oecologia* 121:537–545.

Stamp, N.E. (1992). Relative susceptibility to predation of two species of caterpillar on plantain. *Oecologia* 92:124–129.

Steiner, A.L. (1984). Observations on the possible use of habitat cues and token stimuli by caterpillar-hunting wasps: *Eudynerus foraminatus* (Hymenoptera: Eumenidae). *Quaestiones Entomologicae* 20:25–34.

Taylor, P.W., and Jackson, R.R. (1999). Habitat-adapted communication in *Trite planiceps*, a New Zealand jumping spider (Aranae: Salticidae). *New Zealand Journal of Zoology* 26:127–154.

Trimble, R.M., and Appleby, M.E. (2004). Comparison of efficacy of programs using insecticide and insecticide plus mating disruption for controlling the obliquebanded leafroller in apple (Lepidoptera: Tortricidae). *Journal of Economic Entomology* 97:518–524.

Vasconcelos, H.L. (1991). Mutualism between *Maieta guianensis* Aubl., a myrmecophytic melastome, and one of its ant inhabitants: Ant protection against insect herbivores. *Oecologia* 87:295–298.

Vishakantaiah, M., and Jagabeesh Babu, C.S. (1981). Bionomics of the turf webworm *Maruca testulalis* (Lepidoptera, Pyralidae). *Mysore Journal of Agricultural Science* 14:529–532.

Wagner, D.L. (2005). *Caterpillars of Eastern North America.* Princeton, NJ: Princeton University Press.

Wagner, M.R., and Raffa, K.F. (1993). *Sawfly Life History Adaptations to Woody Plants.* New York: Academic Press.

Walz, A.M., and Whitham, T.G. (1997). Plant development affects arthropod community structure: Opposing impacts of species removal. *Ecology* 78:2133–2144.

Weiss, M.R. (2003). Good housekeeping: Why do shelter-dwelling caterpillars fling their frass? *Ecology Letters* 6:361–370.

———. (2006). Defecation behavior and ecology of insects. *Annual Review of Entomology* 51:635–661.

Weiss, M.R., Lind, E.M., Jones, M.T., Long, J.D., and Maupin, J.L. (2003). Uniformity of leaf shelter construction by early-instar larvae of *Epargyreus clarus* (Hesperiidae), the silver-spotted skipper. *Journal of Insect Behavior* 16:465–480.

Weiss, M.R., Wilson, E.E., and Castellanos, I. (2004). Predatory wasps learn to overcome the shelter defences of their larval prey. *Animal Behavior* 68:45–54.

Wilby, A., Shachak, M., and Boeken, B. (2001). Integration of ecosystem engineering and trophic effects of herbivores. *Oikos* 92:436–444.

Willmer, P.G. (1980). The effects of a fluctuating environment on the water relations of larval Lepidoptera. *Ecol. Entomology* 5:271–292.

———. (1982). Microclimate and the environmental physiology of insects. *Advances in Insect Physiology* 16:1–57.

Wright, J.P., and Jones, C.G. (2006). The concept of organisms as ecosystem engineers ten years on: Progress, limitations, and challenges. *BioScience* 56:203–209.

Wright, J.P., Jones, C.G., and Flecker, A.S. (2002). An ecosystem engineer, the beaver, increases species richness at the landscape scale. *Oecologia* 132:96–101.

Zalucki, M.P., Clarke, A.R., and Malcolm, S.B. (2002). Ecology and behavior of first instar Lepidoptera. *Annual Review of Entomology* 47:361–393.

7

CARPOBROTUS AS A CASE STUDY OF THE COMPLEXITIES OF SPECIES IMPACTS

Nicole Molinari, Carla D'Antonio,
and George Thomson

7.1 ⚊ INTRODUCTION

A long-held interest among ecologists has been to understand traits that characterize species that have dramatic effects on ecosystem structure or functioning. From the plant perspective, Hans Jenny, in his classic work on soil formation, included "the organism," and in particular, plants, as one of the state factors affecting ecosystem development (1958). Likewise early plant ecologists such as Cowles and Clements recognized the important role that individual plant species play in ameliorating abiotic stresses and thus affecting rates or patterns of succession (Cowles 1911, Clements 1916). Hence the general importance of individual plants to shaping ecosystem development has long been recognized. Nonetheless, the context specificity of plant species effects and the mechanisms through which they are exerted was largely unstudied until the rise in interest in invasive, non-native species beginning in the early 1990s. Since that time many examples of plant species effects have been documented and several reviews written (e.g., Vitousek 1990, D'Antonio and Vitousek 1992, Mack and D'Antonio 1998, Ehrenfeld 2003, Levine et al. 2003, D'Antonio and Corbin 2003, D'Antonio and Hobbie 2005), including attention to the conceptual problem of defining types of impacts, and how and when they arise (e.g., Chapin et al. 1996, Parker et al. 1999). These latter authors stress that a species impact arises from a combination of its abundance and its per capita impacts. A species where individuals have a small per capita effect can still have a large impact if the species becomes very abundant. An example of such a species is a fire-promoting annual grass like *Bromus tectorum*: Individuals are diminutive in size, but together their horizontal continuity and slightly less decomposable tissue mass (relative to native annuals) creates a fine fuel bed that has dramatically changed sagebrush steppe ecosystems through the spread of lightning-ignited fires. By contrast, a species where individuals have a large per capita effect due to some unique feature may not have to be abundant numerically to have a notable impact. This is likely common for introduced top predators. For plants this condition can result when an alien species performs a novel function in its introduced range, such as a nitrogen-fixing tree invading early successional sites previously lacking them (Vitousek et al. 1987). Even then it is not clear at what point effects are observed beyond the immediate sphere of an individual plant.

Simultaneous with this rise in interest in species invasions has been a rise in interest in the topic of individual species as ecosystem engineers (Jones et al. 1994, 1997). Generally speaking, species that are considered ecosystem engineers directly or indirectly alter biotic and abiotic materials within an environment. Autogenic engineers change the physical structure through their own body form while allogenic engineers indirectly transform the environment through their effects on important ecosystem processes (Jones et al. 1994). Numerous species may fall under these broad categories, but those species that are necessary to maintain valued ecosystem structure and functioning should be of primary concern to managers and conservation biologists. Likewise, species whose invasion causes dramatic negative changes in important or valued ecosystem properties should also be of conservation concern, particularly species that create positive feedbacks that are difficult to disrupt or reverse. Those invader species that alter large-scale ecosystem structure and function have been referred to as "transformer species" (Richardson et al. 2000a) and these may also be considered to be ecosystem engineers under a broad definition of engineering. Whether particular pathways of engineering are more common among invasive non-native species is not known. Crooks (2002) reviewed invasive species impacts in light of ecosystem engineering and concluded that many allogenic engineer-invaders have unique traits that are responsible for major changes such as alterations in fire regimes (e.g., fire-enhancing grasses; D'Antonio and Vitousek 1992, D'Antonio 2000) or large increases in soil nitrogen pools

(Vitousek et al. 1987, Stock et al. 1995, Yelenik et al. 2004). Some of these invaders, such as fire-enhancing grasses, create positive feedbacks that further promote the altered ecosystem state. Crooks's discussion of unique traits is identical to the earlier suggestion of Chapin et al. (1996) that "discrete trait invaders" would have larger effects than "continuous trait invaders"—that is, species with discretely different traits would have effects that are manifested more rapidly and intensely than species with traits that overlapped with residents. Whether all of these discrete trait invaders qualify as engineers depends on the breadth of definition of this concept. More critical from our perspective is whether impacts are predictable, cause substantial changes in the living conditions for other species, and are readily reversible.

Despite widely cited examples that support the hypothesis that trait uniqueness is critical to impact, very few studies of plant invaders actually compare species impacts across a range of environments where residents vary in their trait values relative to invaders or where invaders vary in their biomass or abundance. Such evaluations would aid in predicting impacts of future invasions and generate an understanding of how rapidly impacts emerge, where they are likely to be most severe, and how hard they are to reverse. Within a single habitat, the impact of a species as an "engineer" could depend on how abundant it is within the site. When the species is uncommon, as in early invasion stages, it may not have a big enough effect to be classified as an engineer, and it may be only when the species completely dominates the site that it can be fully categorized as an engineer.

Most definitions of ecosystem engineering involve descriptions of species that alter habitat structure and abiotic features of the environment in some way. An unanswered question is when might an invader simplify versus diversify environmental heterogeneity? Crooks (2002) points out that some invaders enhance species diversity while others decrease it, and this may be the result of whether the invader adds structure or simplifies structural complexity and homogenizes resource availability. Similarly, comparison of an invader's ability to alter ecosystem properties across habitats allows insight into the conditions under which an invader might facilitate further spread (ala Simberloff and Von Holle 1999, Richardson et al. 2000b) versus reduce the likelihood of further spread.

Here we present a discussion of one invasive plant species that we believe can act as both an autogenic and an allogenic engineer potentially influencing resident species composition and ecosystem processes through any of several pathways. Our goals are: (1) to demonstrate how a single species can have effects through multiple pathways, some of which are ecosystem engineering; (2) to explore the context specificity of impact; and (3) to stimulate discussion about the strength and reversibility of various impacts. Rather than demonstrating that one species is definitively an ecosystem engineer, we aim to demonstrate the usefulness of invasive species as tools for studying questions about the relationship between abundance and impact, invader and resident relative traits, and reversibility of different types of impacts.

7.2 ● CARPOBROTUS AS AN ECOSYSTEM ENGINEER

SPECIES BACKGROUND

Carpobrotus edulis (highway iceplant, hottentot fig) is a succulent, mat-forming perennial plant native to South Africa that grows almost like a vine, spreading from an initially central stem and able to overgrow adjacent plants growing low to the ground (Figure 7.1). Stems root at the nodes as they grow and individual branches can live independently if severed from the initial main stem. Individual plants can live for decades. Stems layer over one another forming mats that can reach up to 30 cm

FIGURE 7.1 Transect through a highly invaded backdune community at Vandenberg Air Force Base, California. The mat forming *C. edulis* can be seen along the entirety of the transect. (See color plate.)

or more in depth and more than 8 m in diameter (D'Antonio, personal observation). The growth form can be thought of as a dense blanket that carpets the soil surface. Stems grow up onto short stature shrubs and subshrubs and eventually completely cover them. Occasionally stems are found blanketing shrubs up to 1 m tall, although this is not typical.

Carpobrotus edulis is considered an invasive weed in California, Australia, France, Spain, and the Balearic Islands of the Mediterranean. It was apparently introduced into California in the early 1900s from South Africa because of its potential to stabilize dune soils. By the 1980s it was recognized as an invasive species of native coastal habitats (e.g., Hoover 1970, Zedler and Scheid 1988). D'Antonio (1990) demonstrated that its indehiscent fruits are eaten by a variety of native vertebrates that inadvertently disperse the seed to new locations.

Controls over the invasion of *C. edulis* into "native" or unmanaged plant communities have been well studied in California. Seedlings of the species are readily consumed by native generalist herbivores, slowing rates of invasion into most habitats (D'Antonio 1993). Establishment probabilities were lowest in coastal scrub sites due to extremely high rates of herbivory by rabbits. Seedlings in dune sites suffer from physiological stress and some herbivory resulting in intermediate establishment probabilities. Seed establishment probabilities are highest in grassland sites where soil disturbance by animals is highest and native vegetation is not competitive once *C. edulis* is established.

We chose to use *C. edulis* as a case study for several reasons. It is an ideal species to assess the impacts of organisms with unique traits, because it has succulent leaves and a mat-forming prostrate growth form that is largely unique in comparison to most other native Californian plant species in the habitats where it is invading. It grows low to the ground, which also contrasts with the generally taller growth forms that dominate several of the habitats where it invades. This is in contrast to many autogenic engineers that appear to introduce taller, more complex structures. Because the species invades a range of coastal habitats and invades relatively slowly, we could measure impact across different habitats at different stages of invasion within any one habitat type. While *C. edulis* has been demonstrated to have negative effects on the growth of neighbors (e.g., D'Antonio and Mahall 1991), mechanisms of impact are not fully understood. From the perspective of engineering, its dense blanket-like form alters the soil surface, the rooting environment, and the chemistry of the soil. At a larger scale, its provision of fruits to animals at a time of year when little else is available (D'Antonio 1990a) could have significant trophic-level impacts. Figure 7.2 presents a schematic of the array of impacts caused by this species. In following text we briefly

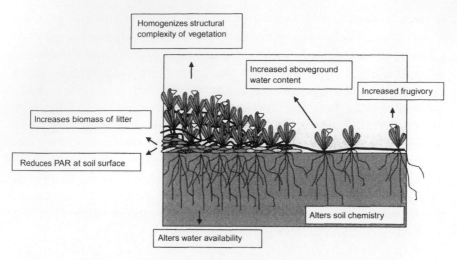

FIGURE 7.2 Conceptual diagram of multiple pathways of impact caused by *C. edulis*.

discuss these, using preliminary and unpublished data to make several points about impacts and the extent to which *C. edulis* functions as an ecosystem engineer.

IMPACTS RELATED TO ABOVEGROUND GROWTH FORM

We hypothesized that along an environmental gradient (dune to coastal sage scrub) *C. edulis* will have differential impacts depending on the community in which it invades. More specifically, we expected environmental modifications to be less extreme in the dune communities because they are dominated by more prostrate growth forms, while impacts should have been greater in shrub-dominated communities such as the coastal sage scrub and chaparral because of the difference in life form. As a first step toward testing this hypothesis, we sampled from two plant community types being invaded by *C. edulis*. We documented vegetation height, species affiliations, light availability at the soil surface, and vegetative and litter biomass at low, intermediate, and high abundances of *C. edulis* in backdune and coastal sage scrub sites at Vandenberg Air Force Base, in Santa Barbara County, California.

Vegetation height is an important component in the determination of a community's structural complexity. The maintenance of vegetative complexity is important to the preservation of the native fauna within a given community. Therefore structural engineering induced by the stature of *C. edulis* may provide insight into the impacts these autogenic changes have on invertebrate, bird, reptile, and mammalian

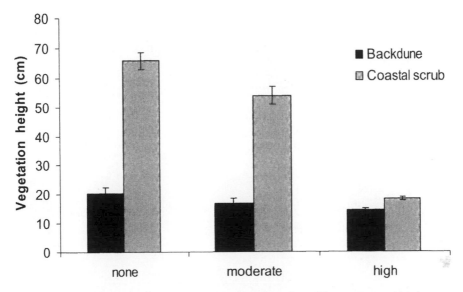

FIGURE 7.3 Average of maximum vegetation heights along 30 m transects. Height was measured every 0.3 m along non-, moderately, and highly invaded backdune and coastal scrub sites (N = 100). Effects of C. edulis invasion were not significant for backdune sites (ANOVA, F = 2.74, P = 0.066). In the coastal scrub, Turkey's post hoc analysis shows significant differences across all levels of invasion (ANOVA, F = 99.0, P < 0.001). Values are means and 1 SE.

communities. In our study, *C. edulis* differentially altered vegetation height and distribution in backdune and coastal sage scrub communities. Backdune communities in this region are typically dominated by low-growing species, such as *Abronia* spp., *Carex pansa* (a low-growing rhizomatous sedge), *Dudleya caespitosa*, and some prostrate shrubs (Holland and Keil 1995). In contrast to the backdune sites, coastal sage scrub is dominated by woody shrubs 1–2 m tall, with little herbaceous understory. We therefore predicted that when *C. edulis* becomes dominant in the coastal sage scrub, it should cause a decline in vegetation stature while this is less likely to be the case in the backdune. The backdune community that we sampled contained low-growing dune vegetation punctuated with scattered shrubs. We found no change in vegetation height across the three levels of *C. edulis* invasion (Figure 7.3) sampled, supporting our prediction that the vertical structure of backdunes will have minimal changes following the invasion of *C. edulis*.

By contrast, we found that the maximum height of vegetation declined dramatically in fully invaded coastal sage scrub sites (Figure 7.3). It did, however, remain the same for noninvaded and partially invaded coastal sage scrub sites. Indeed, our qualitative observations suggest that *C.*

TABLE 7.1 A point intercept was employed along an invasion gradient in back-dune and coastal scrub communities. Hits at the soil surface were recorded every 0.3 meters along a 30-meter transect in uninvaded, moderately invaded, and highly invaded sites.

	Bare Ground	Native Litter	*C. edulis* Litter
Backdune			
None	58	41	0
Moderate	19	55	26
High	4	0	96
Coastal scrub			
None	5	91	0
Moderate	0	20	82
High	0	0	100

edulis actually may increase the structural complexity of sites where it dominates only partially. This occurs through the addition of a low-growing succulent understory to an otherwise shrub-dominated community with occasional erect herbs. By contrast, coastal scrub sites completely dominated by *C. edulis* had almost completely lost their shrub component. D'Antonio (1993) and Vila and D'Antonio (1998a) both suggest that invasion of *C. edulis* into shrublands is a slow process due to herbivory by resident rabbits. Yet it proceeds as *C. edulis* plants find spatial or temporal refuges from herbivory. Fire can trigger the loss of shrubs and speed conversion to full *C. edulis* dominance, at which point the community has very low structural complexity and does not revert to a shrubland.

Additional engineering brought about by alterations in horizontal vegetation cover may also affect organisms, including other plant species. For example, alterations in vegetative cover may have profound influences on seedling recruitment because thick matted vegetation can function as a barrier to seed germination or to seedlings being able to get their roots down into mineral soil. Because we are particularly interested in how *C. edulis* alters the soil surface, we performed point intercept readings across a 30 m transect in uninvaded, moderately, and highly invaded sites. We found that in the backdune there was a decrease from roughly 60% bare ground in noninvaded sites to 4% in invaded sites (Table 7.1). We also found that highly invaded backdunes had a reduction in litter deposition by native species (Table 7.1), but had a significant increase in total litter as a result of increased accumulation under mats of *C. edulis* (Table 7.1 and Figure 7.4). This finding is most likely a

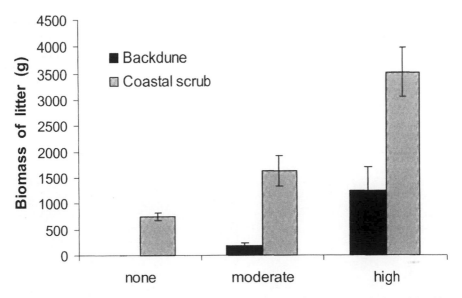

FIGURE 7.4 Biomass of dry litter along noninvaded, moderately invaded, and highly invaded backdune and coastal scrub transects. A minimum of three 1 m × 1 m plots were sub-sampled per site. Post hoc analysis determined that litter was significantly greater in highly invaded backdune (ANOVA, F = 4.65, P < 0.05) and coastal scrub sites (ANOVA, F = 19.85, P < 0.01). Values are means and 1 SE.

result of highly invaded sites having a significant reduction in the abundance of native species coupled (Figure 7.5) with a 15-fold increase in total community biomass that accompanies *C. edulis* invasion (wet biomass (g) = 417 ± 190 SE in noninvaded and 6124 ± 905 SE in highly invaded sites).

In contrast to dune sites, noninvaded coastal sage scrub sites were characterized by having very little bare ground and an abundance of native species litter (Table 7.1). As coastal scrub sites were invaded by *C. edulis* there was a decrease in native litter from 90% to 20% in intermediate invaded sites, which was further reduced to zero in the fully invaded site (Table 7.1). While it might appear superficially that the shift in dominance from native litter to *C. edulis* litter is just a matter of replacement with little effect on seedling establishment, in fact the biomass of *C. edulis* litter is quite different from the biomass of native litter (in pre-invasion condition). There was a 450% increase in the weight of litter per meter^2 between noninvaded and invaded sites (Figure 7.4). This litter tended to carpet the soil in a 5–10 cm layer and could be a physical barrier to seed germination or establishment.

Vegetative cover generated by dense *C. edulis* mats significantly altered light availability at the soil surface of both backdune and coastal scrub

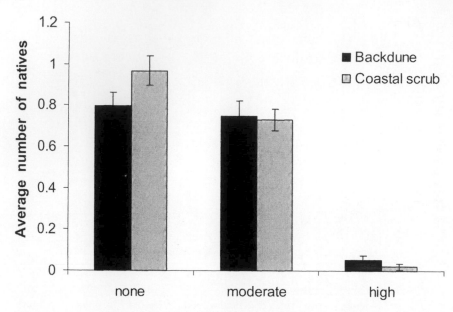

FIGURE 7.5 Average number of native plant species hit along noninvaded, moderately invaded, and highly invaded backdune and coastal scrub transects. Line intercept method was employed, and the total number of native species hit every 0.3 m along the 30 m transect (N = 100) was recorded. Post hoc comparisons show that native species were significantly reduced in highly invaded backdune sites (ANOVA, F = 47.18, P < 0.001). In addition, there was a significant decrease in native species across all levels of invasion in coastal scrub sites (ANOVA, F = 92.62, P < 0.001). Values are means and 1 SE.

communities (Figure 7.6). For example, highly invaded backdune sites had roughly a 40% decrease in photosynthetic active radiation (PAR) at the soil surface when compared to noninvaded sites (Figure 7.6), potentially making it difficult for seedlings of any species to become established or for plants that resprout from basal meristems to regrow each year. The shrub canopy in the "native" coastal sage scrub site resulted in substantially lower light levels at the soil surface than in backdune sites (Figure 7.6). Nonetheless, the coastal sage scrub canopy is not impenetrable to light, and patches of PAR are present but variable at the soil surface. As a result, similar to dune sites, invaded coastal sage scrub experienced a significant decrease in light availability at the soil surface (Figure 7.6) due to the continuous vegetative cover in invaded *C. edulis* sites.

Our investigations suggest that *C. edulis* effectively simplifies communities, through its ability both to reduce vegetation height and to homogenize horizontal vegetative cover. Even though vertical structure

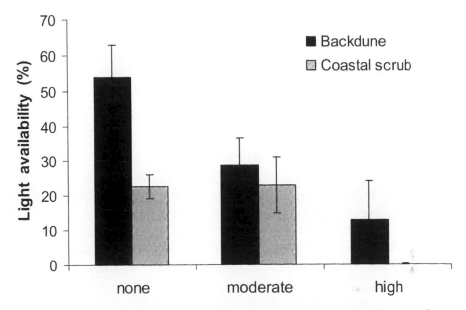

FIGURE 7.6 Percent of available photosynthetic active radiation (PAR) at the soil surface. Measurements were taken every 3 m along a 30 m transect in noninvaded, moderately invaded, and highly invaded backdune and coastal scrub site (N = 10). Post hoc analysis shows a significant difference between noninvaded and highly invaded backdune sites (ANOVA, F = 5.09, P < 0.05) and significantly less light availability in highly invaded coastal scrub sites (ANOVA, F = 4.62, P < 0.05). Values are means and 1 SE.

in the dune community was not affected by *C. edulis* invasion, we found that horizontal cover, light availability, and native species richness were decreased along highly invaded backdune and coastal sage scrub transects. This environmental homogenization may happen along different invasion trajectories in the backdune and coastal sage scrub. Although *C. edulis* seedlings are susceptible to physiological stress in the backdune and young plants grow slowly there (D'Antonio 1993), loss of open space will be a strict function of invasion and growth rates. Invasion into coastal scrub sites will proceed very slowly because of intense herbivory by native generalist herbivores (D'Antonio 1993, Villa and D'Antonio 1998a), thus creating a long period of time when coastal scrub communities may be more structurally diverse.

Regardless of temporary enhancement to the coastal scrub community, however, the fate of the coastal sage scrub in the face of *C. edulis* invasion is ultimately similar to that of the backdune: reduced structural and physiognomic heterogeneity due to invasion. A trigger such as fire that reduces the coastal sage scrub shrubs while not affecting iceplant will likely push the coastal sage scrub community more rapidly into full *C. edulis* dominance. Iceplant is unaffected by fire because of its highly succulent nature and close conformity to the soil surface. The inability of native plant species to regenerate in highly invaded *C. edulis* communities may result from the dense mat-forming nature of *C. edulis*, its effects on light resources, and its access to the soil surface.

BELOWGROUND IMPACTS OF *C. EDULIS*

Carpobrotus edulis, like several other dune-growing species such as European beach grass (*Ammophila arenaria*), was introduced into California to stabilize sand dunes. Its ability to stabilize dunes is a result of its dense fibrous rooting in the upper meter of the soil profile. Most native dune species in California are deeply rooted without dense fibrous roots immediately below the surface. While careful comparisons of *C. edulis* and co-occurring California native species rooting patterns have not been made, we do know that in some habitats *C. edulis* has extremely dense fibrous roots that are fundamentally different in density, biomass per unit area, and distribution than some of the California coastal shrubs (D'Antonio and Mahall 1991). These authors showed changes in rooting pattern in two native shrub species due to *C. edulis* invasion with apparently detrimental effects on water availability to the shrubs. This in itself may simply be viewed as interference competition for a limiting resource, but there may be other structural changes to the belowground environment due to this dense rooting pattern that qualify this species as a "belowground" engineer.

D'Antonio (1990b) documented that *C. edulis* can alter soil chemical properties with the most pronounced and consistent effects being on soil pH and calcium. Sampling four different community types, she found that *C. edulis* can dramatically reduce soil pH in grassland, coastal scrub, and maritime chaparral ecosystems (Figure 7.7) but found very little evidence of a consistent pH effect in dune sites. These data therefore suggest that changes to soil chemistry are habitat specific. The mechanism through which pH is altered is unknown and may be related to salt uptake by growing plants, H^+ ion exudation to balance ammonium or other cation uptake or production of organic acids during decomposition of the abundant litter that accumulates under *C. edulis*. In support of this latter hypothesis D'Antonio (1990b) shows a small but significant decline in pH could be induced by piling iceplant litter onto a grassland soil. Understanding this mechanism is important because it would provide insight into the situations in which *Carpobrotus* will have strong effects and also the potential for reversal of those effects. Further sampling across a broad range of sites in coastal California with varying levels of *Carpobrotus* abundance suggests that the alteration of soil

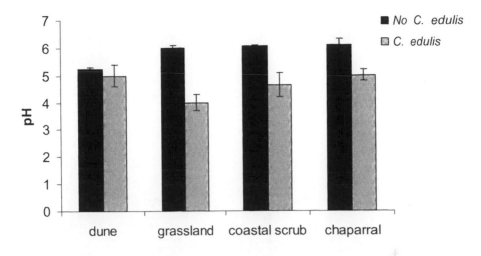

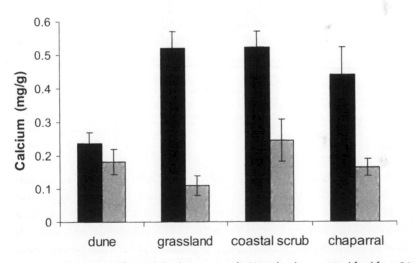

FIGURE 7.7 Impacts of *C. edulis* clones on soil pH and calcium. Modified from D'Antonio (1990b). Soil cores were taken (10 cm deep) from under 12 clones of *C. edulis* spread out across a large area of each vegetation type and compared with soils from under "native" vegetation 1–2 m away from the *C. edulis* plants. Only *C. edulis* plants more than 3 m in diameter were sampled. Soil samples were air dried and then pH measured in a 2:1 water-to-soil slurry. Calcium was measured in mg/g soil after ammonium acetate extraction of air dried soils. For both pH and calcium, differences between under and away from *C. edulis* were significant at $P < 0.001$ using two-way ANOVAs, except for at the backdune sites where $P > 0.16$.

chemistry by *C. edulis* is somewhat proportional to abundance (Figure 7.8). In sites where cover of *Carpobrotus* was higher, soil pH was generally lower. In France, Suehs (2005) found that *C. edulis* had a significant acidifying effect on soil pH.

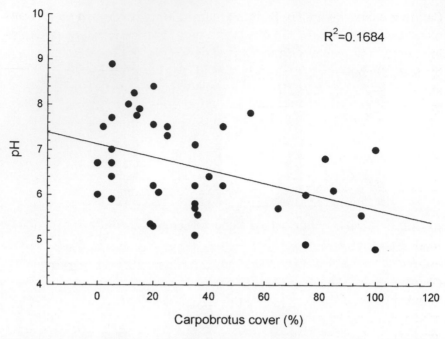

FIGURE 7.8 Relationship between *Carpobrotus* cover and soil pH across sites in coastal California. Figure modified from Albert (1995). Regression: R square = 0.1684, P = 0.009, N = 39.

In addition to altering soil pH, D'Antonio (1990b) found that *Carpobrotus* caused significant declines in soil calcium (Figure 7.7) at the acidified sites and a weak but significant overall increase in sodium across all sites combined (data not shown). All sites showed more sodium under *C. edulis* (average = 0.116 mg/g soil under versus 0.096 away), but variability was high among samples and sites so that the positive effect of *C. edulis* on sodium is seen only when all sites are considered together. It is not known whether the decreased levels of calcium observed would affect the ability of other species to grow in these soils although the percentage reduction is high (approximately 70%). D'Antonio (1990b) followed soil cation levels after *C. edulis* removal in a grassland site where *C. edulis* had caused a fivefold reduction in soil Ca^{++}. After 31 months, levels in removal plots were unchanged suggesting this soil property is slow to recover. Calcium is deposited in coastal sites through marine aerosol deposition or dust (e.g., Schlesinger and Hasey 1980, Chadwick et al. 1999). Hence it could be slowly replaced after *C. edulis* is removed from the site although this did not appear to happen over the 2-plus years of monitoring. Schlesinger and Hasey (1980) found deposition of

Ca^{++} in a chaparral area in Santa Barbara County was much greater on plant leaves than away from plants and perhaps the lack of leaf material in *C. edulis* removal plots impedes the deposition of elements back into the soil. The buildup of sodium in soils under *C. edulis* is relatively slight and it is not known if levels can become high enough as to inhibit other species.

7.3 ● DISCUSSION

CARPOBROTUS EFFECTS ON ABOVEGROUND HETEROGENEITY

In a review of habitat heterogeneity, Tews et al. (2004) found 75 studies positively linking structural complexity and biodiversity. In addition, they pose that the scale at which heterogeneity is measured plays an integral role in its affect on biodiversity of particular groups of species (Tews et al. 2004) because the scales at which microfauna and macrofauna operate (e.g., home range and dispersal) are variable. *Carpobrotus edulis* can exert impacts as an autogenic engineer at multiple scales. At local scales, *C. edulis* homogenizes architectural complexity, which may result in local extirpation of taxa via competitive exclusion of other plant species that function as autogenic engineers. In the case of *C. edulis*, its low growth form may exclude fauna that require large stature plants for nesting sites or shelter. However, micro- and macro-invertebrates may utilize the physical growth form of *C. edulis* and therefore may be locally abundant in highly invaded patches. At the landscape scale, patches of invaded habitat increase the heterogeneity of the regional community. Based on the work presented by Tews et al. (2004) we expect that this increased large-scale heterogeneity may be coupled with increased biodiversity. Although in theory increased heterogeneity should increase biodiversity, the fact remains that low-growing, dense architecture and dense litter of *C. edulis* creates patches of land seemingly uninhabitable by other species. This uninhabitable matrix can be considered a novel type of autogenic engineering when viewed from the landscape scale.

At local scales, the introduction of a plant species into new habitats can have three consequences on the structural heterogeneity of the vegetation. First, it might be assumed that, when the invader is similar in growth form to the native plant species, the invader will have minimal impacts as an autogenic ecosystem engineer. However, Gjerde and Saetersdal (1997) found that bird diversity was highest in mosaics of native pine forests mixed with nonnative spruce, but was lowest in pure stands of nonnative spruce. The mechanisms driving avian communities to prefer pure stands of pine over non-native spruce may be due to alterations of trophic interactions associated with reductions in native pine species, rather than architectural alterations reducing habitat and breeding sites. We suspect that the per capita impact of invaders with similar growth form as the resident community may be minimal in comparison to invasive species with unique growth forms that alter both the physical environment and trophic interactions.

The introduction of unique growth forms can alter vegetation structure by either increasing or decreasing the height of the vegetation. For example, the addition of the non-native tree, *Melaleuca quinquenervia*, adds a woody component to an otherwise low-growing sawgrass wetland in Florida. During moderate levels of invasion, species richness of wildlife increased in the wetlands (mainly driven by avian communities, but also by mammals and macro-invertebrates) (O'Hare and Dalrymple 1997). However, in monospecific stands of *M. quinquenervia* there are reductions in fish, macro-invertebrate, and mammalian abundance (Ceilley et al. 2005), indicating that monospecific stands decrease goods and services. The mechanism through which *M. quinquenervia* disrupts the functioning of sawgrass wetlands is not fully understood, but its impacts are most likely a combination of habitat engineering and its ability to break down trophic interactions. Lastly, there are few examples of invasive species that decrease the height of vegetation as does *C. edulis*. In coastal scrub sites we found that highly invaded communities were transformed from shrublands to low-growing mats of *C. edulis*. The loss of the shrub component in the coastal scrub will surely have impacts on habitat use by other organisms and on trophic interactions, as well as biophysical effects on the local microclimate.

In addition to altering the vegetation structure, a species like *C. edulis* can affect higher trophic levels by providing food or pollen resources that supplement those otherwise available. *Carpobrotus edulis* has large, showy flowers that are open from March to June. Hybridization with another introduced congener, *C. chilensis*, can further extend the flowering season (Vila et al. 1998). The dense nature of *Carpobrotus* flowers and the long season of flowering relative to most native species (personal observation) suggest that the presence of *Carpobrotus* in California could supplement the wide range of native generalist pollinators that appear to use it (Vila et al. 1998). Whether this in turn disrupts native plant–pollinator interactions in California is not known. *C. edulis* has been shown in the Balearic Islands to disrupt the pollination success of native plant species, although this depended on native composition, insect abundance, and the abundance of *C. edulis* (Moragues and Raveset 2005). The eventual homogenization of pollen resources as *Carpobrotus* takes over a site could lead to loss of specialist insect pollinators or herbivores.

Carpobrotus edulis may also exert impacts on native mammal populations by dramatically altering the quantity and quality of food supply. While the fruiting of many coastal species in California has similar phenology as *C. edulis*, the native species typically have dry pappus seeds that likely result in a lower nutritional value. By contrast, *C. edulis* has a fleshy fruit with 80% water content and high energy content (Vila and D'Antonio 1998c). Native mammals such as brush rabbits (*Sylvilagus bachmanii*), mule deer (*Odocoileus hemionus*), ground squirrels (*Spermophilus beecheyi*), and black-tailed jackrabbits (*Lepus californicus*) are common in these habitats. The period when *C. edulis* fruits are ripe (typically summer and early fall) are the driest times of year in this Mediterranean climate, and the provision of fruits to these mammals may supplement populations at a time that is otherwise physiologically stressful for these species. D'Antonio (1990) and Vila and D'Antonio (1998c) observed >90% fruit harvesting rates in *C. edulis* in California during summer and fall. Increased mammal populations could accelerate the loss of other plant species as *Carpobrotus* invades. In addition, increases in frugivorous mammal populations may exert strong trophic alterations via increased numbers of higher predators. The quantity and quality of fruit produced by *C. edulis* are distinct in comparison to native plant species and therefore may have dramatic impacts directly on frugivorous mammals, as well as indirect effects on primary producers and predators.

Overall, there are two emerging themes. First, autogenic engineers will most likely have the greatest impacts on communities in which they are structurally unique. This is not to say that spruce invasion into pine forests and *C. edulis* invasion into backdunes will not have large impacts on these systems; rather, they will most likely affect these systems by altering trophic interactions. We suspect that invasions that introduce a novel growth form into a resident community (such as *M. quinquenervia* into wetland or *C. edulis* into shrub-dominated communities) will have more severe impacts, because not only is there degradation of trophic interactions between native species, but there also is the minimization of shelter and breeding sites brought about through autogenic engineering. Second, as invasions progress, the ecosystem level effects change. In moderately invaded sites there may be a temporary increase in community complexity, resulting in a temporary increase in biodiversity. However, as the invasion progresses, the end product is quite often a monoculture that may facilitate only a few species (such as *Carex pansa* [see following text] or small herbivores for *C. edulis*) but create a net loss of biodiversity through homogenization of the environment resulting in resource loss for multiple species.

ABIOTIC STRESS AND ENGINEERING

In previous studies, D'Antonio (1993) and Vila and D'Antonio (1998a, 1998b) demonstrate that backdune sites are more stressful locations for *C. edulis* growth and establishment than are more stabilized and better vegetated sites such as the coastal scrub site studied here or coastal grassland sites. Dunes typically have more open soil, less stable soil, and greater wind and water stress than more stabilized soil formations such as those that support chaparral or coastal sage scrub. Nonetheless, *Carpobrotus* tolerates the stress of the dune environment well and ultimately can reach almost continuous cover in such sites as demonstrated by our transect through the invaded backdune and through the extensive surveys of Albert (1995).

Numerous studies have validated the importance of facilitation in abiotically stressful environments including dunes (e.g., Lortie and Turkington 2002, Lortie and Callaway 2006). Theoretically, then, a stress-tolerant species like *C. edulis* could facilitate the establishment of other species by ameliorating the abiotic stresses in the local habitat. This would occur if its positive effects on the physical environment outweighed biological interactions such as competition for soil resources that might also occur. In contrast to the dune environment, the coastal sage scrub site has more soil organic matter and higher vegetative cover, and it should be less subject to wind and water stress. As a result, the potential importance of facilitation through amelioration of abiotic stress by *C. edulis* should be low at this site.

We found some evidence for facilitative effects of *C. edulis* in this dune site. This evidence consisted of an increase by 8 cm of the average height of the sedge *Carex pansa*, which was often grazed close to the soil surface in the absence of *C. edulis* (Figure 7.9). *Carex pansa* was the most commonly encountered native backdune species at our study sites. This rhizomatous spreading species appears to tolerate growing under some *C. edulis* and may perhaps benefit by escaping herbivory in its presence. However this species was in low abundance in the completely invaded dune site (Figure 7.9). Unfortunately we do not know if it was never there or if it succumbs to competition when *C. edulis* cover gets high. Like *C. edulis*, *C. pansa* proliferates most of its roots in the upper half meter of soil and so in the long run it may compete more directly with *C. edulis* for root space than does the more deeply rooted *E. ericoides*, which persisted even in the heavily invaded dune. We also found slightly higher

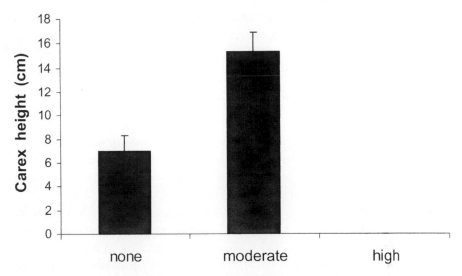

FIGURE 7.9 Average height of *Carex pansa* along a 30m transect in a noninvaded, moderately invaded, and highly invaded backdune. Post hoc analysis shows differences in *C. pansa* height across the invasion gradient (ANOVA, F = 12.44, P < 0.01, $N_{(none)}$ = 18, $N_{(moderate)}$ = 31, $N_{(high)}$ = 0). Values are means and 1 SE.

abundance of *Dudleya caespitosa* along the partially invaded transect compared with either of the other dune transects. We cannot rule out a pre-existing difference between the sites but the sites are in very close proximity and have the same slope and aspect. Also D'Antonio (unpublished) found that heavy herbivory by deer occurred on *Dudleya* when *C. edulis* was removed from around it, suggesting a protective function offered by *C. edulis*. Despite the lack of rigorous data relevant to this issue, we feel that invasive species entering a range of abiotic environments offer the opportunity to explore when ecosystem engineers might act as facilitators via their effects on the harshness of the environment. In this case, the modulating effect of the invader on the soil surface was overridden by either the volume or chemistry of the accumulated litter material.

CARPOBROTUS AS A SOIL CHEMISTRY ENGINEER

Since the rise in interest in impacts of invasive species, many studies have documented that non-native species can alter soil properties including total and available nitrogen pools, soil carbon, soil salinity, and soil microbial communities (Vivrette and Muller 1977, Ehrenfeld 2003, D'Antonio and Corbin 2003, Hawkes et al. 2006, Batten et al. 2006). Few of these changes have been definitively linked to altered species composition and ecosystem development. Likewise the effect of *C. edulis* on the soil is a clear and potentially biogeochemically important plant-driven change, but its implications for long-term successional change or ecosystem function change are less clear. It has not been demonstrated that the changes to pH or cations are creating positive feedbacks that favor further *C. edulis* growth. *Carpobrotus edulis* patches can persist almost unchanged for decades (D'Antonio, personal observation), but the relative contributions of soil effects, litter buildup, outright shading by the *C. edulis* canopy, or competition for water and nutrients in causing declines to natives has yet to be determined.

A close relative of *Carpobrotus edulis, Mesembryanthemum crystallinum* (crystalline iceplant) was one of the first plant species to be demonstrated to have a strong effect on soil chemistry and in particular soil salinity (Vivrette and Muller 1977). Unlike *C. edulis,* however, *M. crystallinum* is an annual species that largely invades annual-dominated ecosystems, and the mechanism through which it alters soil chemistry and ecosystem dynamics is relatively straightforward. After germination in the fall, *M. crystallinum* plants concentrate salts from throughout the soil profile into live leaves thereby reducing soil salinity in the vicinity of their roots. But when plants cease to grow and then senesce in the summer and fall, the salts in the tissues are deposited on the soil surface resulting in very high salinity levels at the soil surface. This in turn inhibits the germination and growth of other potentially co-occurring species and thereby controls community dynamics (Vivrette and Muller 1977). Whether this qualifies as allelopathy or engineering is semantic, but it is a clear example of a positive feedback whereby an invader alters soil chemistry in a way that favors its own persistence. By contrast, *C. edulis*'s effect on soil chemistry, although dramatic in some systems, may not be the main pathway through which it alters future plant community development. D'Antonio and Mahall (1991) demonstrated that *C. edulis* competes for water with native shrub species, and several investigators have documented reduced diversity of native species under *C. edulis* (Suehs 2005, Albert 1995). The extremely dense litter layer could easily inhibit germination of other species, which may be a more important initial impact than soil chemistry changes. Management removal or decomposition could reduce the impact of litter on germination of other species, while soil chemistry changes may take longer to reverse.

Alterations to soil pH are likely to be important to local ecosystem biogeochemistry after removal of iceplant. Soil pH is a fundamental soil characteristic that affects the solubility of ions in the soil solution and therefore nutrient availability to plant growth (Sposito 1989). In order for an acidified soil to become more basic, there must be an input of alkaline cations, and although this can come in from atmospheric deposition, it may take many years to replace lost cations and for pH to return to more typical levels. While the change to soil chemistry induced by *C. edulis* is dramatic, it proves difficult to pinpoint as the cause of long-term changes in the successional trajectory and development of these sites.

7.4 ⬤ CONCLUSIONS

Carpobrotus edulis, like many plant invaders, can affect aspects of ecosystem structure and functioning in multiple ways. Its success in taking over these environments is a function of both its ability to become abundant via seed dispersal and escape from herbivory, and its ability to grow into a dense, low-growing mat once established. This essentially two-dimensional blanket changes soil surface properties, litter and moisture distributions, and ultimately the physical continuity and height of vegetation across sites. Nonetheless, its impacts do not occur immediately and they are different depending on the state of the invasion and the community being invaded. Such context-dependent variation in species impacts are probably the rule but have rarely been documented.

Species such as *C. edulis* are of conservation concern because very few resident species appear to co-exist with them and they have the potential to affect other trophic levels. *Carpobrotus edulis* was introduced because of its ability to engineer the soil, yet its affect on soil stability is probably not the reason why it is having such a large impact on resident species. Its impacts do, however, likely relate to the very dense rooting structure it creates and its influence on the soil environment, as well as its ability to modify aboveground structure. If such an invader is removed, its impacts will not be as readily reversed as those of an invader whose impacts are largely the result of aboveground structures. Such comparative studies of invader impacts and reversibility of impacts could prove insightful to our understanding ecosystem development in a conservation and restoration context.

REFERENCES

Albert, M. (1995). *Morphological variation and habitat associations within the Carpobrotus species complex in coastal California*. Ph.D. dissertation, University of California, Berkeley.

Batten, K., Scow, K., Davies, K.F., and Harrison, S.P. (2006). Two invasive plants alter soil microbial community composition in serpentine grasslands. *Biological Invasions* 8:217–230.

Ceilley, D.W., Buckner, G.G., Schmid, J.R., and Smith, B.W. (2005). A survey of the effects of invasive exotic vegetation on wetland functions: Aquatic fauna and wildlife. Charlotte National Estuary Program. Accessed Novemeber 4, 2006 at www.chnep.org/Grants/R&R_reports/InvasiveExoticPlantImpactsonWetlandFunction_CSWF.pdf.

Chadwick, O.A., Derry, L.A., Vitousek, P.M., Huebert, B.J., and Hedin, L.O. (1999). Changing sources of nutrients during four million years of ecosystem development. *Nature* 397:491–497.

Chapin, F.S., Reynolds, H., D'Antonio, C.M., and Eckhart, V. (1996). The functional role of species in terrestrial ecosystems. In *Global Change in Terrestrial Ecosystems*, B. Walker and W. Steffen, Eds. Cambridge: Cambridge University Press, pp. 403–428.

Clements, F.R. (1916). Plant succession: An analysis of the development of vegetation. Washington, D.C.: Carnagie Institute of Washington.

Cowles, H. (1911). The causes of vegetation cycles. *Botanical Gazette* 51:161–183.

Crooks, J. (2002). Characterizing ecosystem-level consequences of biological invasions: The role of ecosystem engineers. *Oikos* 97:153–166.

D'Antonio, C.M. (1990a). Seed production and dispersal in the non-native, invasive succulent *Carpobrotus edulis* L. (Aizoaceae) in coastal strand communities of central California. *Journal of Applied Ecology* 27:693–702.

——. (1990b). *Invasion and dominance of coastal plant communities by the introduced succulent, Carpobrotus edulis*. Ph.D. dissertation, University of California, Santa Barbara.

——. (1993). Mechanisms controlling invasion of coastal plant communities by the alien succulent, *Carpobrotus edulis*. *Ecology* 74:83–95.

——. (2000). Fire, plant invasions and global changes. In *Invasive Species in a Changing World*, H. Mooney and R. Hobbs, Eds. Covela: Island Press, pp. 65–94.

D'Antonio, C.M., and Mahall, B. (1991). Root overlap and interference for soil moisture between an invasive succulent and two native shrub species. *American Journal of Botany* 78:885–894.

D'Antonio, C.M., and Vitousek, P. (1992). Biological invasions by exotic grasses, the grass-fire cycle and global change. *Annual Review of Ecology and Systematics* 23:63–88.

D'Antonio, C.M., Odion, D., and Tyler, C. (1993). Invasion of maritime chaparral by the introduced succulent, *Carpobrotus edulis*: The roles of fire and herbivory. *Oecologia* 95:14–21.

D'Antonio, C.M., and Corbin, J.D. (2003). Effects of plant invaders on nutrient cycling: Using models to explore the link between invasion and development of species effects. In *Models in Ecosystem Science*, C.D. Canham, J.J. Cole, and W.K. Lauenroth, Eds. Princeton, NJ: Princeton University Press, pp. 363–384.

D'Antonio, C.M., and Hobbie, S. (2005). Plant Species Effects on Ecosystem Processes: Insights from Invasive Species. In *Insights from Invasive Species*, D. Sax, J. Stackowich, and S. Gaines, Eds. Sunderland, MA: Sinauer.

Ehrenfeld, J. (2003). Effects of exotic plants invasions on soil nutrient cycling processes. *Ecosystems* 6:503–523.

Gjerde, I., and Saetersdal, M. (1997). Effects on avian diversity of introducing spruce *Picea* spp. plantations in the native pine *Pinus sylvestris* forests of western Norway. *Biological Conservation* 79:241–250.

Hawkes, C., Belnap J., D'Antonio, C., and Firestone, M.K. (2006). Arbuscular mycorrhizal assemblages in native plant roots change in the presence of invasive exotic grasses. *Plant and Soil* 281:369–380.

Holland, V.L., and Keil, D.J. (1995). *California Vegetation.* Dubuque, IA: Kendall/Hunt Publishing Company.

Hoover, R.F. (1970). *The Vascular Plants of San Luis Obispo County, California.* Berkeley, CA: University of California Press, pp. 1–350.

Jenny, H. (1958). Role of the plant factor in the pedogenic function. *Ecology* 39:5–15.

Jones, C., Lawton, J., and Shachak, M. (1997) Positive and negative effects of organisms as physical ecosystem engineers. *Ecology* 78:1946–1957.

Jones, C.G., Lawton, J.H., and Shachak, M. (1994). Organisms as ecosystem engineers *Oikos* 69:373–386.

Levine, J.M., Vilà, M., D'Antonio, C.M., Dukes, J.S., Grigulis, K., and Lavorel, S. (2003). Mechanisms underlying the impacts of exotic plant invasions. *Proceedings of the Royal Society of London* 270:775–781.

Lortie, C.J., and Turkington, R. (2002). The effect of initial seed density on the structure of a desert plant community. *Journal of Ecology* 90:435–445.

Lortie, C.J., and Callaway, R.M. (2005). Re-analysis of meta-analysis: Support for the stress gradient hypothesis. *Journal of Ecology* 94:7–13.

Mack, M., and D'Antonio, C.M. (1998). Impacts of biological invasions on disturbance regimes. *Trends in Ecology and Evolution* 13:195–198.

Moragues, E., and Traveset, A. (2005). Effect of *Carpobrotus* spp. on the pollination success of native plant species of the Balearic Islands. *Biological Conservation* 122:611–619.

O'Hare, N.K., and Dalrymple, G.H. (1997). Wildlife in southern Everglades wetlands invaded by melaleuca (Melaleuca quinquenervia). *Bulletin of the Florida Museum of Natural History* 41:1–68.

Parker, I., Simberloff, D., Lonsdale, W., Goodell, K., Wonham, M., Kareiva, P., Williamson, M., Von Holle, B., Moyle, P., Byers, J., and Goldwasser, L. (1999). Impact: Toward a framework for understanding the ecological effects of invaders. *Biological Invasions* 1:3–19.

Richardson, D., Pysek, P., Rejmanek, M., Barbour, M., Panetta, F., and West, C. (2000a) Naturalization and invasion of alien plants: Concepts and definitions. *Diversity and Distributions* 6:93–107.

Richardson, D.M., Allsopp, N., D'Antonio, C.M., Milton, S., and Rejmanek, M. (2000b). Plant invasions—the role of mutualisms. *Biological Reviews* 75:65–93.

Schlesinger, W.M., and Hasey, M.M. (1980). The nutrient content of precipitation, dry fallout, and intercepted aerosols in the chaparral of southern California. *American Midland Naturalist* 103:114–122.

Simberloff, D., and Von Holle, B. (1999). Positive interactions of nonindigenous species. *Biological Invasions* 1:21–32.

Sposito, G. (1989). *The Chemistry of Soils*. New York: Oxford University Press.

Stock, W., Weinand, K.T., and Baker, A.C. (1995). Impacts of invading N2-fixing Acacia species on patterns of nutrient cycling in two Cape ecosystems: Evidence from soil incubation studies and 15N natural abundance values. *Oecologia* 101:375–382.

Suehs, C.M. (2005). Ecological and evolutionary factors influencing the invasion of *Carpobrotus* spp (Aizoceae). In *The Mediterranean region*. Ph.D. dissertation, Universite Paul Cezanne Aix-Marseille III.

Tews, J., Brose, U., Grimm, V., Tielborger, K., Wichman, M., Schwager, M., and Jeltsch, F. (2004). Animal species diversity driven by habitat heterogeneity/diversity: The importance of keystone structures. *Journal of Biogeography* 31:79–92.

Vila, M., and D'Antonio, C.M. (1998a). Fitness of invasive *Carpobrotus* (Aizoaceae) in coastal California. *Ecoscience* 5:191–199.

——. (1998b). Hybrid vigor for clonal growth in *Carpobrotus* in coastal California. *Ecological Applications* 8:1196–1205.

——. (1998). Fruit choice and seed dispersal of invasive vs non-invasive morphotypes of *Carpobrotus* (Aizoaceae) in Coastal California. *Ecology* 79:1053–1060.

Vila, M., Weber, E., and D'Antonio, C.M. (1998). Flowering and mating system in hybridizing *Carpobrotus* (Aizoaceae) in coastal California. *Canadian Journal of Botany* 76:1–6.

Vitousek, P.M. (1990). Biological invasions and ecosystem processes: Towards an integration of population biology and ecosystem ecology. *Oikos* 57:7–13.

Vitousek, P.M., Walker, L.R., Whiteaker, L.D., Mueller-Dombois, D., and Matson, P.A. (1987). Biological invasion by *Myrica faya* alters ecosystem development in Hawaii. *Science* 238:802–804.

Vivrette, N., and Muller, C.H. (1977). Mechanism of invasion and dominance of coastal grasslands by *Mesembryanthemum crystallinum*. *Ecological Monographs* 47:301–318.

Yelenik, S.G., Stock, W.D., and Richardson, D.M. (2004). Ecosystem level impacts of invasive *Acacia saligna* in the South African fynbos. *Restoration Ecology* 12:44–51.

Zedler, P.H., and Scheid, G.A. (1988). Invasion of *Carpobrotus edulis* and *Salix lasiolepis* after fire in a coastal chaparral site in Santa Barbara County, California. *Madroño* 35:196–201.

8

ECOSYSTEM ENGINEERING IN THE FOSSIL RECORD: EARLY EXAMPLES FROM THE CAMBRIAN PERIOD

Katherine N. Marenco and David J. Bottjer

8.1 — INTRODUCTION

Ecosystem engineering is a concept that had, until recently, been applied only to modern environments, in which biological interactions can be observed directly. Modern examples of ecosystem engineering have been described from a diverse array of habitats and at a range of scales, as the preceding chapters demonstrate. Only in the past few years, however, have paleoecologists taken notice of the ecosystem engineering concept and begun to identify examples from the fossil record (e.g., Curran and Martin 2003; Gibert and Netto 2006; Hasiotis 2001; Marenco and Bottjer in press; Nicholson and Bottjer 2004, 2005; Parras and Casadio 2006). Although the identification of ancient ecosystem engineers can often be facilitated by comparisons with modern analogues (e.g., burrowing behavior in modern and Pleistocene decapod crustaceans; Curran and Martin 2003), the task is invariably challenging because considerable ecological information is lost during the processes by which living organisms and their surroundings become preserved. Despite the obstacles presented by the fossil record, searching for ancient examples of ecosystem engineering is worthwhile because it helps to improve our understanding of ecological relationships and evolutionary trends throughout the history of life. In this chapter, we discuss the approaches to and challenges of identifying ecosystem engineering in the fossil record, beginning with paleocommunity reconstruction, and we describe two examples from the Cambrian Period (ca. 542–500 million years ago) that are among the earliest-known instances of ecosystem engineering by metazoans.

8.2 ⬤ PALEOCOMMUNITY RECONSTRUCTION

In modern environments, it is possible to directly observe and document the activities of organisms, the effects of those activities on the distribution of resources, and in turn, the impact of changes in resource supply on the ecosystem as a whole. These observations permit the identification of modern ecosystem engineers (Jones et al. 1994). When examining the fossil record for evidence of ancient ecosystem engineering, paleoecologists must use the limited information that is preserved within rocks to reconstruct paleocommunities and, in turn, interactions among community members.

The first task in this process is to determine the type of environment in which the rocks were deposited, whether terrestrial, marine, or transitional. This is best accomplished by examining the rocks for sediment characteristics and sedimentary structures that reflect environment-specific physical processes and for fossil organisms that inhabited a limited range of environments. Echinoderms, for example, are known to have lived almost exclusively in marine settings since their appearance over 500 million years ago (Brusca and Brusca 2003).

Second, the fossils preserved within a rock unit must be identified. Metazoan fossils occur in two primary forms: *body fossils*, the physical remains of anatomical structures; and *trace fossils*, structures that were generated by organisms in the course of their activities (e.g., Bromley 1996). Paleoecological data that can be obtained directly from body fossils include estimates of community diversity, relative abundances of species or groups, and occupation of *morphospace*, or the set of theoretically possible body plans (morphotypes) (e.g., Thomas et al. 2000). Life habits of community members can be inferred from body fossils using *functional morphology*, in which modeling and comparisons between analogous body structures in modern and ancient organisms facilitate the interpretation of fossil behavior (e.g., Brenchley and Harper 1998). Trace fossils are useful indicators of the range of benthic activities that took place within ancient environments. In most cases, trace fossils are the only preserved evidence for the presence of soft-bodied (nonskeletonized) organisms within a rock unit, unless exceptional conditions at the time of deposition permitted the preservation of soft tissues (Bottjer et al. 2002).

Third, the physical condition of the fossils and their distribution within the rock unit must be taken into account. For example, a marine rock unit that contains abundant, randomly oriented shells is likely to represent a *shell bed*, a dense accumulation of wave-transported and often fragmentary shell material, rather than a community of high population density that was preserved in place. Determining whether marine organisms were transported away from their habitat prior to preservation is more difficult when studying ancient *soft substrate* environments, such as muddy seafloors, than *hard substrate* environments, such as reefs and carbonate *hardgrounds*, in which many benthic organisms lived permanently or semipermanently attached to hard surfaces. In most instances of hardgrounds in the fossil record, surface-attaching benthic organisms are preserved in life position, providing opportunities for the study of spatial relationships among many members of the benthic community (Taylor and Wilson 2003). Although reefs generally require more reconstruction due to their tendency to break apart prior to preservation, evidence of original spatial relationships typically is found in well-preserved fossil reefs, and this can provide insights into reef community structure (e.g., Wood 1999).

Throughout the process of paleocommunity reconstruction, caution must be taken to avoid the misinterpretation of evidence obtained from rocks and fossils. One of the more troublesome factors to take into account is the rate of sediment accumulation versus the rates of erosion and bioturbation in a given environment. In modern environments, sediment accumulation rates rarely remain constant for extended periods of time. A meter-thick unit of rock may in one location represent 5 million years of slow sediment accumulation whereas in another area it may represent merely 500,000 years, if sediment accumulation is rapid. In addition, the rate of erosion may temporarily exceed the rate of sediment accumulation, leaving a gap, or *hiatus*, in the rock record. Bioturbation intensity often appears greater during periods of slow sediment accumulation, when the seafloor experiences prolonged exposure to benthic activity. In addition, benthic organisms that engage in vertical burrowing may transport younger material down into older layers of sediment, and vice versa. Thus, fluctuations in rates of erosion, bioturbation, and sediment accumulation may lead to *time-averaging* of body fossils, or the adjacent preservation of organisms that did not coexist in life (e.g., Brenchley and Harper 1998).

A second complication to be dealt with during paleocommunity reconstruction is the incompleteness of the fossil record, even after hiatuses and time-averaging have been taken into consideration. As is the case in modern environments, many ancient organisms were soft-bodied, or nonskeletonized. Soft tissues are much more susceptible to decay than skeletal components and are not preserved unless decay is inhibited through exceptional circumstances. Thus, most fossil-bearing rock units contain body fossils only of skeletonized organisms. Trace fossils are excellent indicators of this "missing" diversity because they record the activities of both skeletonized and soft-bodied organisms.

8.3 ● IDENTIFYING ECOSYSTEM ENGINEERS IN THE FOSSIL RECORD

Jones and colleagues (1994) define two distinct categories of ecosystem engineers: *autogenic engineers*, whose biogenic structures (living and dead) change the environment; and *allogenic engineers*, whose activities alter the "physical state" of pre-existing materials, thereby changing the environment. Ecosystem engineers and the products of their engineering vary in their potential to be preserved in the fossil record.

Autogenic engineers of hard structures, such as reefs, can readily be recognized in the fossil record because their often-substantial engineering products have high preservation potential, their original shape can usually be reconstructed, and in most cases they reflect the taxonomic affinities of the engineers that built them. Such "hard substrates" often are preserved with other organisms still attached, in life position. This type of preservation facilitates interpretation of inter-species relationships and paleocommunity trophic structure.

Some autogenic engineers are soft-bodied, such as aquatic plants (macrophytes) and most sponges. Aquatic plants may grow densely in bodies of fresh water. In doing so, they may affect the environment, for instance by changing the amount of light that reaches the bottom (Carpenter and Lodge 1986). However, they are unlikely to be preserved as fossils unless anoxic bottom conditions, generated by stagnation of the water column, inhibit the decay of plant material (e.g., Brenchley and Harper 1998). Sponges alter fluid flow in marine environments, through both their passive filtration systems and their physical presence on the seafloor. Some sponges also create nutrient-rich habitats for fish and other animals (Saito et al. 2003). Sponge construction, which typically consists of soft tissue surrounding a matrix of unarticulated skeletal elements or "spicules" (e.g., Brusca and Brusca 2003), is not conducive to complete fossilization. Only the spicules are commonly preserved, and these often become scattered and mixed in with sediment grains, leaving no record of the sponge's original structure or life position. Rare examples of spiculate sponges that are preserved intact have been found in sedimentary deposits, such as the Lower Cambrian Chengjiang Biota of southern China, that appear to have formed under exceptional circumstances (Hou et al. 2004, Xiao et al. 2005). Thus, physical evidence for the influence of macrophytes or sponges on benthic communities usually is absent or limited to indirect sources such as scattered spicules, except in rare cases.

Allogenic engineers that construct macroscopic burrows (e.g., fiddler crabs; Bertness 1985) or mounds (e.g., thalassinidean shrimp; Ziebis et al. 1996) or graze on hard surfaces (e.g., periwinkles; Bertness 1984) may have their activities recorded as trace fossils. However, a problem may arise if the allogenic engineer is a soft-bodied organism, such as a polychaete worm. In such a case, the engineering activity itself may be identifiable from trace fossils, but the identity of the engineer will likely remain unknown. The reverse of this situation is also possible: An engineer may have a preservable skeleton but produce an ephemeral structure (e.g., skeletonized diatoms produce mucilaginous mats; Winterwerp and van Kesteren 2004). If no modern analogues for such an allogenic engineer are known, then its engineering behavior may never come to light.

An additional complication is that the quantity and depth of bioturbation in marine environments have increased throughout the past approximately 540 million years concurrently with the gradual rise in benthic biodiversity (Ausich and Bottjer 1982, Droser and Bottjer 1993). Through this time interval, as bioturbation structures began to extend to greater depths and occur in greater densities within the sediment, individual burrows and tunnels became obliterated, and the resulting sediment and sedimentary rocks are left with a homogeneous appearance. Trace fossils left by earlier seafloor communities often are later "overprinted" by a different set of structures as environmental conditions change (e.g., Orr 1994). Thus, it may be difficult to discern the preserved work of a single allogenic engineer from that of any of several hundred other benthic bioturbators within a rock unit.

Allogenic engineers that either are microscopic (e.g., meiofauna, zooplankton) or engage in engineering activities that do not result in the production of physical structures (e.g., chemical effects), or both, can be very difficult to identify in the fossil record. Zooplankton concentrate organic matter into fecal pellets, which assist in the vertical transport of material to the seafloor (e.g., Dunbar and Berger 1981). Although individual pellets were not preserved under normal conditions, the rise of zooplankton in ancient oceans is reflected in marine rocks by a change in the $\delta^{13}C$ values of preserved organic matter (Logan and Butterfield 1998). In most cases, however, biochemical engineering effects are too subtle to be recorded in rocks and fossils. If such effects are, in fact, recorded, they may mistakenly be attributed to abiotic causes.

Incomplete fossil preservation of animals and their behavior precludes the identification of more than a fraction of the ecosystem engineers that once existed. At the same time, many well-preserved examples of ecosystem engineering in the fossil record have yet to be recognized. In the next section, we describe and present evidence for two of the earliest examples of metazoan allogenic and autogenic engineering in the history of life.

8.4 ✸ SETTING THE STAGE: THE CAMBRIAN PERIOD

The Cambrian Period (ca. 542–500 million years ago) was an important time of transition in ecological and evolutionary history. Mineralized skeletons and skeletal elements, such as "small shelly fossils" and sponge spicules, appeared in the earliest Cambrian but did not become widespread and diverse until the end of the Cambrian (e.g., Brasier et al. 1997, Brasier and Hewitt 1979). A wide variety of soft-bodied fossils have been described from the exceptionally preserved Early Cambrian Chengjiang Biota in southern China, suggesting that nonmineralized metazoans constituted a substantial component of Early Cambrian benthic communities (Hou et al. 2004). Biomineralizing organisms, with predator- and pressure-resistant skeletons, were capable of occupying a greater range of niches than their soft-bodied counterparts, and this competitive advantage allowed the populations of such organisms to expand into a variety of marine environments (e.g., Vermeij 1989). Paralleling the trend toward widespread biomineralization among metazoans was the rapid diversification of metazoan body plans known as the *Cambrian explosion* (e.g., Conway Morris 2006, Marshall 2006, Thomas et al. 2000). Metazoan body plans in the earliest Cambrian were commonly simple and limited to few types, whereas by the latest Cambrian, most of the biological "architecture" considered characteristic of the major metazoan groups had already become established (Sepkoski 1979, Thomas et al. 2000).

Rocks that represent Early Cambrian shallow marine environments below tidal range typically contain limited disruption of sedimentary layers, which reflects a lack of vertically oriented bioturbation, and common *microbially mediated sedimentary structures* in siliciclastic facies (Hagadorn and Bottjer 1997) (Figure 8.1). Microbially mediated sedimentary structures are thought to represent the effects of sediment

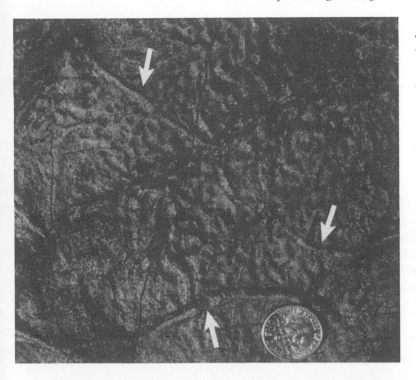

FIGURE 8.1 Characteristic features of Lower Cambrian rocks. (Left) Positive x-radiograph of a vertically sectioned sedimentary rock sample from the Lower Cambrian Campito Formation showing finely laminated sediment that was not disrupted by vertical bioturbation. Scale bar = 1 cm. (Right) Horizontal (bedding plane) exposure of a Lower Cambrian sedimentary rock showing "wrinkle structures" (microbially mediated sedimentary structures), which are thought to have formed due to the compaction of a seafloor microbial mat. Superimposed on the wrinkle structures are examples of the simple horizontal trace fossil *Planolites* (arrows). (Although they are sedimentary structures, trace fossils are classified using biological nomenclature.) *Planolites* likely represents the work of a shallow-burrowing organism that may have been feeding on nutrients associated with the microbial mat. (See color plate.)

binding by microbial mats (Hagadorn and Bottjer 1997, 1999; Noffke et al. 1996), namely cohesive sediment behavior (Schieber 1999). Support for this interpretation comes from observations of modern microbial mats (e.g., Gerdes et al. 1993, Hagadorn and Bottjer 1999). The surfaces of many such modern mats strikingly resemble the strange features preserved in Lower Cambrian rocks, including "wrinkle structures" (Hagadorn and Bottjer 1997) (Figure 8.1), "elephant skin" (Gehling 1999), "domal structures" (Schieber 1999), and "syneresis cracks" (Pflüger 1999).

In Lower Cambrian carbonate rocks deposited primarily in shallow water, microbial structures are common (e.g., Rowland and Shapiro 2002). *Microbialites*, structures that formed through precipitation of carbonate in the presence of (and often triggered by) benthic microbial communities, first appeared approximately 3.5 billion years ago (Wood 1999). Prior to the earliest Cambrian, the dominant form of microbialite was the *stromatolite*, a laminated structure produced primarily by photosynthetic cyanobacteria (Wood 1999). *Thrombolites*, nonlaminated microbialites with "clotted" textures, appeared in the Neoproterozoic (ca. 1000–542 million years ago) but did not become abundant until the earliest Cambrian (Wood 1999). The first true reefs were constructed by microbial communities in the Neoproterozoic, and stromatolite–thrombolite reefs persisted into the earliest Cambrian (e.g., Rowland and Shapiro 2002). The rise of metazoan reefs in the Early Cambrian likely contributed to the decline of microbialites in many shallow marine environments (Zhuravlev 2001).

Studies to date have shown that microbially mediated sedimentary structures and microbialites are common in rocks of the Neoproterozoic and Early Cambrian and are comparatively scarce in younger rocks (e.g., Gehling 1999; Hagadorn and Bottjer 1997, 1999). This implies that most seafloor sediments in the Neoproterozoic through Early Cambrian were bound together by microbial filaments, making them firmer and more cohesive than those of the modern oceans (e.g., Gehling 1999, Hagadorn and Bottjer 1997). These "matgrounds" would likely have been difficult or impossible for benthic metazoans to penetrate, and the combination of ubiquitous mats and a lack of infaunal bioturbation would have prevented aeration of the sediment, allowing an oxic–anoxic boundary to develop in the sediment close to the seafloor surface (McIlroy and Logan 1999). As a result, all metazoan activity likely took place on the top surfaces of mats, within mats, or immediately beneath them.

Seilacher (1999) proposed four guilds to characterize the categories of metazoan activity that took place in late Neoproterozoic shallow marine

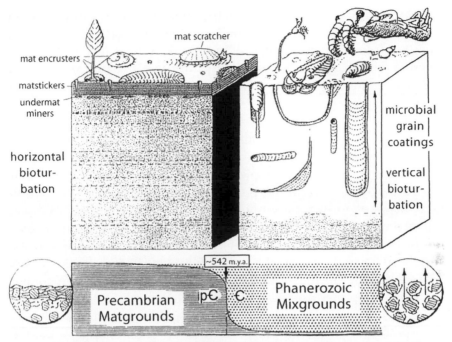

FIGURE 8.2 Illustration of typical Precambrian "matground" seafloors and post-Cambrian "mixground" seafloors with their associated communities. Matgrounds supported a specialized community of "mat scratchers," "mat encrusters," "mat stickers," and "undermat miners" (Seilacher 1999). Post-Cambrian mixgrounds are characterized by a diverse community of organisms that was active both on the seafloor surface and within the sediment. Modified from Seilacher (1999).

benthic communities (Figure 8.2). These are "mat encrusters," organisms that lived permanently attached to the mat surface; "mat scratchers," mobile organisms that scavenged or hunted for food on the surface of the mat without damaging it; "mat stickers," suspension feeders that used conical shells to maintain an upright orientation in the surface of the mat; and "undermat miners," burrowers that tunneled directly beneath the mat and fed on detritus from the layers above (Seilacher 1999). These mat-associated lifestyles persisted into the Early Cambrian but gradually disappeared from open marine environments along with the microbial mats themselves (Dornbos and Bottjer 2001).

8.5 ▬ EARLY METAZOAN ALLOGENIC ENGINEERS

The absence of vertical bioturbation in siliciclastic rocks deposited during the Neoproterozoic and Early Cambrian indicates that conditions beneath the seafloor surface may have been unfavorable for metazoan activity (Bottjer et al. 2000). Although limited food resources within the sediment may have provided little incentive for organisms to burrow infaunally, considerable evidence, including the presence of abundant microbially mediated sedimentary structures, suggests that physical and adaptive limitations were primarily responsible for restricting benthic organisms to epifaunal habitats (e.g., Bottjer et al. 2000).

The fossil record of bioturbation exhibits a prominent trend over time toward increasing trace fossil complexity, density, and penetration depth beneath the seafloor (Ausich and Bottjer 1982, Droser and Bottjer 1993). The earliest-known macroscopic trace fossils are found in rocks that were deposited during the late Neoproterozoic (e.g., Jensen 2003, Jensen et al. 2006). Most of these early biogenic structures consist of simple, bilaterally symmetrical, horizontal forms that likely represent the activities of soft-bodied vermiform organisms on or just beneath the seafloor surface or beneath microbial mats (Collins et al. 2000, Valentine 1995). Trace fossils do not begin to exhibit a vertically oriented component until the Neoproterozoic–Cambrian boundary (ca. 542 million years ago), when *Treptichnus pedum*, a trace fossil that consists of a series of shallow scoop-like marks, appears in rocks representing shallow marine environments (e.g., Droser et al. 1999, Gehling et al. 2001) (Figure 8.3). Although deeply vertical burrows occurred in nearshore and shoreface environments in the earliest Cambrian ("*Skolithos* piperock"; Droser 1991), shallow burrow structures with little or no verticality were the dominant form of bioturbation in subtidal environments until the Middle to Late Cambrian (Bottjer et al. 2000) (Figure 8.3). The gradual increase in bioturbation depth in shallow marine environments from the Early to Late Cambrian has been demonstrated by Droser (1987) and Droser and Bottjer (1988, 1989). Rocks that were deposited in shallow marine settings of the Late Cambrian through the Modern display a very different set of characteristics from those representing Early Cambrian seafloors, including visibly disrupted sedimentary layers, common vertical burrows that may overprint earlier bioturbation structures, and absent microbially mediated sedimentary structures (Bottjer et al. 2000) (Figure 8.3). Thus, a transition occurred during the Cambrian Period between seafloors that were characterized by primarily horizontal bioturbation and extensive microbial mats (as reflected by the abundance of microbially mediated sedimentary structures and dearth of vertical sediment disruption in Lower Cambrian rocks) and those that were characterized by extensive vertical bioturbation and absent microbial mats.

Seilacher and Pflüger (1994) proposed the *agronomic revolution* hypothesis to explain how and why this transition in seafloor conditions and benthic behavior took place. According to this hypothesis, benthic

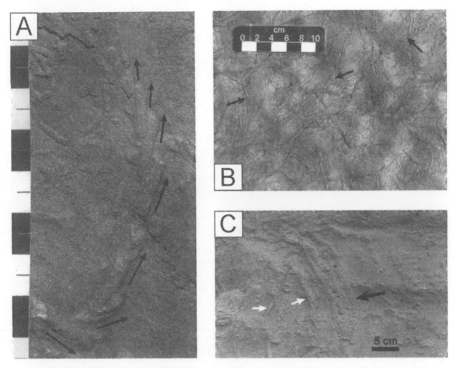

FIGURE 8.3 Trace fossils preserved in Lower Cambrian (A, B) and Eocene (ca. 56–34 million year old) (C) rocks. (A) *Treptichnus pedum*, the first trace fossil to exhibit a vertically oriented component, preserved upside down on the bottom of a Lower Cambrian sedimentary rock unit. The nested lobes (arrows) of the trace likely represent systematic probing of the seafloor sediment by a priapulid-like deposit-feeding organism. (B) A Lower Cambrian bedding plane surface that contains abundant horizontal trace fossils (*Planolites*; arrows). (C) Abundant vertically oriented trace fossils in Eocene exposures near San Diego, CA. *Conostichus* (black arrows), a large, lobe-shaped burrow, is produced by anemones and other stationary benthic suspension feeders during sediment influx. *Ophiomorpha* (white arrows), a deep mud-lined burrow, is produced by many types of benthic suspension-feeding crustaceans, which require stable semipermanent dwellings. (See color plate.)

metazoans acquired evolutionary adaptations during the Cambrian explosion that allowed them to burrow vertically into matgrounds. Bioturbation depth and intensity increased, eventually disrupting the layered structure of the microbial mats and increasing the water and oxygen content of the seafloor sediment. Mat development was relegated to marginal environments in the wake of the agronomic revolution, and the seafloor took on characteristics more typical of post-Cambrian marine settings, such as improved nutrient distribution and an indistinct water-sediment boundary (Bottjer et al. 2000) (Figure 8.1).

The ecological and evolutionary effects of the agronomic revolution are reflected in the record of body and trace fossils and have collectively been termed the *Cambrian substrate revolution* (Bottjer et al. 2000). Among the more significant of these effects were those felt by the mat-ground community (Seilacher 1999). Mat scratchers and undermat miners were better equipped than the other guilds for adjusting to new seafloor conditions because their mobile lifestyles allowed them to reposition themselves in response to changes in oxygen and substrate consistency (Bottjer et al. 2000). However, mat scratchers were adapted to living and feeding on cohesive sediment surfaces, and the disappearance of such surfaces from open marine environments forced many species to migrate into more restricted areas where hard substrates were common, such as rocky coastlines and the deep ocean (Bottjer et al. 2000). Mat encrusters and mat stickers faced a greater challenge due to their specialized sessile lifestyles. Lacking a means of migrating to more suitable environments, many of these groups evolved stems or direct attachment mechanisms that allowed them to utilize the limited hard surfaces that were available in shallow marine settings (Bottjer et al. 2000). Not all such groups were successful, however. The mat-sticking helicoplacoid echinoderms, for example, did not adapt to the new substrate conditions and became extinct before the end of the Cambrian (Bottjer et al. 2000; Dornbos and Bottjer 2000, 2001).

The agronomic and Cambrian substrate revolutions together represent the earliest-known instance of allogenic ecosystem engineering by metazoans in the history of life. With increasing depth and intensity of bioturbation, benthic metazoans brought about a dramatic change in shallow subtidal seafloors of the Early Cambrian, supplanting microbes as the dominant biotic influence on many seafloor conditions and making available to other members of the community a variety of previously inaccessible resources and ecological niches (Bottjer et al. 2000, Dornbos et al. 2004) (Figure 8.4). The transformative effects of bioturbation have been recognized in a wide variety of modern ecosystems as well (e.g., Meysman et al. 2006).

In this case of allogenic ecosystem engineering, as in some of the examples discussed earlier, the engineers themselves were not necessarily preserved, but the impact of the engineering activity can easily be recognized in rocks. Efforts to identify the ecosystem engineer(s) of the agronomic revolution are in their early stages, although soft-bodied metazoans are likely candidates based on their abundance in exceptionally preserved deposits such as the Chengjiang Biota (Hou et al. 2004). Given the scarcity of preserved soft tissues in the fossil record, studying the distribution and abundance of trace fossils in Lower Cambrian rocks

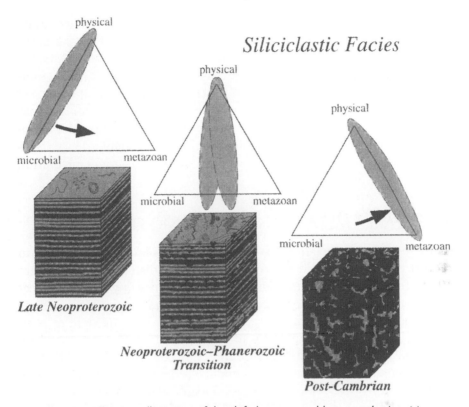

FIGURE 8.4 A schematic illustration of the shift that occurred between the Late Neoproterozoic and post-Cambrian in the dominant processes that controlled seafloor conditions. As indicated in the triangular diagrams, Late Neoproterozoic seafloor conditions were controlled by physical and microbial processes. Microbial influence gradually decreased during the Neoproterozoic–Phanerozoic transition, when metazoan bioturbation became more abundant and disruptive. In the post-Cambrian, abundant and extensive bioturbation by metazoans was the primary factor, in addition to physical processes, that governed seafloor conditions. Modified from Bottjer et al. (2000).

may be the best way to determine the role soft-bodied organisms may have played in engineering Early Cambrian ecosystems. A study of Lower Cambrian shallow marine rocks in eastern California demonstrates that the simple horizontal trace fossil *Planolites*, likely the product of shallow burrowing by soft-bodied vermiform metazoans, was the most abundant type of bioturbation present on horizontal "bedding plane" surfaces throughout the rocks examined (Marenco and Bottjer in press). The proliferation of *Planolites* burrows on Early Cambrian seafloors likely reflects the presence of a steady nutrient supply, generated by widespread microbial mats, that likely was capable of sustaining a diverse matground community. Other evidence for the existence of such a community in these Lower Cambrian rock units includes the common occurrence of *Volborthella*, a small enigmatic Cambrian fossil interpreted as the skeleton of a matground-adapted animal (e.g., Seilacher 1999), and shell casts and molds of possible linguliform brachiopods, which may have been adapted to life in the low-oxygen conditions promoted by microbial mats (Bailey et al. 2006). Despite the abundance of horizontal bioturbation, microbial activity was likely still the dominant factor influencing substrate conditions in these particular Early Cambrian shallow marine environments prior to the agronomic revolution (Bailey et al. 2006).

8.6 ⬡ EARLY METAZOAN AUTOGENIC ENGINEERS

As mentioned in preceding text, the earliest reefs known from the fossil record were constructed entirely by microorganisms (e.g., Grotzinger 1989). These microbial reefs, with frameworks consisting entirely of stromatolitic and thrombolitic fabrics, were dominant in marine settings until the Early Cambrian. The transition toward a new style of reef-building was gradual, beginning in the Neoproterozoic and extending 10–15 million years into the Cambrian Period (Copper 2001). In the Late Neoproterozoic (ca. 550 million years ago), the first metazoans to construct calcium carbonate skeletons appeared (Grotzinger et al. 1995). These include two possible cnidarian forms: *Cloudina*, a tube-building organism (Grant 1990); and *Namacalathus*, a goblet-shaped organism (Grotzinger et al. 2000). Although fossil evidence suggests that these and other early skeletonized animals commonly lived within microbial reefs or constructed small "thickets" and mounds independently, their skeletons were not substantial enough to constitute a primary reef framework (Wood 1999). It was not until approximately 530 million years ago that metazoans began to play a more significant role in reef construction.

Archaeocyath sponges were the first skeletonized metazoan components of Early Cambrian reefs (Copper 2001, Wood 1999). These animals lacked spicules, having instead a calcified skeleton with a complex internal structure (Wood 1999). Typical archaeocyath skeletons were cone- or cup-shaped with double or single walls constructed of calcite (Copper 2001). In double-walled forms, the two walls commonly were joined together by septa; this septate region, the intervallum, likely housed soft tissue (Wood 1999). Archaeocyaths were solitary or colonial, and their skeletal morphologies varied widely from single cones to branching or sheet-like forms (Copper 2001) (Figure 8.5). Archaeocyath skeletons, significantly more robust than those of earlier reef-associated metazoans,

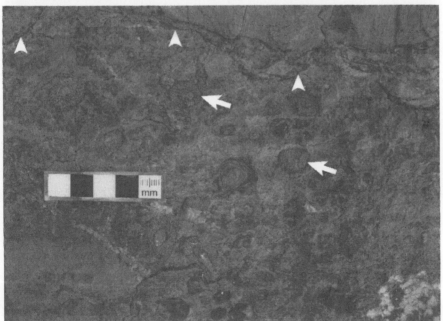

FIGURE 8.5 Close-up views of an archaeocyath–calcimicrobial reef, Stewart's Mill, NV. (Top) A branching archaeocyath sponge, which appears to have been preserved in life position. Surrounding the archaeocyath is carbonate sediment that may have accumulated slowly while the animal was alive, allowing it to be preserved in place. Scale bar = 1 cm. (Bottom) Accumulated fragments of skeletal material, including that of archaeocyaths (arrows), and microbial structures surrounded by carbonate sediment. The sharp boundary in the upper portion of the photograph (arrowheads) likely represents the floor of a reef cavity, which would have harbored organisms that were specially adapted to life in these cryptic settings. Photo courtesy of Matthew Clapham. (See color plate.)

not only enhanced their preservation potential but likely also helped the animals deter potential encrusters and competitors (Zhuravlev 2001). Although they rarely grew to more than 20 cm in height (Wood 1999), archaeocyaths were capable of constructing substantial reef frameworks (Zhuravlev 2001). Unlike later reef-building organisms, however, archaeo-cyaths required the assistance of calcified microorganisms to build reefs (Rowland and Shapiro 2002, Wood 1999).

Calcified microorganisms, or *calcimicrobes*, likely were cyanobacteria that were preserved by the precipitation of calcium carbonate around their extracellular sheaths (Wood 1999). These "skeletal" microorgan-isms rose to prominence as reef builders in the Late Neoproterozoic, constructing mounds on the order of several hundred meters thick and one kilometer wide (e.g., Aitken 1989). Calcimicrobes differ from stro-matolites in their growth morphology, which commonly is clumpy or shrublike (Wood 1999), and probably also in their mode of carbonate precipitation, which is unknown but may have been influenced by envi-ronmental factors (Copper 2001). The three main categories of calcimi-crobes are *Renalcis*, a globular form comprised of clumps or clots of fine-grained calcite; *Epiphyton*, a shrublike colonial form; and *Girvanella*, a sheetlike or crustlike form (Wood 1999).

The presence of well-established, abundant calcimicrobes in earliest Cambrian seafloor communities facilitated the reef-building success of archaeocyaths 10 to 15 million years later (Wood 1999). Calcimicrobes appear to have become cemented in calcium carbonate during active reef growth (e.g., Kruse et al. 1995), which would have lent added strength to any of their associated structures. In high-energy shallow-water envi-ronments, calcimicrobial crusts likely stabilized the seafloor sediment, allowing archaeocyaths to become established (Wood 1999). The pres-ence of complexly intergrown calcimicrobes and archaeocyaths in fossil reefs suggests that calcimicrobes strengthened archaeocyath frame-works at later stages of reef growth (e.g., Zhuravlev 2001). Thus, calcimi-crobes served as non-metazoan autogenic engineers by facilitating the growth of the earliest substantial metazoan reefs.

As reef builders, archaeocyaths were autogenic ecosystem engineers in their own right. The growth of archaeocyath–calcimicrobial reefs expanded benthic ecospace on Early Cambrian seafloors. In addition to increasing available surface area for organism attachment, these reefs promoted diversification of the benthic community by dividing the habitat into open-surface and cryptic (cavity) settings and increasing the number of energy-dependent microhabitats through varied topography (Wood 1999, Zhuravlev 2001) (Figure 8.5). Among the groups that

colonized archaeocyath–calcimicrobial reefs were brachiopods, echinoderms, gastropods, and trilobites, many of which evolved reef-specific
adaptations (Zhuravlev 2001). For example, the trilobite genus *Gior-
danella* became specialized as a reef-dwelling stationary suspension-
feeder (Zhuravlev 2001). Other groups, such as sponges and metazoan
microburrowers, became specialized inhabitants of reef cavities (Kobluk
1988, Wood 1999). Thus, in archaeocyath–calcimicrobial reefs, we have
a well-preserved multistage example of early autogenic engineering,
which likely contributed to the Cambrian explosion of marine animal
diversity.

8.7 ⬤ CONCLUSIONS

Recognizing ancient examples of ecosystem engineering in the fossil
record is challenging due to the loss of primary ecological information
that occurs during preservation. Problems such as time-averaging, fluctuating sediment accumulation rates, and preferential preservation of
skeletonized organisms can hamper paleoecological investigations. Evidence for engineering behavior, or for the presence of engineers themselves, may be impossible to obtain from the fossil record unless
exceptional conditions prevailed at the time of preservation. Autogenic
engineers, which altered the environment through their biogenic structures, are generally more apparent in the fossil record than allogenic
engineers, which altered the environment through the transformation
of pre-existing materials; this "bias" may become more apparent as the
study of ancient ecosystem engineering progresses.

The Early Cambrian agronomic revolution and development of
archaeocyath–calcimicrobial reefs are two of the earliest examples of
allogenic and autogenic engineering in the history of life. By expanding
benthic ecospace, these instances of engineering had broad ecological
and evolutionary effects. Erwin (2005) argues that the construction of
new ecological niches is essential if organisms' genetic *inventions* are to
become successful *innovations* that persist in communities through
time. The development of new niches via the agronomic revolution and
the expansion of reefs likely helped facilitate the Cambrian explosion of
marine innovations.

The search for ancient ecosystem engineers is in its early stages, but
it promises to greatly improve our understanding of community ecology
over broad timescales. Paleoecologists must continue to refine and build
upon current strategies for identifying examples of ancient ecosystem
engineering in the fossil record.

ACKNOWLEDGMENTS

This manuscript benefited considerably from the suggestions of two anonymous reviewers. K. Marenco is grateful to the following for providing research support: Sigma Xi, the Paleontological Society, the Geological Society of America, the Evolving Earth Foundation, and the University of Southern California Department of Earth Sciences.

REFERENCES

Aitken, J.D. (1989). Giant "algal" reefs, Middle/Upper Proterozoic Little Dal Group (>770, <1200 Ma), Mackenzie Mountains, N.W.T. Canada. In *Reefs: Canada and Adjacent Areas*, Vol. 13. H.H.J. Geldsetzer, N.P. James, and G.E. Tebbutt, Eds. Calgary, Alberta, Canada: Canadian Society of Petroleum Geologists, Memoir 13, pp. 13–23.

Ausich, W.I., and Bottjer, D.J. (1982). Tiering in suspension-feeding communities on soft substrata throughout the Phanerozoic. *Science* 216:173–174.

Bailey, J.V., Corsetti, F.A., Bottjer, D.J., and Marenco, K.N. (2006). Microbially mediated environmental influences on metazoan colonization of matground ecosystems: Evidence from the Lower Cambrian Harkless Formation. *Palaios* 21:215–226.

Bertness, M.D. (1984). Habitat and community modification by an introduced herbivorous snail. *Ecology* 65:370–381.

——. (1985). Fiddler crab regulation of *Spartina alterniflora* production on a New England salt marsh. *Ecology* 66:1042–1055.

Bottjer, D.J., Etter, W., Hagadorn, J.W., and Tang, C.M. (2002). *Exceptional Fossil Preservation: A Unique View on the Evolution of Marine Life*. New York: Columbia University Press.

Bottjer, D.J., Hagadorn, J.W., and Dornbos, S.Q. (2000). The Cambrian substrate revolution. *GSA Today* 10:1–7.

Brasier, M.D., Green, O., and Shields, G. (1997). Ediacarian sponge spicule clusters from southwestern Mongolia and the origins of the Cambrian fauna. *Geology* 25:303–306.

Brasier, M.D., and Hewitt, R.A. (1979). Environmental setting of fossiliferous rocks from the uppermost Proterozoic–Lower Cambrian of central England. *Palaeogeography, Palaeoclimatology, Palaeoecology* 27:35–57.

Brenchley, P.J., and Harper, D.A.T. (1998). *Palaeoecology: Ecosystems, Environments and Evolution*. London: Chapman and Hall.

Bromley, R.G. (1996). *Trace Fossils: Biology, Taphonomy, and Applications*. London: Chapman and Hall.

Brusca, R.C., and Brusca, G.J. (2003). *Invertebrates*. Sunderland, MA: Sinauer Associates.

Carpenter, S.R., and Lodge, D.M. (1986). Effects of submerged macrophytes on ecosystem processes. *Aquatic Botany* 26:341–370.

Collins, A.G., Lipps, J.H., and Valentine, J.W. (2000). Modern mucociliary creeping trails and the bodyplans of Neoproterozoic trace-makers. *Paleobiology* 26:47–55.

Conway Morris, S. (2006). Darwin's dilemma: The realities of the Cambrian 'explosion.' *Philosophical Transactions of the Royal Society of London, Series B* 361:1069–1083.

Copper, P. (2001). Evolution, radiations, and extinctions in Proterozoic to mid-Paleozoic reefs. In *The History and Sedimentology of Ancient Reef Systems,* G.D. Stanley, Jr., Ed. New York: Kluwer Academic/Plenum Publishers, pp. 89–119.

Curran, H.A., and Martin, A.J. (2003). Complex decapod burrows and ecological relationships in modern and Pleistocene intertidal carbonate environments, San Salvador Island, Bahamas. *Palaeogeography, Palaeoclimatology, Palaeoecology* 192:229–245.

Dornbos, S.Q., and Bottjer, D.J. (2000). Evolutionary paleoecology of the earliest echinoderms: Helicoplacoids and the Cambrian substrate revolution. *Geology* 28:839–842.

——. (2001). Taphonomy and environmental distribution of helicoplacoid echinoderms. *Palaios* 16:197–204.

Dornbos, S.Q., Bottjer, D.J., and Chen, J.-Y. (2004). Evidence for seafloor microbial mats and associated metazoan lifestyles in Lower Cambrian phosphorites of Southwest China. *Lethaia* 37:127–137.

Droser, M.L. (1987). *Trends in extent and depth of bioturbation in Great Basin Precambrian-Ordovician strata, California, Nevada and Utah.* Ph.D. dissertation, University of Southern California, Los Angeles.

——. (1991). Ichnofabric of the Paleozoic *Skolithos* ichnofacies and the nature and distribution of *Skolithos* piperock. *Palaios* 6:316–325.

Droser, M.L., and Bottjer, D.J. (1988). Trends in extent and depth of bioturbation in Cambrian carbonate marine environments, western United States. *Geology* 16:233–236.

——. (1989). Ordovician increase in extent and depth of bioturbation: Implications for understanding early Paleozoic ecospace utilization. *Geology* 17:850–852.

——. (1993). Trends and patterns of Phanerozoic ichnofabrics. *Annual Review of Earth and Planetary Sciences* 21:205–225.

Droser, M.L., Gehling, J.G., and Jensen, S. (1999). When the worm turned: Concordance of Early Cambrian ichnofabric and trace-fossil record in siliciclastic rocks of South Australia. *Geology* 27:625–628.

Dunbar, R.B., and Berger, W.H. (1981). Fecal pellet flux to modern bottom sediment of Santa Barbara Basin (California) based on sediment trapping. *GSA Bulletin* 92:212–218.

Erwin, D.H. (2005). The origin of animal body plans. In *Evolving Form and Function: Fossils and Development: Proceedings of a symposium honoring Adolf Seilacher for his contributions to paleontology, in celebration of his 80th birthday.* D.E.G. Briggs, Ed. New Haven: Peabody Museum of Natural History, Yale University, pp. 67–80.

Gehling, J.G. (1999). Microbial mats in terminal Proterozoic siliciclastics: Ediacaran death masks. *Palaios* 14:40–57.

Gehling, J.G., Jensen, S., Droser, M.L., Myrow, P.M., and Narbonne, G.M. (2001). Burrowing below the basal Cambrian GSSP, Fortune Head, Newfoundland. *Geological Magazine* 138:213–218.

Gerdes, G., Claes, M., Dunajtschik-Piewak, K., Riege, H., Krumbein, W.E., and Reineck, H.E. (1993). Contribution of microbial mats to sedimentary surface structures. *Facies* 29:61–74.

Gibert, J.M. de, and Netto, R.G. (2006). Commensal worm traces and possible juvenile thalassinidean burrows associated with *Ophiomorpha nodosa*, Pleistocene, southern Brazil. *Palaeogeography, Palaeoclimatology, Palaeoecology* 230:70–84.

Grant, S.W.F. (1990). Shell structure and distribution of *Cloudina*, a potential index fossil for the terminal Proterozoic. *American Journal of Science* 290A:261–294.

Grotzinger, J.P. (1989). Introduction to Precambrian reefs. In *Reefs, Canada and Adjacent Areas*, H.H.J. Geldsetzer, N.P. James, and G.E. Tebbutt, Eds. Calgary, Alberta, Canada: Canadian Society of Petroleum Geologists, Memoir 13, pp. 9–12.

Grotzinger, J.P., Bowring, S.A., Saylor, B.Z., and Kaufman, A.J. (1995). Biostratigraphic and geochronological constraints on early animal evolution. *Science* 270:598–604.

Grotzinger, J.P., Watters, W.A., and Knoll, A.H. (2000). Calcified metazoans in thrombolite-stromatolite reefs of the terminal Proterozoic Nama Group, Namibia. *Paleobiology* 26:334–359.

Hagadorn, J.W., and Bottjer, D.J. (1997). Wrinkle structures: Microbially mediated sedimentary structures common in subtidal siliciclastic settings at the Proterozoic-Phanerozoic transition. *Geology* 25:1047–1050.

———. (1999). Restriction of a Late Neoproterozoic biotope: Suspect-microbial structures and trace fossils at the Vendian-Cambrian transition. *Palaios* 14:73–85.

Hasiotis, S.T. (2001). "Traces" of hidden biodiversity in Paleosols: Examples from Phanerozoic terrestrial deposits. *Geological Society of America Abstracts With Programs* 33:336.

Hou, X.-G., Aldridge, R.J., Bengstrom, J., Siveter, D.J., and Feng, X.-H. (2004). *The Cambrian Fossils of Chengjiang, China.* London: Blackwell Science Ltd.

Jensen, S. (2003). The Proterozoic and Earliest Cambrian trace fossil record: Patterns, problems and perspectives. *Integrative and Comparative Biology* 43:219–228.

Jensen, S., Droser, M.L., and Gehling, J.G. (2006). A critical look at the Ediacaran trace fossil record. In *Neoproterozoic Geobiology and Paleobiology*, S. Xiao and A.J. Kaufman, Eds. Dordrecht, The Netherlands: Springer, pp. 115–157.

Jones, C.G., Lawton, J.H., and Shachak, M. (1994). Organisms as ecosystem engineers. *Oikos* 69:373–386.

Kobluk, D.R. (1988). Pre-Cenozoic fossil record of cryptobionts and their presence in early reefs and mounds. *Palaios* 3:243–250.

Kruse, P.D., Zhuravlev, A.Y., and James, N.P. (1995). Primordial metazoan-calcimicrobial reefs: Tommotian (Early Cambrian) of the Siberian platform. *Palaios* 10:291–321.

Logan, G.A., and Butterfield, N.J. (1998). Plankton dynamics, organic geochemistry, and the Precambrian-Cambrian transition. *Geological Society of America Abstracts With Programs* 30:147.

Marenco, K.N., and Bottjer, D.J. (in press). The importance of *Planolites* in the Cambrian substrate revolution. *Palaeogeography, Palaeoclimatology, Palaeoecology*.

Marshall, C.R. (2006). Explaining the Cambrian "Explosion" of animals. *Annual Review of Earth and Planetary Sciences* 34:355–384.

McIlroy, D., and Logan, G.A. (1999). The impact of bioturbation on infaunal ecology and evolution during the Proterozoic-Cambrian transition. *Palaios* 14:58–72.

Meysman, F.J.R., Middelburg, J.J., and Heip, C.H.R. (2006). Bioturbation: A fresh look at Darwin's last idea. *Trends in Ecology and Evolution* 21:688–695.

Nicholson, K.A., and Bottjer, D.J. (2004). Ecosystem engineers during the Cambrian Explosion: Trace fossil record from the Lower Cambrian of eastern California. *Geological Society of America Abstracts With Programs* 36:522.

——. (2005). Lower Cambrian trace fossils of eastern California: Engineering an ecological revolution. *Geological Society of America Abstracts With Programs* 37:486.

Noffke, N., Gerdes, G., Klenke, T., and Krumbein, W.E. (1996). Microbially induced sedimentary structures: Examples from modern sediments of siliciclastic tidal flats. *Zentralblatt für Geologie und Paläontologie* 1995:307–316.

Orr, P.J. (1994). Trace fossil tiering within event beds and preservation of frozen profiles: An example from the Lower Carboniferous of Menorca. *Palaios* 9:202–210.

Parras, A., and Casadio, S. (2006). The oyster *Crassostrea? hatcheri* (Ortmann, 1897), a physical ecosystem engineer from the upper Oligocene–lower Miocene of Patagonia, southern Argentina. *Palaios* 21:168–186.

Pflüger, F. (1999). Matground structures and redox facies. *Palaios* 14:25–39.

Rowland, S.M., and Shapiro, R.S. (2002). Reef patterns and environmental influences in the Cambrian and earliest Ordovician. In *Phanerozoic Reef Patterns*, W. Kiessling, E. Flügel, and J. Golonka, Eds. Tulsa, OK: Society for Sedimentary Geology, SEPM Special Publication No. 72. pp. 95–128.

Saito, T., Uchida, I., and Takeda, M. (2003). Skeletal growth of the deep-sea hexactinellid sponge *Euplectella oweni*, and host selection by the symbiotic shrimp *Spongicola japonica* (Crustacea: Decapoda: Spongicolidae). *Journal of Zoology (London)* 258:521–529.

Schieber, J. (1999). Microbial mats in terrigenous clastics: The challenge of identification in the rock record. *Palaios* 14:3–12.

Seilacher, A. (1999). Biomat-related lifestyles in the Precambrian. *Palaios* 14:86–93.

Seilacher, A., and Pflüger, F. (1994). From biomats to benthic agriculture: A biohistoric revolution. In *Biostabilization of Sediments*, W.E. Krumbein, D.M. Paterson, and L.J. Stal, Eds. Oldenburg, Germany: Bibliotheks und Informationssystem der Carl von Ossietzky Universität, pp. 97–105.

Sepkoski, J.J., Jr. (1979). A kinetic model of Phanerozoic taxonomic diversity, II: Early Phanerozoic families and multiple equilibria. *Paleobiology* 5:222–251.

Taylor, P.D., and Wilson, M.A. (2003). Palaeoecology and evolution of marine hard substrate communities. *Earth-Science Reviews* 62:1–103.

Thomas, R.D.K., Shearman, R.M., and Stewart, G.W. (2000). Evolutionary exploitation of design options by the first animals with hard skeletons. *Science* 288:1239–1242.

Valentine, J.W. (1995). Late Precambrian bilaterians: Grades and clades. In *Tempo and Mode in Evolution: Genetics and Paleontology 50 Years After Simpson*, W.M. Fitch and F.J. Ayala, Eds. Washington, D.C.: National Academy Press.

Vermeij, G.J. (1989). The origin of skeletons. *Palaios* 4:585–589.

Winterwerp, J.C., and van Kesteren, W.G.M. (2004). *Introduction to the Physics of Cohesive Sediment in the Marine Environment.* Amsterdam: Elsevier B.V.

Wood, R. (1999). *Reef Evolution.* Oxford: Oxford University Press.

Xiao, S., Hu, J., Yuan, X., Parsley, R.L., and Cao, R. (2005). Articulated sponges from the Lower Cambrian Hetang Formation in southern Anhui, South China: Their age and implications for the early evolution of sponges. *Palaeogeography, Palaeoclimatology, Palaeoecology* 220:89–117.

Zhuravlev, A.Y. (2001). Paleoecology of Cambrian reef ecosystems. In *The History and Sedimentology of Ancient Reef Systems,* G.D. Stanley, Jr., Ed. New York: Kluwer Academic/Plenum Publishers, pp. 121–157.

Ziebis, W., Forster, S., Huettel, M., and Jorgensen, B.B. (1996). Complex burrows of the mud shrimp *Callianassa truncata* and their geochemical impact in the sea-bed. *Nature* 382:619–622.

9

HABITAT CONVERSION ASSOCIATED WITH BIOERODING MARINE ISOPODS

Theresa Sinicrope Talley and Jeffrey A. Crooks

9.1 ⬤ INTRODUCTION

The biotic consequences of ecosystem engineering are typically complex and depend on many factors, including interactions among different engineers within a system and spatio–temporal scales. Such complexities demonstrate the richness of engineering-related activities as well as the value of having one overarching context with which to view seemingly disparate interactions. Among the most extreme effects of engineers (or any species) is the complete biogenic transformation of one habitat type to another. When this occurs via the removal of physical structure, claims are often made that "habitat has been lost" and the species that effect this change are labeled "habitat destroyers." However, the consequences of habitat conversion activities are often not this simple, and depend on the frame of reference and spatial scales under consideration.

In nearshore marine systems, engineers such as vascular plants and macroalgae are often the dominant structural forms. The biotic communities associated with these autotrophs, such as salt marshes, mangroves, seagrass beds, and kelp forests, are typically distinct from and more diverse than communities that occupy sediments not occupied by these engineers (Crooks 2002 and references therein). Species that cause the disappearance of these structural elements through engineering, therefore, can cause dramatic shifts in resident biotic assemblages. Bioeroders, such as crustaceans, echinoderms, molluscs, polychaete worms, sponges, and fish, are examples of such engineers because their drilling, rasping, scraping, and boring activities break down otherwise solid substrates. Some of the most important bioeroders in marine systems are the marine pill bugs (Crustacea: Isopoda). Sphaeromatid and limnoriid isopods can burrow into a wide variety of substrates, which can lead to cascading ecological consequences. The economic effects of their activities can also be substantial, including the decay of wooden structures such as pier pilings and docks (Miller 1926, Ray 1959).

In this chapter, we will examine engineering activities and their consequences for three types of marine isopods, *Sphaeroma quoianum*, *S. terebrans*, and *Limnoria* spp., on the biogenic marine habitats formed by marsh plants, mangroves, and kelp (respectively). We will focus on the case of *S. quoianum*, drawing from published (Talley et al. 2001) and previously unpublished (Levin et al.) data, and compare this example with the two other taxa. Each of these isopods has been anthropogenically spread around the world. Although these species represent conservation concerns, they also provide an opportunity for ecological insight afforded by the study of biological invasions (Vitousek 1990, Crooks 2002). These isopods often perform their bioerosive activities in multi-engineer systems, with the plants and kelp creating biogenic structure and the isopods removing it. Sometimes the activities may act directly on the physical environment, such as *S. quoianum* burrowing into unvegetated banks. However, the actual mechanisms by which they cause this loss of structure differ. Their activities include burrowing into the substrate into which plants grow (*S. quoianum*); direct, but nontrophic burrowing into the plant tissue (*S. terebrans*); and burrowing into tissue associated with feeding activities (*Limnoria* spp.) (Figure 9.1). For each taxon, we will consider the ecosystem engineering context of these activities, and discuss how the loss of vegetated habitat can actually lead to creation of habitat on two different scales. We will also identify gaps in knowledge that are needed to provide a more complete picture of the role of these species in ecosystems.

9.2 • SPHAEROMA QUOIANUM

S. quoianum (= *S. quoyanum*, *S. pentodon*) lives intertidally (Riegel 1959, Morris et al. 1980) and burrows into a variety of available soft substrates including peat, mud, and soft rock. It also bores into floating material

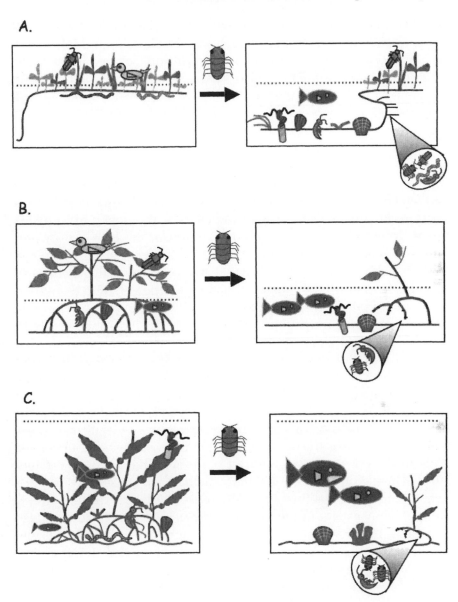

FIGURE 9.1 Schematic diagrams showing the conversion of a (A) salt marsh, (B) mangrove, and (C) kelp forest to open, unvegetated systems initiated by the burrowing activities of isopods. The burrows of the isopods also create fine-scale habitat for burrowing–dwelling organisms. Burrows are depicted as small black lines, and the water surface by a blue dotted line. Despite the different systems, note the similarity of players and processes—all systems contain an allogenic engineering (isopod) whose burrowing activities create fine-scale habitat and remove a second, autogenic engineer leading to the conversion of habitat from one state to another. See text for a full explanation. (See color plate.)

such as wood or Styrofoam (Barrows 1919, Miller 1926, Abbott 1940, Higgins 1956, Carlton 1979), making it susceptible to human-influenced dispersal. *Sphaeroma quoianum* likely was introduced to western North America from Australasia by ships from Australia coming to California for the gold rush in the mid- to late-1800s (Carlton 1979). It was first reported in San Francisco Bay in 1893 and, since the turn of the twentieth century, has spread to several bays, from Bahia de San Quintín in Baja California, Mexico, to Yaquina Bay, Oregon (Carlton 1979, T. Davidson personal communication). Although this isopod can burrow into a variety of available substrata, its preferred habitats in this range are the peat and mud banks of tidal creeks and marshes (Figure 9.1).

Relatively little is known about the effect of the isopod in its native range. Where it has been introduced, however, the high-density and intensive burrowing activity of *S. quoianum* have been observed to weaken mud and clay banks of salt marsh edges, thus making them more susceptible to erosion by wave action or creek flow than in uninvaded areas, even in the presence of native burrowers (Carlton 1979, Josselyn 1983, Nichols and Pamatmat 1988). Carlton (1979) estimates that in some areas of San Francisco Bay, tens of meters or more of marsh edge had been lost since the introduction of this isopod at the end of the nineteenth century, and that this species was likely one of the most important agents of shoreline erosion (e.g., Figures 9.2 and 9.3).

To excavate the burrows in which it lives and eats, *S. quoianum* faces headfirst into the substrate and creates a current with its pleopods that passes forward over the dorsum, down in front of the head, under the body to the posterior end, and out of the forming burrow (Rotramel 1975). It breaks off substrate with its mandibles and releases the particles near its midline so that they are washed out of the burrow by the current, without being caught in the feeding brushes on the first and third legs (Rotramel 1975). *S. quoianum* does not appear to feed on its burrow substrate, and is thought to filter-feed by using a similar current as that used for burrowing. The action of the pleopods brings water into the burrow, over the isopod, and through the brushes on the front legs, thereby trapping particles. The pleopods occasionally stop beating and the isopod cleans the brushes with its maxillipeds (Rotramel 1975).

In order to confirm that *S. quoianum* does not eat plant material found within the mud banks, we used stable isotope analysis to eliminate these as food sources. In assessing trophic relationships using stable isotope analysis, a shift of about 1 unit higher (heavier) of $\delta^{13}C$ and 3 units higher (heavier) of $\delta^{15}N$ occurs from lower to higher trophic levels (Peterson et al. 1984). The results indicate that *S. quoianum* ($\delta C_{13} = -17.4 \pm 0.4$, $\delta N_{15} = 9.2 \pm 0.40$) is likely not eating pickleweed, *Sarcocornia perennis*

FIGURE 9.2 Extreme undercutting of salt marsh surface along the bay front of Corte Madera marsh, San Francisco Bay, December 1998. The burrowing activity of *Sphaeroma quoianum* into vertical marsh banks loosens sediments causing increased localized erosion and undercutting. Photo credit: T.S. Talley. (See color plate.)

(= *Salicornia virginica*), the dominant plant found in the marsh plain above burrows ($\delta C_{13} = -24.5 \pm 1.3$, $\delta N_{15} = 4.5 \pm 1.6$), or the belowground organic material ($\delta C_{13} = -21.9 \pm 1.8$, $\delta N_{15} = 8.9 \pm 0.1$) found in the vicinity of its burrows.

The effects of *S. quoianum* on the abiotic environment and subsequent abiotic interactions have been quantitatively explored only over fine scales (cm–m) in San Francisco Bay and San Diego Bay, California (Talley et al. 2001). Throughout this region, *S. quoianum* prefers to burrow into vertical, firm peat banks, such as those occupied by *S. perennis*, compared with the softer, more sloping banks occupied by Pacific cordgrass, *Spartina foliosa* (Talley et al. 2001). Burrow casts and X-rays revealed that anastomosing burrow networks of the isopod contained

FIGURE 9.3 Erosion of salt marsh bank on the bay front of Corte Madera marsh, San Francisco Bay, December 1998. Extensive undercutting results in the breakage of large chunks of the marsh surface and subsequent loss of vegetated salt marsh. Photo credit: T.S. Talley. (See color plate.)

excavations that were generally somewhat horizontal and averaged 0.6 cm width and 2.2 cm length, with a maximum length of nearly 6 cm. When present, burrows occupied 3–15% of the volume of the outer 5 cm of marsh bank sediments (Talley et al. 2001). Densities of *S. quoianum* burrows were highly correlated with densities of the isopod itself, but the burrows were more persistent than isopod, with visually assessed densities remaining similar throughout the year of the study even though densities of the isopod were up to 13× lower in the winter (Talley et al. 2001). The explicit consideration of this persistence of an organism's structural modification beyond the life span of the organism itself is an advantage of the engineering concept—as opposed to consideration of traditional biotic interactions (e.g., predator–prey relationships), engineering readily accounts for direct effects that outlast the species that created them (Hastings et al. 2007).

Patches of high *S. quoianum* burrow density reduced local (several cm) bank sediment stability, or shear strength, by an average of 2–4× (in both bays), with the largest reductions occurring in firmer sediments. Supporting these findings were the results of an enclosure experiment, which revealed that the presence of this isopod caused 2.4× more wet sediment to be lost from enclosures (10 cm diameter × 10 cm depth) than was lost from enclosures without isopods (Talley et al. 2001). Accordingly, areas of bank with intermediate and high, but not low, burrow

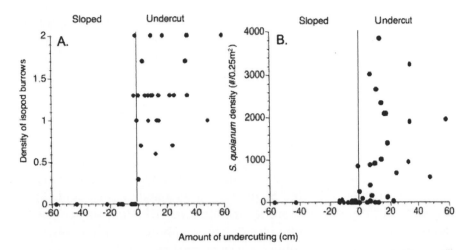

FIGURE 9.4 Relationship between undercutting of marsh banks and (A) *Sphaeroma quoianum* burrow density and (B) the density of *S. quoianum* individuals. Undercutting was measured as the largest distance from a vertical pole set on end on the creek bottom or tidal flat surface and aligned parallel to and touching the top surface of the bank (i.e., a true vertical bank distance would be ~0 cm). Burrow density was assessed visually, and ranked on a scale of 0 (low) to 2 (many). Data were collected in July 1998 from San Diego Bay and San Francisco Bay, n = 45 sampling stations (see Talley et al. 2001 for more information). Data demonstrate a clear threshold, with few burrows or individuals on nonundercut banks and more burrows and individuals on undercut ones.

densities tended to have the most undercut bank faces (Figure 9.4A). Also, the density of isopods themselves were significantly correlated with bank undercutting (p = 0.0006, $F_{1,43}$ = 14, R^2 = 0.25; Figure 9.4B).

The isopod-induced bank erosion and subsequent undercutting led to the eventual collapse of sections of bank (Figure 9.2). Measures of marsh edge lateral loss averaged 15 ± 3 cm year^{-1} in San Diego and 27 ± 7 cm year^{-1} in San Francisco, with losses as high as 112 cm year^{-1} (Talley et al. 2001). Losses of over 60 cm occurred when sections of marsh bank (overhangs) fell to the surface below (Figure 9.3). Loss due to slumping was most common on the bay front marsh edge in San Francisco Bay where wave energy appeared to be highest due to large boat and ferry traffic (Talley et al. 2001; Figure 9.3).

Salt marsh plant composition and cover atop banks that were suitable for the isopod did not appear to differ with *S. quoianum* density (Levin et al. unpublished data), illustrating either the lack of an effect or a lag in response due to the long-lived nature of most of the marsh plants, which consisted of perennial succulents and grasses (e.g., *S. perennis,*

Batis maritima, Jaumea carnosa, Distichlis spicata, and *Spartina foliosa).* These species are themselves ecosystem engineers, as they provide aboveground and belowground physical structure affecting abiotic conditions such as hydrodynamics, shading, sediment deposition, and bank stabilization (e.g., Leonard and Luther 1995, Hacker and Bertness 1999, Cavatorta et al. 2003). While isopod-induced horizontal loss of marsh reduces habitat area for these species and the biota associated with them, there would be a gain in the mudflat and creek-bottom habitat created as a result of marsh loss (Figure 9.1A). Although not quantified, logically this habitat conversion would tend to favor a different suite of species than that supported by the vegetated marsh. For example, one might expect transitions from a predominance of organisms with terrestrial traits such as insects, oligochaetes, pulmonate snails, mammals, and songbirds to those with marine traits such as polychaetes, bivalves, fish, and migratory shorebirds (e.g., Szedlmayer and Able 1996, Levin et al. 1998, Talley personal observation; Figure 9.1A).

Habitat also appeared to be created at smaller spatio–temporal scales, through the increase in the structural complexity of marsh banks (Figure 9.1A). Surveys were conducted in both bays to determine the effects that the invasion of this isopod was having on the associated marsh benthic community (Levin et al. unpublished data). Benthic samples were taken equidistantly from vertical banks along each reach of salt marsh banks used in the study, for a total of nearly 50 15 cm diameter × 5 cm deep cores per bay. Each core was washed after preservation through 0.3 mm mesh, and all macrofauna (fauna ≥0.3 mm) were identified and enumerated (estimates do not include the presence or abundance of *S. quoianum* or its often-present commensal isopod, *Iais californica).* For the benthic macrofauna, higher densities of *S. quoianum* were associated with greater species richness ($R^2 = 0.21$, $P < 0.001$, $F_{1,51} = 13.5$) and higher total numbers of fauna ($R^2 = 0.15$, $P = 0.003$, $F_{1,52} = 9.4$) in San Francisco Bay, but not in San Diego Bay ($p \geq 0.66$). Densities of several taxa increased with higher *S. quoianum* numbers. Other peracarid crustaceans had the strongest association in both bays (SF: $R^2 = 0.30$, $P < 0.001$, $F_{1,52} = 22.6$; SD: $R^2 = 0.40$, $P < 0.001$, $F_{1,45} = 30.1$), and, in San Francisco Bay, densities of most insect, polychaete, and enchytraeid oligochaete taxa also increased in association with *S. quoianum* ($R^2 = 0.11 - 0.16$, $P = 0.003 - 0.01$, $F_{1,52} = 6.6 - 9.9$). The taxa in both bays that tended to decline with increased *S. quoianum* were those normally associated with anaerobic conditions, such as tubificid oligochaetes and capitellid polychaetes ($R^2 = 0.10 - 0.37$, $P = <0.001 - 0.04$, $F = 4.5 - 29.7$). In these descriptive studies, the individual effects on the benthos by the isopod, its burrows, and the resulting abiotic alterations were not separated, but could be by performing future experiments using artificial burrows and/or abiotic manipulations in order to separate biotic from abiotic effects of the species.

Overall, *S. quoianum* has the potential to dramatically alter wetland ecosystems (Carlton 1979, Talley et al. 2001). Locally, the increase in habitat heterogeneity created by isopod burrows can facilitate some suites of resident biota (while inhibiting others). At larger spatio–temporal scales, the isopod can convert high-elevation, vegetated salt marsh into low-elevation, unvegetated flats. The quantitative assessment of both the rate and the biotic consequences of this conversion require additional study, however. This work should focus on decoupling isopod-induced erosion rates from background rates due to both natural and other anthropogenic factors (e.g., boat wake). Quantitative assessment of isopod effects on the larger complex of vegetated marsh and unvegetated flats will require explicit sampling of these two distinct habitat types, which will often necessitate different sampling strategies and comparisons across taxa not typically considered together (e.g., birds vs. fish).

Despite these uncertainties, the general effects of *S. quoianum* invasion are clear enough to warrant conservation concern and management action. Although we argue that "habitat" is not truly being lost, just being converted, this transformation from vegetated marsh to unvegetated flats within invaded areas is largely undesirable. On the West Coast of the United States, much salt marsh habitat has been lost to human development and remaining areas are often encroached upon, no longer able to migrate upland and compensate for loss due to sea level rise and waterfront erosion. Yet, these marshes support a number of endangered and threatened species (Mitsch and Gosselink 2000). A principal goal for *S. quoianum* management should be to limit the spread of this species to uninfested systems, which appears chiefly to occur through human-aided transport on wood or man-made materials. The potential for this species to be transported in plant material being used for marsh restoration efforts also warrants concern. In terms of restoration design, within infested areas it will also be beneficial to recognize the factors that limit the engineering activity of the isopod, such as bank slope, and where possible create systems that will not promote extensive isopod burrowing or accelerated erosion rates. Assessments of invasion risk could also be considered with appropriate landscape planning of proportions of habitat type that would account for future habitat conversion.

9.3 ⬤ *SPHAEROMA TEREBRANS*

Sphaeroma terebrans is a cosmopolitan species of tropical and subtropical waters. It is thought to be native to the Indo-Pacific, and to have been spread in wooden ships to other locations around the world, such as Florida (U.S.) (Rehm and Humm 1973, Carlton and Ruckelshaus 1997, Brooks and Bell 2005), Central and South America (Ellison and Farnsworth 1990), and East Africa (Svavarsson et al. 2002). Like *S. quoianum*, it is a burrowing species that appears to create burrows primarily for living space and not for direct access to food (John 1971). Unlike *S. quoianum*, however, it burrows directly into plant tissue—the exposed woody prop roots of mangrove trees (Figure 9.1B). It is typically most abundant in the prop roots at the lower extent of mangrove forests, and densities decrease at sites with higher tidal elevations (Svavarsson et al. 2002).

There are several multispecies, scale-dependent contexts for considering engineering in relation to the activities of *S. terebrans* in mangals. As with salt marsh plants, mangroves are engineers that structure the environment through production of vegetative cover on intertidal sediments. Mangals increase shading and modify hydrodynamics, as well as provide habitat in the form of structural complexity, supratidal canopy structure, and root surfaces as sites for faunal attachment and burrowing (Lugo and Snedaker 1974). It appears that the engineering effects of mangroves may locally outweigh direct trophic inputs from the mangrove plants, which are relatively inedible (Newell et al. 1995).

The burrowing activities of *S. terebrans*, like that of *S. quoianum*, represent allogenic engineering that can be considered on (at least) two different scales. At the broadest scale, there has been considerable debate about the effects of *S. terebrans*, especially in Florida (e.g., Rehm and Humm 1973, Simberloff et al. 1978, Ribi 1981), where it is thought to have been introduced. Some workers suggest that the boring activities of the isopod cause loss of mangrove habitat through aerial prop root damage and death resulting in the toppling of trees (Rehm and Humm 1973, Ribi 1981, Perry 1988, Perry and Brusca 1989, Ellison and Farnsworth 1990, Svavarsson et al. 2002). Others, however, indicate that damage caused by *S. terebrans* may in fact increase the amount of root branching, perhaps benefiting the mangrove (Simberloff et al. 1978). More recently, it has been argued that factors such as productivity (Carlton and Ruckelshaus 1997, Olafsson 1998) and the healing vs. branching response of the plant (Brooks and Bell 2002) must be examined. Descriptive work in East Africa has indicated, however, that the isopod can limit the lower intertidal extent of mangal through higher infection rates and burrow densities, and higher subsequent rates of root death and tumbling of trees in the lower intertidal (Svavarsson et al. 2002). In general it is possible that *S. terebrans* has dramatically affected the distribution of mangroves wherever it has invaded (Carlton and Ruckelshaus 1997).

In this sense, then, *S. terebrans* again has the potential to undo the engineering effects of mangroves and convert complex mangal habitat back to relatively simple tidal flat (Figure 9.1B), albeit sometimes with the added structure of downed trees (e.g., Svavarsson et al. 2002). While there have been no known studies directly comparing assemblages of bare vs. mangrove-vegetated tidal sediments as a result of *Sphaeroma* activity, first principles (e.g., an understanding of the role of habitat complexity in shaping assemblages) and comparisons of mangrove vs. mudflat ecosystems (e.g., Sheridan 1997) suggest rather obvious outcomes. When mangroves are destroyed, mangal-associated species, especially those associated with the forest canopy and roots, will be replaced by species associated with more open and/or marine conditions (Figure 9.1B). This will likely result in a loss of overall diversity, which is why the loss of mangroves due to *S. terebrans* in Florida has been labeled an "ecocatastrophe" (Rehm and Huhm 1973). However, as Enright (1974) points out, the natural spread of mangroves into tidal flats could equally well be viewed as ecocatastrophic from a marine perspective (although the status of *S. terebrans* as an exotic was not yet known). This highlights the need to consider larger perspectives and that the loss of one habitat type will yield another.

The burrowing activities of *S. terebrans* likely also have smaller-scale effects comparable to *S. quoianum* (Figure 9.1B). *S. terebrans* holes have been reported to support other fauna (Brooks and Bell 2005), including congeners (*S. quadridentatum*), which appear to benefit from the *S. terebrans* adults within the burrows (Thiel 2000). Interestingly, it has been suggested that cover of other epibionts, such as colonial ascidians and sponges (Perry 1988, Ellison and Farnsworth 1990, Ellison et al. 1996), and burrowing molluscan shipworms (John 1971) inhibit the burrowing activities of the isopod through the physical or chemical modification of the root, which could also be viewed as ecosystem engineering.

As with *S. quoianum*, there are serious conservation implications associated with *S. terebrans*, especially if it is exotic throughout much of its range and has shifted what we perceive to be the natural distributional limits of mangroves (Carlton and Ruckelshaus 1997). Although eradication of *S. terebrans* is virtually impossible, management of mangroves should take *S. terebrans* distribution and impacts into account. For example, mangrove restoration is difficult in lower intertidal areas where plants are subject to isopod attack, so protection of mature trees in this zone is especially critical (Svavarsson et al. 2000). Ensuring the presence of beneficial engineers such as diverse native root fouling communities may also limit attack by *S. terebrans* (e.g., Ellison and Farnsworth 1990).

9.4 ● *LIMNORIA* SPP.

Isopods in the genus *Limnoria* are cosmopolitan, with two of the wood-boring species (*L. tripunctata* and *L. quadripunctata*) often inadvertently transported and, therefore, introduced throughout the world (Carlton 1989, Williamson et al. 2002). Unlike sphaeromatids, the limnorid isopods (*Limnoria* spp.), or gribbles, consume their burrowing substrate, which consists of wood, seagrass, and algal material (Johnson and Menzies 1956, Cookson and Lorenti 2001, Thiel 2003). These species are best known for their abilities as bioeroders of wooden structures, and the damage associated with the activities of these and other bioeroders (such as shipworms) can be substantial (e.g., Kofoid 1921, Reish et al. 1980). Given the economic implications of the engineering activities of such species, a considerable field related to prevention of marine biodeterioration has arisen (e.g., Costlow and Tipper 1984).

Although relatively little is known about the ecological effects of limnorid isopods, like the sphaeromatids, they are ecosystem engineers able to cause habitat change by interacting with other engineers at several spatial scales. Two noninvasive species, *Limnoria algarum* and *L. chilensis*, can be found burrowing in kelp (Barrales and Lobbas 1975, Thiel 2003), where their activities can weaken holdfasts, leading to the loss of entire plants and contributing to or even instigating natural cycles of kelp bed die back, especially in high exposure areas (Barrales and Lobbas 1975, North 1979; Figure 9.1C). Kelps are important engineers in marine systems; their canopies and holdfasts influence water-flow and sedimentation rates (Eckman et al. 1989), as well as provide habitat for a variety of species, including relatively sessile invertebrates and more-mobile fish (Foster and Schiel 1985). The loss of these plants therefore may transform canopied forest to open seafloor (Figure 9.1C), although again the magnitude and direct effects of this loss have not been quantified. In general terms, though, species composition in open or deforested areas compared with forested areas will differ dramatically (e.g., Graham 2004). Loosening kelps from the seafloor will also produce kelp paddies, or rafts, which serve as habitat and transport mechanisms for a variety of fish and invertebrates (Hobday 2000). At finer scales, habitat is created for the organisms that inhabit burrows, including arthropods and juvenile *Limnoria* (Sleeter and Coull 1973, El-Shanshoury et al. 1994, Thiel 2003; Figure 9.1C).

As with the other two examples, there are applied implications for the engineering activities of limnoriid isopods. As mentioned previously, the burrowing activities of the invasive, wood-boring gribbles can lead to substantial economic costs associated with damage to wooden structures such as piers, docks, and boats (Costlow and Tipper 1984). The conservation implications of the algal-boring isopods are less clear, as they are not thought to be non-native species. Nonetheless, effective kelp forest management should take into account the engineering activities of these organisms, including adopting regional perspectives to kelp patch distribution that would consider potential refugia from attack (e.g., Holyoak 2000) and proportions of habitat necessary to account for habitat conversion.

9.5 ➾ LESSONS AND IMPLICATIONS

The central focus of much of the research on these isopods, the destruction of vegetated communities, correctly reflects the high conservation priority placed on these biogenic habitats, especially in light of the propensity of these crustacean bioeroders to be invasive. However, in order to more fully characterize the roles of these species in affecting available habitat, broader perspectives are needed. The preceding examples, for instance, highlight the importance of considering different scales and contexts in assessments of the effects of ecosystem engineers. Although such views typically have not been addressed for these isopods (and many other habitat converters), they would be achieved by landscape-level considerations (e.g., Wright et al. 2002, Talley in press). Such studies would be potentially complicated in that they necessarily encompass multiple and often distinct habitat types and associated biota, which may require different types of sampling tools, approaches, and expertise. This is a likely cause of the general tendency to focus attention on the loss of one habitat type with less explicit consideration of the habitat that replaces it.

These isopod examples also demonstrate the power of the engineering concept to reveal commonalities in ecosystem responses to species with seemingly distinct activities. Habitat conversion by *Limnoria* arises primarily from feeding on plant tissue, whereas *S. quoianum* has little direct influence on the plants whose habitat it ultimately destroys. The engineering construct, of one organism affecting other organisms via changes to the abiotic environment, irrespective of whether or not these changes are associated with trophic interactions (Jones et al. 1994, 1997), neatly captures the multifaceted nature of these interactions. In fact, trophic relationships *per se* (e.g., in terms of flows of material and energy through food webs), would not be sufficient to describe the ecosystem-level changes caused by the activities of these organisms (although they would clearly be important for understanding other ecological dynamics). The concept of ecosystem engineering also helps identify gaps in current knowledge that require further attention, such as how available habitat changes across the landscape, and offers explanations for outcomes when purely trophic explanations are not sufficient. Utilizing the principles of engineering and addressing the full breadth of species-induced changes to habitat will improve our ability to make ecological generalizations and predictions, as well as provide information necessary for applied efforts such as restoration and invasions management.

ACKNOWLEDGMENTS

We thank Lisa A. Levin for permission to use our previously unpublished data collected during research funded by the National Oceanographic and Atmospheric Administration's National Sea Grant College, Project number R/CZ-150 NOAA, grant number NA66RG0477 administered by California Sea Grant. The views expressed herein are those of the authors and do not necessarily reflect the views of NOAA or any of its subagencies. We thank the NCEAS working group on Modeling Ecosystem Engineers for discussions about the ecosystem engineer concept. Helpful comments on this chapter were provided by Kim Cuddington, Jeb Byers, and Timothy Davidson, who also provided information on the northern extent of *S. quoianum*.

REFERENCES

Abbott, C.H. (1940). Shore isopods: Niches occupied, and degrees of transition toward land life with special reference to the family Lygididae. In *Proceedings of the 6th Pacific Science Congress*, Vol. 3. Berkeley: University of California Press, pp. 505–511.

Barrales, H.L., and Lobban, C.S. (1975). The comparative ecology of *Macrocystis pyrifera*, with emphasis on the forests of Chubut, Argentina. *Journal of Ecology* 63:657–677.

Barrows, A.L. (1919). The occurrence of a rock-boring isopod along the shore of San Francisco Bay, California. *University of California Publications in Zoology* 19:299–316.

Brooks, R.A., and Bell, S.S. (2002). Mangrove response to attack by a root boring isopod: Root repair versus architectural modification. *Marine Ecology Progress Series* 231:85–90.

Brooks, R.A., and Bell, S.S. (2005). The distribution and abundance of *Sphaeroma terebrans*, a wood-boring isopod of red mangrove (*Rhizophera mangle*) habitat within Tampa Bay. *Bulletin of Marine Science* 76:27–46.

Carlton, J.T. (1979). *History, biogeography, and ecology of the introduced marine and estuarine invertebrates of the Pacific Coast of North America*. Ph.D. dissertation, University of California, Davis.

Carlton, J.T. (1989). Man's role in changing the face of the ocean: Biological invasions and implications for conservation of near-shore environments. *Conservation Biology* 3:265–273.

Carlton, J.T., and Ruckelshaus, M.H. (1997). Nonindigenous marine invertebrates and algae. In *Strangers in Paradise: Impact and Management of Nonindigenous Species in Florida*, D. Simberloff, D.C. Schmitz, and T.C. Brown, Eds. Washington, D.C.: Island Press, pp. 87–210.

Cavatorta, J.R., Johnston, M., Hopkinson, C., and Valentine, V. (2003). Patterns of sedimentation in a salt marsh-dominated estuary. *Biological Bulletin* 205:239–241.

Cookson, L.J., and Lorenti, M. (2001). A new species of limnoriid seagrass borer (Isopoda) from the Mediterranean. *Crustaceana* 74:339–346.

Costlow, J.D., and Tipper, R.C. (1984). *Marine Biodeterioration: An Interdisciplinary Study*. Annapolis, Maryland: Naval Institute Press.

Crooks, J.A. (2002). Characterizing the consequences of invasions: The role of introduced ecosystem engineers. *Oikos* 97:153–166.

Eckman, J.E., Duggins, D.O., and Sewell, A.T. (1989). Ecology of understory kelp environments. 1. Effects of kelps on flow and particle transport near the bottom. *Journal of Experimental Marine Biology and Ecology* 129:173–187.

Ellison, A.M., and Farnsworth, E.J. (1990). The ecology of Belizean mangrove-root fouling communities: I. Epibenthic fauna are barriers to isopod attack of red mangrove roots. *Journal of Experimental Marine Biology and Ecology* 142:91–104.

Ellison, A.M., Farnsworth, E.J., and Twilley, R.R. (1996). Facultative mutualism between red mangroves and root-fouling sponges in Belizean mangal. *Ecology* 77:2431–2444.

El-Shanshoury, A.R., Mona, M.H., Shoukr, F.A., and El-Bossery, A.M. (1994). The enumeration and characterization of bacteria and fungi associated with marine wood-boring isopods, and the ability of these microorganisms to digest cellulose and wood. *Marine Biology* 119:321–326.

Enright, J.T. (1974). Mangroves, isopods and ecosystem. *Science* 183:1038.

Foster, M., and Schiel, D.R. (1985). *The Ecology of Giant Kelp Forests in California: A Community Profile*. Washington, D.C.: U.S. Fish and Wildlife Service Biological Report 85(7.2).

Graham, M. (2004). Effects of local deforestation on the diversity and structure of Southern California giant kelp forest food webs. *Ecosystems* 7:341–357.

Hacker, S.D., and Bertness, M.D. (1999). Experimental evidence for factors maintaining plant species diversity in a New England salt marsh. *Ecology* 80:2064–2073.

Hastings, A., Byers, J.E., Crooks, J.A., Cuddington, K., Jones, C.G., Lambrinos, J.G., Talley, T.S., and Wilson, W.G. (2007). Ecosystem engineers in space and time. *Ecological Letters* 10:153–164.

Higgins, C.G. (1956). Rock-boring isopods. *Geological Society of America Bulletin* 67:1170.

Hobday, A. (2000). Persistence and transport of fauna on drifting kelp (*Macrocystis pyrifera* (L.) C. Agardh) rafts in the Southern California bight. *Journal of Experimental Marine Biology and Ecology* 253:75–96.

Holyoak, M. (2000). Habitat patch arrangement and metapopulation persistence of predators and prey. *American Naturalist* 156:378–389.

John, P.A. (1971) Reaction of *Sphaeroma terebrans* Bate toother sedentary organisms infesting the wood. *Zoologischer Anzeiger* 186:126–136.

Johnson, M.W., and Menzies, R.J. (1956). The migratory habits of the marine gribble *Limnoria tripunctata* Menzies in San Diego Harbor, California. *Biological Bulletin* 110:54–68.

Jones, C.G., Lawton, J.H., and Shachak, M. (1994). Organisms as ecosystem engineers. *Oikos* 689:373–386.

Jones, C.G., Lawton, J.H., and Shachak, M. (1997). Positive and negative effects of organisms as physical ecosystem engineers. *Ecology* 78:1946–1957.

Josselyn, M. (1983). *The Ecology of San Francisco Bay Tidal Marshes: A Community Profile.* Washington, D.C.: U.S. Fish and Wildlife Service Biological Report 82(23).

Kofoid, C.A. (1921). The marine borers of the San Francisco Bay region. In *Report on the San Francisco Bay Marine Piling Survey,* C.A. Kofoid, Ed. San Francisco, CA.: San Francisco Bay Marine Piling Committee, pp. 23–61.

Leonard, L.A., and Luther, M.E. (1995). Flow hydrodynamics in tidal marsh canopies. *Limnology and Oceanography* 40:1474–1484.

Levin, L.A., Talley, T.S., and Hewitt, J. (1998). Macrobenthos of *Spartina foliosa* (Pacific cordgrass) salt marshes in Southern California: Community structure and comparison to a Pacific mudflat and a *Spartina alterniflora* (Atlantic smooth cordgrass) marsh. *Estuaries* 21:129–144.

Lugo, A.E., and Snedaker, S.C. (1974). The ecology of mangroves. *Annual Review of Ecology and Systematics* 5:39–64.

Miller, R. (1926). Ecological relations of marine wood-boring organisms in San Francisco Bay. *Ecology* 7:247–254.

Mitsch, W.J., and Gosselink, J.G. (2000). *Wetlands,* 3rd ed. New York: John Wiley and Sons.

Morris, R.H., Abbott, D.P., and Haderlie, E.C. (1980). *Intertidal Invertebrates of California.* Stanford: Stanford University Press.

Newell R.I.E., Marshall, N., Sasekumar, A., and Chong, V.C. (1995). Relative importance of benthic microalgae, phytoplankton, and mangroves as sources of nutrition for penaeid prawns and other coastal invertebrates from Malaysia. *Marine Biology* 123:595–606.

Nichols, F.H., and Pamatmat, M.M. (1988). *The Ecology of Soft-Bottom Benthos of San Francisco Bay: A Community Profile.* Washington, D.C.: U.S. Fish and Wildlife Service Biological Report 85(7.23).

North, W.J. (1979). Adverse factors affecting giant kelp and associated seaweeds. *Experientia* 35:445–447.

Olafsson, E. (1998). Are wood-boring isopods a real threat to the well-being of mangrove forests? *Ambio* 27:760–761.

Perry, D.M. (1988). Effects of associated fauna on growth and productivity in the red mangrove. *Ecology* 69:1064–1075.

Perry, D.M., and Brusca, R.C. (1989). Effects of the root-boring isopod *Sphaeroma peruvianum* on red mangrove forests. *Marine Ecology Progress Series* 57:287–292.

Peterson, B.J., Howarth, R.W., and Garritt, R.H. (1984). Multiple stable isotopes used to trace the flow of organic matter in estuarine food webs. *Science* 227:1361–1363.

Ray, D.L. (1959). *Marine Boring and Fouling Organisms*. Seattle, WA: University of Washington Press.

Rehm, A., and Humm, H.J. (1973). *Sphaeroma terebrans*: A threat to the mangroves of southwestern Florida. Science 182:173–174.

Reish D.J., Soule, D.F., and Soule, J.D. (1980). The benthic biological conditions of Los Angeles–Long Beach Harbors: Results of 28 years of investigations and monitoring. *Helgolander Meeresuntersuchungen* 34:193–205.

Ribi, G. (1981). Does the wood-boring isopod *Sphaeroma terebrans* benefit red mangroves (*Rhizophora mangle*)? *Bulletin of Marine Science* 31:925–928.

Riegel, J.A. (1959). Some aspects of osmoregulation in two species of sphaeromid isopod Crustacea. *Biological Bulletin* 116:272–284.

Rotramel, G. (1975). Filter feeding by the marine boring isopod *Sphaeroma quoyanum* H. Milne Edwards, 1840 (Isopoda: Sphaeromatidae). *Crustaceana* 28:7–10.

Sheridan, P. (1997). Benthos of adjacent mangrove, seagrass and non-vegetated habitats in Rookery Bay, Florida, USA. *Estuarine, Coastal and Shelf Science* 44:455–469.

Simberloff, D., Brown, B.J., and Lowrie, S. (1978). Isopod and insect root borers may benefit Florida mangroves. *Science* 201:630–632.

Sleeter, T.D., and Coull, B.C. (1973). Invertebrates associated with the marine wood boring isopod, *Limnoria tripunctata*. *Oecologia* 13:97–102.

Svavarsson, J., Osore, M.K.W., and Ólafsson, E. (2002). Does the wood-borer *Sphaeroma terebrans* (Crustacea) shape the distribution of the mangrove *Rhizophora mucronata*? *Ambio* 31:574–579.

Szedlmayer S.T., and Able, K.W. (1996). Patterns of seasonal availability and habitat use by fishes and decapod crustaceans in a southern New Jersey Estuary. *Estuaries* 19:697–709.

Talley, T.S. (in press). Which spatial heterogeneity framework? Consequences for conclusions about patchy population distributions. *Ecology*.

Talley, T.S., Crooks, J.A., and Levin, L.A. (2001). Habitat utilization and alteration by the burrowing isopod, *Sphaeroma quoyanum*, in California salt marshes. *Marine Biology* 138:561–573.

Thiel, M. (2000). Juvenile *Sphaeroma quadridentatum* invading female-offspring groups of *Sphaeroma terebrans*. *Journal of Natural History* 34:737 745.

Thiel, M. (2003). Reproductive biology of *Limnoria chilensis*: Another boring peracarid species with extended parental care. *Journal of Natural History* 37:1713–1726.

Vitousek, P.M. (1990). Biological invasions and ecosystem processes: Towards an integration of population biology and ecosystem studies. *Oikos* 57:7–13.

Williamson, A.T., Bax, N.J., Gonzalez, E., and Geeves, W. (Eds). (2002). *Development of a Regional Risk Management Framework for APEC Economies for Use in the Control and Prevention of Introduced Marine Pests.* APEC Marine Resource Conservation Working Group, Final Report. Accessed at www.marine.csiro.au/crimp/reports/APEC_Report.pdf.

Wright, J.P., Jones, C.G., and Flecker, A.S. (2002). An ecosystem engineer, the beaver, increases species richness at the landscape scale. *Oecologia* 1332:96–101.

10

SYNTHESIS: LESSONS FROM DISPARATE ECOSYSTEM ENGINEERS

James E. Byers

The chapters in this section illustrate that the precise effects of ecosystem engineers can be highly system specific, but the ecosystem engineering concept reveals commonalities in engineering-related processes. The intricacies of an insect tying leaves together (Lill and Marquis, Ch 6) and an isopod collapsing salt marsh banks (Talley and Crooks, Ch 9) can readily be viewed as distinctly disparate examples, yet both have community-level effects initiated by alteration of physical structure. The idiosyncratic details of these examples are certainly important in their own right for providing insight into individual systems; however, examining a diversity of examples provides unique opportunities for gaining general insights and unifying theories. Here I draw out five major messages that are reflected in these chapters and evaluate some implications for future directions for the study of ecosystem engineers.

First, one distinct benefit in considering many different ecosystem engineers in side-by-side case studies is the identification of the unique advantages that different systems may offer for examining different lines of research questions. For example, the shelter-building insects described by Lill and Marquis (Ch 6) and the soil-tilling earthworms described by Lavelle (Ch 5) clearly alter the physical structure of habitat in important ways, but are often overlooked because of their small size. It is the small size of these engineers, however, that makes them easy to manipulate and replicate in experiments. Because most ecosystem engineering studies are observational, systems such as these may provide valuable insight into the mechanisms behind engineering outcomes. Similarly, some systems are more heavily influenced by broad physical forces than others allowing for examination of the interactions between engineering and external, often larger-scale processes. Wave energy, for instance, influenced the engineering potential of bioeroding isopods (Talley and Crooks, Ch 9).

Second, and perhaps most notably, the various examples in the chapters underscore that the temporal scale of the engineering and the persistence of the engineered aspects differ greatly between systems (Hastings et al. 2007). In temperate regions the structural changes of leaf tiers are cast aside when leaves are shed by deciduous trees every autumn. In contrast, the chemical and salinity changes to soil deposited by iceplant often persist for years even after the plant itself is removed (Molinari et al., Ch 7). As Molinari et al. further emphasize, differences in the spatial scale of ecosystem engineering can also be apparent. On a small scale, invasive ecosystem engineers can exact great physical changes resulting in lower (Molinari et al., Ch 7) or higher (Talley and Crooks, Ch 9) species richness. If the engineering skews the environment heavily enough, higher richness could especially be due to an increase in exotic species. At larger scales, a mosaic of engineered and unengineered habitat is likely in many cases to lead to high regional-scale species richness due to enhanced habitat heterogeneity. However, in extreme cases of engineering, like iceplant, where almost all species were excluded underneath it, low species richness can still result at large scales. Thus, although we see a common thread of engineers altering the physical environment and enhancing environmental heterogeneity, the resultant community effects are determined largely from the scale at which environmental heterogeneity affects biodiversity for a particular group of species as well as the baseline richness of unmodified habitats (Tews et al. 2004, Wright et al. 2006).

Third, physical, structural modification remains one of the most clearcut examples of ecosystem engineering. Such modification is easily identified and has obvious effects on subsequent biotic interactions within a community. For example the effect of certain earthworm species to mesh soil particles into solid macroaggregated structures has direct consequences for nutrient distributions to plants. In other cases, like the bioeroding isopods, the structural modification may be so drastic that a habitat is completely converted to another habitat type.

Although any changes to the abiotic environment could be thought of as engineering, if such changes occur due to trophic, assimilatory, or even competitive purposes they may be better characterized with existing ecological terminology and frameworks like energy flow, metabolism, or allelopathy. For example, a filter-feeding mollusc could increase water clarity by removing plankton or sediment from the water column. Although the effect on water clarity may be the same, the removal of plankton is trophic while sediment removal is engineering. Placing the emphasis of ecosystem engineering on the process (filtration of sediment) as opposed to the consequence (water clarity) is important because it helps to indicate which ecological theories (e.g., ecosystem engineering, food webs) might be most applicable in a given instance. In this instance, dynamic feedbacks of the predator feeding on its planktonic prey and subsequent community-level consequences will surely differ from those arising from interactions of predators and nonliving sediment particles. What makes ecologists' task both difficult and compelling is that species may often be influential due to a mixture of engineering and biotic interactions. However, Talley and Crooks (Ch 9) make a clear case that, from a management perspective, the bioeroding isopods are important mostly in nontrophic ways. Thus an explicit ecosystem engineering framework in and of itself would be particularly helpful to management applications in this system.

An emphasis on the processes behind ecosystem engineering can lead to some grey areas. In particular, it can sometimes be difficult to categorize chemical changes to an ecosystem. For example, are chemical inputs by iceplant into the soil best examined with an engineering framework or with alleopathy or Lotka-Volterra competition models? Ultimately the distinction between ecosystem engineering and biotic interactions that yield similar environmental effects (like filtration or alleopathy) may depend on the perspective and needs of the practitioner and which framework is easiest and most efficient to apply. In the case of iceplant, the clearest examples of chemical engineering may be through its spatially and temporally extended abiotic influence via inorganic chemicals (e.g., salt). Legacy effects of salt or chemicals that persist after an ecosystem engineer is removed might also be effectively framed as ecosystem engineering since there is no intentional competitive target of these lingering abiotic changes.

Fourth, for most ecologists who deal with contemporary systems, the paleontologic examples of Marenco and Bottjer (Ch 8) depicting some of the earliest forms of engineering are intriguing. Specifically, the soft-bottom bioturbating–aerating species they describe opened up a new third dimension of habitat for marine infauna. By providing a broader temporal view, such paleontological evidence provocatively implies that ecosystem engineering may have important ramifications for evolutionary processes, particularly the appearance of novel functional groups of organisms. That is, if ecosystem engineers facilitate use of a completely novel habitat, they can catalyze new modes of life. It would be a tantalizing exercise to try to identify the explosive radiations of species throughout time and determine how many may have been attributable to novel ecosystem engineers facilitating expansion into previously uninhabited ecological niches. Such novel ecosystem engineers that began engineering in a new way or in a new habitat that physics alone could not engineer effectively may have been critical catalysts in the radiation of lifestyles and life-forms.

Fifth, two of the chapters in this section (Molinari et al., Ch 7 and Talley and Crooks, Ch 9) dealt predominantly with ecosystem engineers in their non-native environments. Although ecosystem engineers typically function as engineers in both their native and introduced environments, when they are introduced to a new environment, ecosystem engineers may become more abundant or we may simply have a tendency to notice the engineering effects more in a place where the effects are novel. Invasive ecosystem engineers will often have unique traits (Crooks 2002), unless they happen to be structurally identical to a native species, e.g., one tree species replacing another. The large community changes that can often occur in an environment where an ecosystem engineer is introduced stem from the fact that the native species are often not adapted to the newly engineered abiotic conditions. Even if native species survive the direct alterations, the abiotic playing field, which provides the context upon which all biotic interactions are dependent, may be severely skewed. These disturbances may therefore erase a native species' prior advantage of local environmental adaptation accrued over evolutionary time, giving non-native species equal or better opportunity to compete their way into the community (Byers 2002). As opposed to direct anthropogenic disturbances, the modification of historic, environmental conditions by introduced ecosystem engineers may be particularly enhanced because, once established, they chronically alter the environment. This is one reason the removal of invasive ecosystem engineers is frequently a top priority in restoration efforts (Byers et al. 2006, Byers in press).

In summary, the scientific literature has an increasing number of clear examples of ecosystem engineers (Wright and Jones 2006). The most convincing of these are cases where engineering effects far outweigh effects from biotic interactions. Burrowing isopods and beavers are certainly part of food webs, but their largest impacts on the communities are through their engineering activities. Even though the effects of ecosystem engineers on their communities can be pervasive and extreme, there is still no widely used, off-the-shelf theoretical approach to study these effects that is analogous to the concepts and models available for studying predation and competition. Mutualism theory is a partial inroad in this direction (Bruno et al. 2003), and some progress has recently been made with explicit ecosystem engineering models (Gurney and Lawton 1996, Cuddington and Hastings 2004; Wilson and Wright, Ch 11, Cuddington and Hastings, Ch 13, Meron et al., Ch 12). Ecosystem engineering, with its dynamic components of organisms affecting physical structure and consequent feedbacks on the engineers and their communities, would benefit from full development of analytical, conceptual, and theoretical approach in ecology (Jones et al. 1994, Gurney and Lawton 1996, Cuddington and Hastings 2004).

Generalizing types of ecosystem engineering would greatly aid such a development of a full theoretical and conceptual treatment because one of the impediments may be that each case of ecosystem engineering has been viewed as idiosyncratic, to be addressed on a case-by-case basis. In describing the paleo explosion of bioturbators–aerators as an important, engineering life form, Marenco and Bottjer (Ch 8) have provided an example of how we could meaningfully categorize engineers according to their functional alterations of the environment. Examples of other major categories of species that share overarching similarities of engineering effects may include the following: flow modifiers, habitat modifiers, and biogeochemical modifiers (Gutierrez et al. 2003). Identifying common, unifying groups of ecosystem engineers is a challenging, yet potentially fruitful pursuit for ecologists (Gutierrez et al. 2003). Because some ecosystem engineers, including the ones in this section, span multiple categories, the category applied may depend on which affected species one cares about. For example, earthworms modify both habitat and nutrient flows. For ground-dwelling insects the habitat modification may likely be the most important aspect, because aggregations and disaggregations of soil structures have a profound influence on certain other belowground species. However, for plants, the worms' role as nutrient distributors is likely to be a large one. In any event, such classification schemes would likely be welcomed by theoreticians seeking to develop general models for particular suites of engineers, or empiricists looking for common patterns across systems. The development of sound classifications is perhaps one of the most important needs to advance a generalized, unified study of ecosystem engineers.

REFERENCES

Bruno, J.F., Stachowicz, J.J., and Bertness, M.D. (2003). Inclusion of facilitation into ecological theory. *Trends in Ecology & Evolution* 18:119–125.

Byers, J.E. (2002). Impact of non-indigenous species on natives enhanced by anthropogenic alteration of selection regimes. *Oikos* 97:449–458.

Byers, J.E. (in press). Invasive animals in marshes: Biological agents of change. In *Salt marshes under global siege*, B.R. Silliman, E.D. Grosholz, and M.D. Bertness, Eds. University of California Press.

Byers, J.E., Cuddington, K., Jones, C.G., Talley, T.S., Hastings, A., Lambrinos, J.G., Crooks, J.A., and Wilson, W.G. (2006). Using ecosystem engineers to restore ecological systems. *Trends in Ecology & Evolution* 21:493–500.

Crooks, J.A. (2002). Characterizing ecosystem-level consequences of biological invasions: The role of ecosystem engineers. *Oikos* 97:153–166.

Cuddintgon, K., and Hastings, A. (2004). Invasive engineers. *Ecological Modelling* 178:335–347.

Gurney, W.S.C., and Lawton, J.H. (1996). The population dynamics of ecosystem engineers. *Oikos,* 76(2):273–283.

Gutierrez, J.L., Jones, C.G., Strayer, D.L., and Iribarne, O.O. (2003). Mollusks as ecosystem engineers: The role of shell production in aquatic habitats. *Oikos* 101:79–90.

Hastings, A., Byers, J.E., Crooks, J., Cuddington, K., Jones, C., Lambrinos, J., Talley, T., and Wilson, W. (2007). Ecosystem engineering in space and time. *Ecology Letters* 10:153–164.

Jones, C.G., Lawton, J.H., and Shachak, M. (1994). Organisms as ecosystem engineers. *Oikos* 69:373–386.

Tews, J., Brose, U., Grimm, V., Tielborger, K., Wichmann, M.C., Schwager, M., and Jeltsch, F. (2004). Animal species diversity driven by habitat heterogeneity/diversity: The importance of keystone structures. *Journal of Biogeography* 31:79–92.

Wright, J.P., and Jones, C.G. (2006). The concept of organisms as ecosystem engineers ten years on: Progress, limitations, and challenges. *Bioscience* 56(3):203–209.

Wright, J.P., Jones, C.G., Boeken, B., and Shachak, M. (2006). Predictability of ecosystem engineering effects on species richness across environmental variability and spatial scales. *Journal of Ecology* 94(4):815–824.

Figure 5.2 Tower-shaped earthworm cast in fallows in Vietnam. This structure is formed after weeks of daily deposition of casts at the top edge of the structure. Note dense colonization by fine roots. Photo P. Jouquet.

FIGURE 7.1 Transect through a highly invaded backdune community at Vandenberg Air Force Base, California. The mat forming *C. edulis* can be seen along the entirety of the transect.

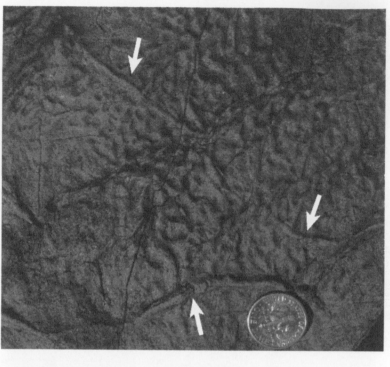

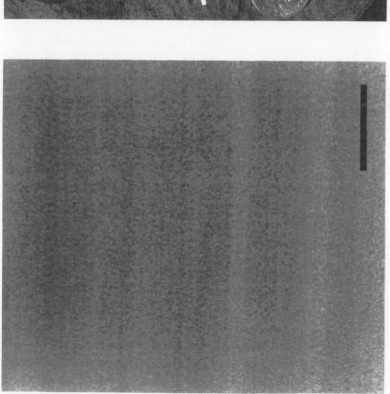

FIGURE 8.1 Characteristic features of Lower Cambrian rocks. (Left) Positive x-radiograph of a vertically sectioned sedimentary rock sample from the Lower Cambrian Campito Formation showing finely laminated sediment that was not disrupted by vertical bioturbation. Scale bar = 1 cm. (Right) Horizontal (bedding plane) exposure of a Lower Cambrian sedimentary rock showing "wrinkle structures" (microbially mediated sedimentary structures), which are thought to have formed due to the compaction of a seafloor microbial mat. Superimposed on the wrinkle structures are examples of the simple horizontal trace fossil *Planolites* (arrows). (Although they are sedimentary structures, trace fossils are classified using biological nomenclature.) *Planolites* likely represents the work of a shallow-burrowing organism that may have been feeding on nutrients associated with the microbial mat.

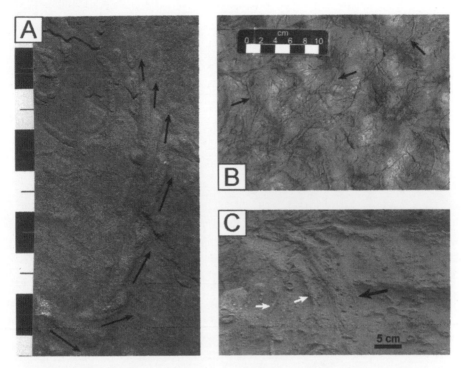

FIGURE 8.3 Trace fossils preserved in Lower Cambrian (A, B) and Eocene (ca. 56–34 million year old) (C) rocks. (A) *Treptichnus pedum*, the first trace fossil to exhibit a vertically oriented component, preserved upside down on the bottom of a Lower Cambrian sedimentary rock unit. The nested lobes (arrows) of the trace likely represent systematic probing of the seafloor sediment by a priapulid-like deposit-feeding organism. (B) A Lower Cambrian bedding plane surface that contains abundant horizontal trace fossils (*Planolites*; arrows). (C) Abundant vertically oriented trace fossils in Eocene exposures near San Diego, CA. *Conostichus* (black arrows), a large, lobe-shaped burrow, is produced by anemones and other stationary benthic suspension feeders during sediment influx. *Ophiomorpha* (white arrows), a deep mud-lined burrow, is produced by many types of benthic suspension-feeding crustaceans, which require stable semipermanent dwellings.

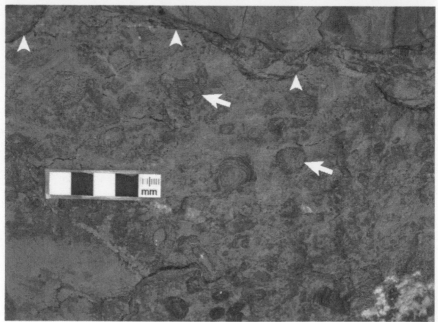

FIGURE 8.5 Close-up views of an archaeocyath–calcimicrobial reef, Stewart's Mill, NV. (Top) A branching archaeocyath sponge, which appears to have been preserved in life position. Surrounding the archaeocyath is carbonate sediment that may have accumulated slowly while the animal was alive, allowing it to be preserved in place. Scale bar = 1 cm. (Bottom) Accumulated fragments of skeletal material, including that of archaeocyaths (arrows), and microbial structures surrounded by carbonate sediment. The sharp boundary in the upper portion of the photograph (arrowheads) likely represents the floor of a reef cavity, which would have harbored organisms that were specially adapted to life in these cryptic settings. Photo courtesy of Matthew Clapham.

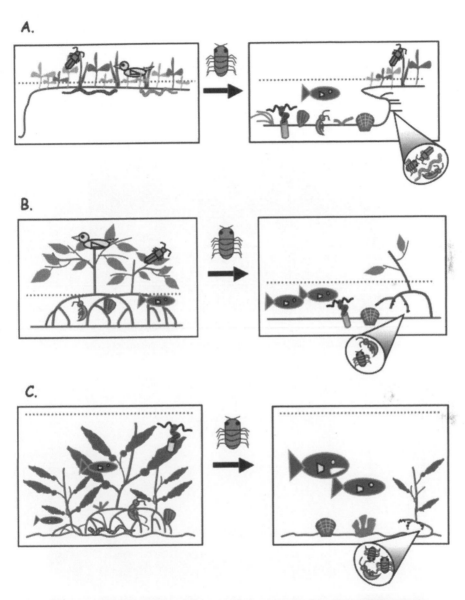

FIGURE 9.1 Schematic diagrams showing the conversion of a (A) salt marsh, (B) mangrove, and (C) kelp forest to open, unvegetated systems initiated by the burrowing activities of isopods. The burrows of the isopods also create fine-scale habitat for burrowing–dwelling organisms. Burrows are depicted as small black lines, and the water surface by a blue dotted line. Despite the different systems, note the similarity of players and processes—all systems contain an allogenic engineering (isopod) whose burrowing activities create fine-scale habitat and remove a second, autogenic engineer leading to the conversion of habitat from one state to another. See text for a full explanation.

FIGURE 9.2 Extreme undercutting of salt marsh surface along the bay front of Corte Madera marsh, San Francisco Bay, December 1998. The burrowing activity of *Sphaeroma quoianum* into vertical marsh banks loosens sediments causing increased localized erosion and undercutting. Photo credit: T.S. Talley.

FIGURE 9.3 Erosion of salt marsh bank on the bay front of Corte Madera marsh, San Francisco Bay, December 1998. Extensive undercutting results in the breakage of large chunks of the marsh surface and subsequent loss of vegetated salt marsh. Photo credit: T.S. Talley.

III

THEORIES AND MODELS

Relatively few models exist that specifically examine the ecosystem engineering concept and its myriad implications for the engineer's population dynamics, not to mention the outcomes for the rest of an ecological community and the environment being modified. This section examines models related to ecosystem engineering. One key aspect of these models is the link between the biotic and abiotic features of an ecosystem. There are two ways that this linkage can take place. First, and perhaps simplest, the link can be "one way," a situation in which an organism modifies the environment and the environmental modifications have subsequent implications for other features or species in the community. These implications can be important in and of themselves when one is interested in, say, the effects of climate change or community disruptions in the presence of an invasive species. Perhaps more interesting from a theoretical point of view, the second situation concerns a feedback between the ecosystem engineer and the consequences of its engineering. In other words, the engineer's activities modify an environmental feature that either directly or indirectly affects the engineer's population dynamics. Here we examine models that represent both possibilities.

11

COMMUNITY RESPONSES TO ENVIRONMENTAL CHANGE: RESULTS OF LOTKA–VOLTERRA COMMUNITY THEORY

William G. Wilson and Justin P. Wright

11.1 ⬤ INTRODUCTION

A principal feature of ecosystem engineering is an alteration of one or more environmental variables, an alteration that potentially has important impacts on the resident community. In some situations the alteration of the environment is so complete that the entire community changes from one habitat type to another, the prototypical example being the beaver (Naiman et al. 1988, Wright et al. 2002). This rather dramatic class of alterations is not the one we address in this chapter. Instead, we are interested in less extreme situations in which the ecosystem engineer has a weaker impact on environmental variables, causing a less extreme impact on the ecological community. Examples include the modification of resource availability by desert shrubs (Boeken et al. 1998, Raffaele and Veblen 1998, Wright et al. 2006), pocket gophers (Williams and Cameron 1986, Inouye et al. 1987), or alpine cushion plants (Badano et al. 2006). Within this scenario, our goal is to understand what ecological factors enhance or reduce the influence of an engineering species on patterns of species diversity in a community.

Within a single trophic-level community model, species-specific growth rates and species interactions typically are represented as fixed constants, imagined as either environmentally dependent, but with a constant environment, or as independent of environmental variation. In either case, explicit environmental dependence is unnecessary except when considering dynamic resources. The single trophic-level community model examined here relaxes this constant environment assumption and, although no explicit representation is made, supposes that there exists a species that modifies the environment and produces a quantitative response in the pairwise interaction strengths and species-specific growth rates in the community.

It is a fair question to ask whether our formulation differs from one designed to study the impact of a general, slow environmental change on an ecological community. Perhaps the best answer is that our formulation is robust to either biotic or abiotic drivers of the environmental change, but does not examine the potential feedback response on the driver. This formulation makes our study relevant beyond ecosystem engineers, applicable to any situation or mechanism causing a change in environmental variables that affects community composition, for example, abiotic interactions or human-induced climate change. Our results are also robust to differing definitions of ecosystem engineer (Wright and Jones 2006, Wilson, Ch 3).

Our approach differs significantly from previous attempts to model the effects of ecosystem engineers on system dynamics that have explicitly accounted for feedbacks between the fitness of the engineering species and the changes that it creates in environmental variables (Gurney and Lawton 1996, Wright et al. 2004). Consideration of these feedback loops provides useful insight into the situations in which the engineer's persistence is dependent upon its engineering activities. Our model does not account for these feedbacks simply because our primary interest is the community-level response of species richness to engineering effects.

We extend the recent Wilson et al.(2003) and Wilson and Lundberg (2004, 2006) community model that yields approximate analytic results for community properties. That framework considers linearized species interactions, a la Lotka–Volterra models, as the mathematical foundation, and then proposes an approximate mean-field solution for many interesting ecological measures. Comparisons with numerical results not subject to the approximations demonstrate the validity of the analytic results. Far and away the greatest benefit of the approximate solution is the explicit connection between community properties and the underlying distribution of species growth rates and interaction strengths.

Here we consider these underlying species-level distributional properties, namely the mean and variance of growth rates and the strength of species interactions, to be dependent on environmental variables. No explicit consideration of the engineer is necessary; instead, its influence on the community is mediated through the environment. Given environmentally driven change in the distributions of species-level growth rates and interaction strengths, we can ask about changes in community-level response variables, namely the number of species persisting in the environment. We can also examine questions such as at what levels of productivity does ecosystem engineering have weaker or stronger effects. Finally, given the model predictions, we examine whether this model analysis fits qualitative arguments made in the literature.

11.2 ▬ LOTKA–VOLTERRA COMMUNITY MODEL

We extend the Lotka–Volterra (LV) community model for interacting species under the assumption that every species in a local community has the potential to interact with any other. Dynamics of closed systems result from species-specific per capita births, $B(\{n\}; E)$, and deaths, $D(\{n\}; E)$, dependent on the set of species densities, $\{n\}$, and the environmental state E. The traditional formulation of an LV model can be viewed as a linear approximation (by Taylor series expansion) of these arbitrary birth and death functions, with an intermediate set of density-independent per capita birth and death rates for species i, $b_i(E)$ and $d_i(E)$ respectively, and density-dependent terms representing the effect of species j on species i's birth and death rates, $\beta_{ij}(E)$ and $\delta_{ij}(E)$. Collecting all of these terms,

$$\frac{dn_i}{dt} = n_i[B_i(\{n_j\}; E) - D_i(\{n_j\}; E)] \tag{1a}$$

$$\approx n_i\left[\left(b_i(E) - \sum_{j=1}^{P}\beta_{ij}(E)n_j\right) - \left(d_i(E) + \sum_{j=1}^{P}\delta_{ij}(E)n_j\right)\right] \tag{1b}$$

$$= n_i\left(K_i(E) - \alpha_{ii}(E)n_i - \sum_{j \neq i}^{P}\alpha_{ij}(E)n_j\right) \tag{1c}$$

reproduces the classic Lotka–Volterra community model (Lotka 1925, Volterra 1926) for a regional pool of P species. The parameter $K_i(E)$ not only represents the maximum per capita growth rate of species i, but, as shown in later text, it also scales the species' equilibrium density. Interaction rates, both intraspecific, $\alpha_{ii}(E)$, and interspecific, $\alpha_{ij}(E)$, also can depend on the environmental state. Each of these parameters, in general, involves combined aspects of both birth and death rates, but isolates density-independent and density-dependent interactions. As just indicated, all of the parameters are implicit functions of the environmental state E, but in the remainder of this manuscript we will drop the explicit dependence. We also assume that the interaction parameters, considered across all species, can be described by a mean and variance, for example $\bar{K}$ and σ_K^2, both dependent on the environmental state, and similarly for the intraspecific interaction parameters, α_{ii}, distributed with mean and variance, α_I and $\sigma_{\alpha I}^2$, and the interspecific interaction parameters, α_{ij}, distributed with mean α_H and variance $\sigma_{\alpha H}^2$, where the H denotes interspecific (or heterospecific) interactions. Note that this description does not require that the parameters are normally distributed; the following calculations simply use only the first two nonzero moments of the distributions, which are the mean and variance. It is not clear that including higher moments would introduce qualitatively new changes to our results.

PSEUDO-EQUILIBRIUM ANALYSIS: TARGET DENSITIES

Our analysis begins by setting the derivatives in Equation 1c to zero, indicating an equilibrium situation. Although it is typically thought that Lotka–Volterra competition models yield rather stable equilibria, van Nes and Scheffer (2004) clearly demonstrate multiple stable equilibria as well as cyclic and chaotic dynamics when interspecific and intraspecific interactions take on overlapping distributions. Although we have also observed these interesting dynamics under appropriate conditions (Wilson unpublished), we have not examined the importance of this issue in detail, in part because the community tends to collapse to a small subset of species with high density (for example, the large $\bar{\alpha}$ limit of Fig. 3 in Wilson et al. 2003). This small species number limit is beyond the range of this target density approach's validity.

Proceeding with the analysis, it is clear that no species density can be negative, for trivial reasons, but there are no mathematical reasons to impose such a restriction. Indeed, at least at the outset of the analysis, allowing negative species densities at equilibrium provides an immediate solution to the steady-state situation,

$$\alpha_{ii}\tilde{n}_i = K_i - \sum_{j\neq i}^{P}\alpha_{ij}\tilde{n}_j, \tag{2}$$

where $\tilde{n}_i$ is the equilibrium density for species i while allowing negative species densities. These values are called "target densities" because they represent the values towards which population densities would move if they were allowed to take on any value. Another way to think of these values is that, if a species is rare and the other species are at their target densities, then the per capita growth rate of the focal species is proportional to $K_i - \sum_{j\neq i}\alpha_{ij}\tilde{n}_j$. This growth rate when rare is also a primary determinant of the target density, making the target density related to the force increasing or decreasing the population density when the species is rare.

The next step in the analysis begins by defining the average target density, $\bar{\tilde{n}}$, and substituting $\tilde{n}_i = \bar{\tilde{n}} + \Delta\tilde{n}_i$, where $\Delta\tilde{n}_i$ represents a small deviation from the average value, and similarly for all of the parameters, into Equation 2. Explicitly,

$$(\bar{\tilde{n}} + \Delta\tilde{n}_i)(\alpha_I + \Delta\alpha_{ii}) = (\bar{K} + \Delta K_i) - \sum_{j\neq i}^{P}(\alpha_H + \Delta\alpha_{ij})(\bar{\tilde{n}} + \Delta\tilde{n}_j) \tag{3}$$

and then expanding all terms gives,

$$\alpha_I\bar{\tilde{n}} + \alpha_I\Delta\tilde{n}_i + \bar{\tilde{n}}\Delta\alpha_{ii} + \Delta\alpha_{ii}\Delta\tilde{n}_i =$$
$$\bar{K} + \Delta K_i - (P-1)\alpha_H\bar{\tilde{n}} - \bar{\tilde{n}}\sum_{j\neq i}^{P}\Delta\alpha_{ij} + \alpha_H\Delta\tilde{n}_i - \sum_{j\neq i}^{P}\Delta\alpha_{ij}\Delta\tilde{n}_j. \tag{4}$$

Equation 4 holds much information on the equilibrium densities of the different species, as well as community-level properties. However, obtaining all this information requires a slightly confusing analysis that hinges on the idea that the sets of terms involving each order of the small deviations must separately be zero. Consider, for example, the case when all of the small deviations are zero. Substituting zero into Equation 4 for all the small deviations leads to Equation 5a. This equation is called the "zeroth order" one because its terms are proportional to the deviations to the zero power. Next, consider deviations so small that the product of any two deviations is negligible. This consideration has no affect on Equation 5a, but yields another equation because the sum of all the first-order terms must cancel in order for the equality Equation 4 to hold.

Likewise, similarly considering the second-order terms provides us with a total of three conditions,

$$\alpha_I \bar{\bar{n}} = \bar{K} - (P-1)\alpha_H \bar{\bar{n}} \tag{5a}$$

$$\alpha_I \Delta \tilde{n}_i + \bar{\bar{n}} \Delta \alpha_{ii} = \Delta K_i - \bar{\bar{n}} \sum_{j \neq i}^{P} \Delta \alpha_{ij} + \alpha_H \Delta \tilde{n}_i \tag{5b}$$

$$\sum_{j \neq i}^{P} \Delta \alpha_{ij} \Delta \tilde{n}_j = 0. \tag{5c}$$

We can simplify the solutions to these equations by defining the overall interaction strength of species i with its community, $\chi_i = \alpha_{ii} + \sum_{j \neq i} \alpha_{ij}$, and its average over all species, $\bar{\chi} = \alpha \chi_I + (P-1)\alpha_H$. With these new identities, we obtain from Equation 5a the average target density,

$$\bar{\tilde{n}} = \frac{\bar{K}}{\bar{\chi}}. \tag{6}$$

Straightforward manipulation of Equation 5b provides each species' deviation,

$$\Delta \tilde{n}_i = \frac{\bar{K}}{\alpha_I - \alpha_H} \left[\frac{\Delta K_i}{\bar{K}} - \frac{\Delta \chi_i}{\bar{\chi}} \right], \tag{7}$$

where $\Delta \chi_i = \Delta \alpha_{ii} + \sum_{j \neq i} \Delta \alpha_{ij}$, leading to the variance, $\sigma_\chi^2 = \sigma_{\alpha I}^2 + (P-1)\sigma_{\alpha H}^2$.
Equation 7 then provides the variance in the target densities,

$$\sigma_{\tilde{n}}^2 = \frac{1}{P} \sum_{i=1}^{P} (\Delta \tilde{n}_i)^2 = \left(\frac{\bar{K}}{\alpha_I - \alpha_H} \right)^2 \left[\frac{\sigma_{\bar{K}}^2}{\bar{K}^2} + \frac{\sigma_{\bar{\chi}}^2}{\bar{\chi}^2} \right]. \tag{8}$$

It is an obvious note to point out that the set of Equation 7 defining the deviations sums to zero as expected, but the less obvious implication of this condition is that the deviations are thus correlated. Equation 5c provides the condition that expresses this fine-scale correlation as something that involves the interaction structure of the community and the target densities. We will not delve into this detailed correlation structure, but it may be of interest elsewhere.

COMMUNITY RESPONSE TO ENVIRONMENTAL CHANGE

Analysis of the model provides insight into the dependence of community-level properties on changes in the underlying model parameters due to the engineer's effect. We first focus on an estimate for species richness based on the supposition that species with negative target densities eventually will be excluded from the system. Given a species pool of size P and assuming a normal distribution of target densities with mean and variance given by Equations 6 and 8, the size of the remaining community, S, is

$$\frac{S}{P} \approx 1 - \frac{\sigma_{\tilde{n}}}{\sqrt{2\pi}\tilde{n}} e^{-\tilde{n}^2/2\sigma_{\tilde{n}}^2}. \tag{9}$$

This approximation results from an integration by parts of the species-density distribution (Mathews and Walker 1970), and the most important aspect of this expression is that S/P increases with increasing $\tilde{n}^2/\sigma_{\tilde{n}}^2$ (Wilson et al. 2003). Thus, the number of species in a community should decrease with increasing interaction strength, $\bar{\chi}$, and variability in interaction strengths or carrying capacity, and increase with increasing $\bar{K}$ (see Equations 6 and 8). This general statement is qualified slightly by the dependence on $\alpha_I - \alpha_H$.

In the following text we consider more detailed implications for species richness, but we can also generate predictions for *productivity*. A terminological aside might be useful at this point. There are three things that could potentially be called productivity, each associated with a distinct biological level. The first thing is the average species growth rate, represented in the model by $\bar{K}$, which is a measure of nutrient levels and is a function of many environmental variables. We will refer to the aggregate of all these environmental variables as *fertility*. Processes that change the fertility level might be called *enrichment* (Wilson and Lundberg 2006), although the two terms might also be used synonymously. The second and third things make a distinction between individual-level mass and population-level biomass. Consider, for example, in our model a species at its equilibrium population density. At this density the population-level biomass is unchanging, yet it is well understood that the model represents population dynamics as the combined result of reproduction, individual growth, and mortality. One possible population-level measure of *productivity* is the standing biomass represented by this equilibrium density, which is the measure of productivity we use here. However, empiricists sometimes focus on the average individual growth, as is sometimes appropriate, measuring this quantity as a biomass production over some time interval and calling it *productivity*. One can likely identify slow-growth and high-growth species that have similar standing biomass, meaning that these two measures of productivity may provide contrasting conclusions that are not reconcilable without further empirical study. Our model makes no predictions concerning the individual-level measure of productivity; indeed, the Lotka–Volterra interaction coefficient α_{ij}, as defined in Equation 1c, confounds the density-dependent growth and death rates, β_{ij} and δ_{ij}, respectively. Unless one assumes that there are no density-dependent loss rates, making predictions regarding individual-level productivity will require a different analysis. In our model the sum of all species densities represents "community productivity," or,

$$\text{Productivity} = \bar{\bar{n}}\frac{S}{P} = \frac{\bar{K}S}{\bar{\chi}P}. \tag{10}$$

This measure should increase with increasing $\bar{K}$ and decrease with increasing interaction strength, $\bar{\chi}$, and variability in interaction strengths or carrying capacity (via Equation 9). The relative importance of these various dependencies is unclear for the most part, however in a two-trophic level community, only the productivity dependence on $\bar{K}$ and the species-richness dependence on σ_k^2 were particularly strong (Wilson and Lundberg 2006).

Equation 9 outlines a complicated dependence of species richness on model parameters. However, we can pursue a more limited but focused examination to understand how an engineer modifying some environmental variable E affects community composition. Given that S/P increases with increasing $\bar{\bar{n}}^2/\sigma_{\bar{n}}^2$, we can differentiate these expressions to understand the general trends as

$$\frac{\partial\left(\dfrac{S}{P}\right)}{\partial E} \propto \frac{\partial\left(\dfrac{\bar{\bar{n}}^2}{\sigma_{\bar{n}}^2}\right)}{\partial E} = \frac{\bar{\bar{n}}^2}{\sigma_{\bar{n}}^2}\left[\frac{1}{\bar{\bar{n}}^2}\frac{\partial\bar{\bar{n}}^2}{\partial E} - \frac{1}{\sigma_{\bar{n}}^2}\frac{\partial\sigma_{\bar{n}}^2}{\partial E}\right]. \tag{11}$$

One result of the approximations taken in Equation 9 is that our interest mostly involves the general trends in community size dependence. With this understanding, it becomes mathematically simpler to examine the scaled derivative that we define as the quantity ΔS,

$$\Delta S \equiv \frac{\sigma_{\bar{n}}^2}{\bar{n}^2} \frac{\partial \left(\frac{\bar{n}^2}{\sigma_{\bar{n}}^2} \right)}{\partial E} = \frac{1}{\bar{n}^2} \frac{\partial \bar{n}^2}{\partial E} - \frac{1}{\sigma_{\bar{n}}^2} \frac{\partial \sigma_{\bar{n}}^2}{\partial E}. \tag{12}$$

The derivatives in Equation 12 can be calculated using Equations 6 and 8 to give

$$\frac{1}{\bar{n}^2} \frac{\partial \bar{n}^2}{\partial E} = \frac{2}{\bar{K}} \frac{\partial \bar{K}}{\partial E} - \frac{2}{\bar{\chi}} \frac{\partial \bar{\chi}}{\partial E} \tag{13a}$$

$$\frac{1}{\sigma_{\bar{n}}^2} \frac{\partial \sigma_{\bar{n}}^2}{\partial E} = \frac{\frac{1}{\bar{K}^2} \frac{\partial \sigma_{\bar{K}}^2}{\partial E} + \frac{1}{\bar{\chi}^2} \frac{\partial \sigma_{\bar{\chi}}^2}{\partial E}}{\frac{\sigma_{\bar{K}}^2}{\bar{K}^2} + \frac{\sigma_{\bar{\chi}}^2}{\bar{\chi}^2}} + \frac{2 \frac{\sigma_{\bar{\chi}}^2}{\bar{\chi}^2}}{\frac{\sigma_{\bar{K}}^2}{\bar{K}^2} + \frac{\sigma_{\bar{\chi}}^2}{\bar{\chi}^2}} \left[\frac{1}{\bar{K}} \frac{\partial \bar{K}}{\partial E} - \frac{1}{\bar{\chi}} \frac{\partial \bar{\chi}}{\partial E} \right]$$

$$- \frac{1}{\alpha_I - \alpha_H} \frac{\partial (\alpha_I - \alpha_H)}{\partial E} \tag{13b}$$

Substituting these expressions into Equation 12 results in a rather complicated expression in need of further simplification. To this end, it seems reasonable, or at least not unreasonable, to assume that $\partial(\alpha_I - \alpha_H)/\partial E = 0$ under the idea that the environmental influence on intra- and inter-specific interactions are similar. We can further define relative changes in each of the important parameters,

$$\Delta K = \frac{1}{\bar{K}} \frac{\partial \bar{K}}{\partial E} \tag{14a}$$

$$\Delta \sigma_{\bar{K}}^2 = \frac{1}{\bar{K}^2} \frac{\partial \sigma_{\bar{K}}^2}{\partial E} \tag{14b}$$

$$\Delta \chi = \frac{1}{\bar{\chi}} \frac{\partial \bar{\chi}}{\partial E} \tag{14c}$$

$$\Delta \sigma_{\chi}^2 = \frac{1}{\bar{\chi}^2} \frac{\partial \sigma_{\chi}^2}{\partial E}. \tag{14d}$$

Our goal will be to examine how the change in the relative measure of species richness, ΔS, depends on the relative engineer-induced changes in the species interaction parameters of interest. Substituting Equations 13 into Equation 12, using the shorthand notation defined by Equations 14, provides the relative importance of the various interaction parameters to changes in species richness,

$$\frac{\Delta S}{\Delta K} = \frac{2\sigma_{\bar{K}}^2 / \bar{K}^2}{\sigma_{\bar{\chi}}^2 / \bar{\chi}^2 + \sigma_{\bar{K}}^2 / \bar{K}^2} \tag{15a}$$

$$\frac{\Delta S}{\Delta K} = -\frac{2\sigma_{\bar{K}}^2 / \bar{K}^2}{\sigma_{\bar{\chi}}^2 / \bar{\chi}^2 + \sigma_{\bar{K}}^2 / \bar{K}^2} \tag{15b}$$

$$\frac{\Delta S}{\Delta \sigma_{\bar{K}}^2} = -\frac{1}{\sigma_{\bar{\chi}}^2 / \bar{\chi}^2 + \sigma_{\bar{K}}^2 / \bar{K}^2} \tag{15c}$$

$$\frac{\Delta S}{\Delta \sigma_{\chi}^2} = -\frac{1}{\sigma_{\bar{\chi}}^2 / \bar{\chi}^2 + \sigma_{\bar{K}}^2 / \bar{K}^2}. \tag{15d}$$

These expressions demonstrate that relative changes in species richness are affected twice as strongly by changes in the average growth rate and interaction strengths as by changes in the variances.

Suppose that environmental change primarily affects fertility, or the community's distribution of maximum growth rates through the parameters $\bar{K}$ and σ_K^2, and affect all other parameters only very weakly. Similar arguments, however, can be made comparing changes in the average growth rate with the parameters governing the distribution of interaction strengths. Possible outcomes for the distributions of species values are shown in Figure 11.1, depicting how engineering can change the two distributional parameters quite independently. As seen by inspection of Equations 15, increases in the mean and variance in K have opposing effects on community richness. Concerning oneself with only growth rate changes, the net effects on community richness might just be the sum of the independent contributions through the growth rate parameters, having a sum of relative species change,

$$\frac{\Delta S}{\Delta K} + \frac{\Delta S}{\Delta \sigma_{\bar{K}}^2} = \frac{2\sigma_{\bar{K}}^2 / \bar{K}^2 - 1}{\sigma_{\bar{\chi}}^2 / \bar{\chi}^2 + \sigma_{\bar{K}}^2 / \bar{K}^2}. \tag{16}$$

There is then a critical value, $2\sigma_{\bar{K}}^2 / \bar{K}^2 = 1$, that serves as the tipping point for when engineering will decrease community richness ($2\sigma_{\bar{K}}^2 / \bar{K}^2 < 1$) or

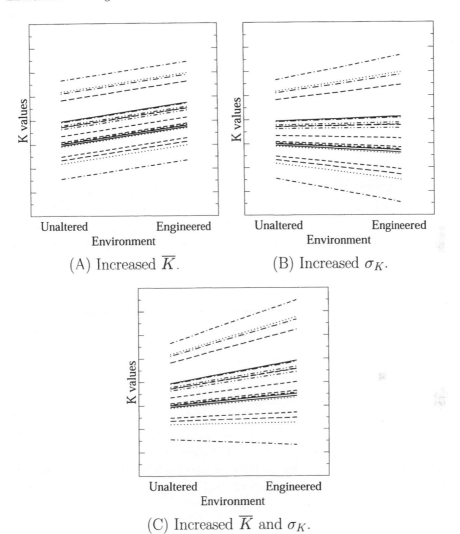

(A) Increased $\overline{K}$.

(B) Increased σ_K.

(C) Increased $\overline{K}$ and σ_K.

Figure 11.1 Possible shifts in species maximum growth rates in an engineered/modified environment. (A) All growth rates increase equally, resulting in an increased $\overline{K}$ but unchanged σ_K. Species richness is expected to increase under this scenario. (B) The average growth rate $\overline{K}$ may stay constant while the standard deviation σ_K increases. Species richness is expected to decrease in this situation. (C) Both the average growth rate $\overline{K}$ and the standard deviation σ_K might increase. The response in species richness depends on which distributional parameter changes most markedly.

increase community richness ($2\sigma_{\bar{K}}^2/\bar{K}^2 > 1$). When the community has relatively homogeneous maximum growth rates, $\sigma_{\bar{K}}^2 \approx 0$, the denominator in Equation 16 takes on its smallest value. Thus, the magnitude of community change is large, and the environmental change induced by ecosystem engineering should have a strong, negative impact on species richness. In contrast, communities with relatively heterogeneous maximum growth rates, or large $\sigma_{\bar{K}}^2$, ought to become more species rich with environmental change. As the community becomes increasingly heterogeneous, relative community change saturates at a value of 2.

11.3 ● DISCUSSION

Our conclusions for the effects of engineering on community properties are relatively straightforward and follow from the results of Wilson et al. (2003). If engineering (or some other process) modifies the environment such that the average maximum growth rate of the species in the community is increased, or in the above representation, $\Delta K > 0$, then the community's species richness increases. On the other hand, if engineering increases any other aspect of species growth, including the average competitive strength, the variance in the maximum growth rate, or the variance in competitive strength, then the community's species richness decreases.

The model presented here envisions an ecological community as a set of interacting species not too far from their equilibrium such that pairwise species interactions can be linearized. Preliminary examinations of a model having nonlinear interactions suggest that the linearization assumption is not critical to the conclusions (Wilson unpublished). We have supposed that the influence of an ecosystem engineer can be collapsed to a "single" environmental variable that alters the mean growth rate, the variation in growth rates, the mean interaction strength, and the variation in interaction strengths. These changes then affect the community, in particular species richness and productivity, defined here as the sum of all species densities.

Wilson and Lundberg (2006) found that in a two-trophic-level community model, increasing $\bar{K}$ (or the mean and/or variance in competition) increased both productivity and species richness at the resource level, whereas increasing $\sigma_{\bar{K}}^2$ increased resource productivity and decreased resource species richness. A hump-shaped form would result from differing relative effects of engineering on the different interaction parameters along their gradient in strength (see also Wilson et al. 2003). This conceptual picture is shown schematically in Figure 11.2. In this situation, environmental change leading to higher fertility could lead to

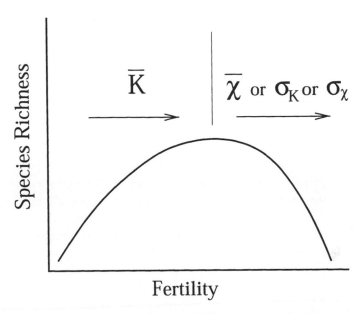

Figure 11.2 General model results demonstrate that species richness in a community increases with increasing $\bar{K}$ and decreases with increasing interaction strength, $\bar{\chi}$, and variability in interaction strengths, σ^2_{χ}, or carrying capacity, σ_K. It is anticipated that at low fertility, enrichment primarily increases $\bar{K}$, whereas at high fertility, enrichment primarily increases $\bar{\chi}$, producing a hump-shaped curve in species richness with fertility.

lower productivity depending on the effects on other species-level parameters.

An earlier conceptual model relating the effects of ecosystem engineering on species richness to the effects that the habitat modification has on primary productivity (Wright and Jones 2004) has results that are potentially consistent with the results of our current model. Wright and Jones (2004) assume a hump-shaped relationship between fertility and species richness (see also Mittelbach et al. 2001). In low-fertility environments on the increasing limb of the fertility–richness relationship, engineers that cause changes in the environment leading to higher fertility should increase richness, while engineers causing a decrease in fertility should decrease richness. Conversely, in high-fertility environments, on the decreasing limb of the fertility–diversity relationship, engineers increasing fertility cause a decrease in richness while those decreasing richness cause an increase in richness.

To compare the two models, we make two assumptions, both of which can be, and in one case have been, tested empirically. The first assumption is that in low-fertility environments, increasing fertility has a stronger effect on growth rates than on the strength of interspecific interactions while the opposite is true in high-fertility environments. In other words, the importance of competition increases as fertility increases. This assumption has theoretical (Grime 1973, 1977) and empirical (Reader 1990, TwolanStrutt and Keddy 1996) support, although there are contrasting theories (Newman 1973, Tilman 1988) that have also been supported by empirical evidence (Wilson and Tilman 1991, 1993; Reader et al. 1994), and a meta-analysis of the intensity of competition across environmental gradients showed mixed results (Goldberg et al. 1999). The second assumption is that species respond similarly to changes in fertility, implying that the change in the variance in growth rates and interaction strengths is small relative to the change in mean values of growth rates and interaction strengths. There are suggestions that interaction strength might be environmentally dependent (Sala and Graham 2002, Navarette and Berlow 2006), but the magnitude of these changes in variance relative to changes in the means are unknown.

If both assumptions hold, then the predictions of the current model and the model proposed in Wright and Jones (2004) are consistent. In relatively low fertility environments, ecosystem engineers that increase fertility cause relatively large increases in mean growth rates ($\bar{K}$) leading to increases in species richness while those that decrease fertility decrease $\bar{K}$ leading to decreases in richness as predicted in Wright and Jones (2004). In high-fertility environments, ecosystem engineers that increase fertility result in relatively large increases in overall interaction strength ($\bar{\chi}$) leading to decreases in richness, while those that decrease fertility decrease $\bar{\chi}$ leading to increases in richness. This again matches the predictions of Wright and Jones (2004). The two models become more difficult to compare directly if these assumptions do not hold.

An important point to be emphasized is the utility of a mechanistic understanding our model provides for interpreting changes in community structure and function. We have indicated how ecosystem engineering, or even any other source of environmental change, alters environmental variables that we collectively call *fertility*. Fertility then alters species-level processes that can be summarized by density-independent growth rate and density-dependent interaction strengths. These alterations are best described as changes in the distributional properties, the mean and variance, of the species-level growth parameters. Our model identifies how community properties, including species richness and productivity, change with changes in these distributional properties. Key empirical information is how all of the distributional parameters change with changes in fertility. Estimating the distribution of interaction strengths and growth rates in diverse, intact natural communities in different environments is clearly a daunting challenge. Yet, designing appropriate experimental communities in which to examine how the distribution of growth rates and interaction strengths change in different environments, thereby testing our assumptions, should be feasible.

ACKNOWLEDGMENTS

Support was provided by the National Center for Ecological Analysis and Synthesis, a center funded by NSF (DEB-0072909), the University of California, and the Santa Barbara campus.

REFERENCES

Badano, E., Jones, C.G., Cavieres, L., and Wright, J.P. (2006). Assessing impacts of ecosystem engineers on community organization: A general approach illustrated by effects of a high-Andean cushion plant. *Oikos* 115:369–385.

Boeken, B., Lipchin, C., Gutterman, Y., and van Rooyen, N. (1998). Annual plant community responses to density of small-scale soil disturbances in the Negev Desert of Israel. *Oecologia* 114:106–117.

Goldberg, D.E., Rajaniemi, T., Gurevitch, J., and Stewart-Oaten, A. (1999). Empirical approaches to quantifying interaction intensity: Competition and facilitation along productivity gradients. *Ecology* 80:1118–1131.

Grime, J.P. (1973). Competitive exclusion in herbaceous vegetation. *Nature* 242: 344–347.

Grime, J.P. (1977). Evidence for existence of 3 primary strategies in plants and its relevance to ecological and evolutionary theory. *American Naturalist* 111: 1169–1194.

Gurney, W.S.C., and Lawton, J.H. (1996). The population dynamics of ecosystem engineers. *Oikos* 76:273–283.

Inouye, R.S., Huntly, N.J., and Tester, J.R. (1987). Pocket gophers (*Geomys bursarius*), vegetation, and soil nitrogen along a successional sere in east central Minnesota. *Oecologia* 72:178–184.

Lotka, A.J. (1925). *Elements of Physical Biology*. Baltimore, MA: Williams and Wilkins.

Mathews, J., and Walker, R.L. (1970). *Mathematical Methods of Physics*, 2nd ed. Menlo Park, CA: Benjamin Cummins Publishing Co.

Mittelbach, G.G., Steiner, C.F., Scheiner, S.M., Gross, K.L., Reynolds, H.L., Waide, R.B., Willig, M.R., Dodson, S.I., and Gough, L. (2001). What is the observed relationship between species richness and productivity? *Ecology* 82:2381–2396.

Naiman, R.J., Johnston, C.A., and Kelley, J.C. (1988). Alteration of North American streams by beaver. *BioScience* 38:753–762.

Navarrete, S.A., and Berlow, E.L. (2006). Variable interaction strengths stabilize marine community pattern. *Ecology Letters* 9:526–536.

Newman, E.I. (1973). Competition and diversity in herbaceous vegetation. *Nature* 244:310.

Raffaele, E., and Veblen, T.T. (1998). Facilitation by nurse shrubs of resprouting behavior in a post-fire shrubland in northern Patagonia, Argentina. *Journal of Vegetation Science* 9:693–698.

Reader, R.J. (1990). Competition constrained by low nutrient supply—An example involving Hieracium-Floribundum Wimm and Grab (Compositae). *Functional Ecology* 4:573–577.

Reader, R.J., Wilson, S.D., Belcher, J.W., Wisheu, I., Keddy, P.A., Tilman, D., Morris, E.C., Grace, J.B., McGraw, J.B., Olff, H., Turkington, R., Klein, E., Leung, Y., Shipley, B., Vanhulst, R., Johansson, M.E., Nilsson, C., Gurevitch, J., Grigulis, K., and Beisner, B.E. (1994). Plant competition in relation to neighbor biomass—An intercontinental study with *Poa pratensis*. *Ecology* 75:1753–1760.

Sala, E., and Graham, M.H. (2002). Community-wide distribution of predator-prey interaction strength in kelp forests. *Proceedings of the National Academy of Science* 99:3678–3683.

Tilman, D. (1988). *Plant Strategies and the Dynamics and Structure of Plant Communities*. Princeton, NJ: Princeton University Press.

TwolanStrutt, L., and Keddy, P.A. (1996). Above- and belowground competition intensity in two contrasting wetland plant communities. *Ecology* 77:259–270.

van Nes, E.H., and Scheffer, M. (2004). Large species shifts triggered by small forces. *American Naturalist* 164:255–266.

Volterra, V. (1926). Fluctuations in the abundance of a species considered mathematically. *Nature* 118:558–560.

Williams, L.R., and Cameron, G.N. (1986). Effects of removal of pocket gophers on a Texas coastal prairie. *American Midland Naturalist* 115:216–224.

Wilson, S.D., and Tilman, D. (1991). Components of plant competition along an experimental gradient of nitrogen availability. *Ecology* 72:1050–1065.

Wilson, S.D., and Tilman, D. (1993). Plant competition and resource availability in response to disturbance and fertilization. *Ecology* 74:599–611.

Wilson, W.G., and Lundberg, P. (2004). Biodiversity and the Lotka–Volterra theory of species interactions: Open systems and the distribution of logarithmic densities. *Proceedings of the Royal Society of London B* 271:1977–1984.

Wilson, W.G., and Lundberg, P. (2006). Non-neutral community dynamics: Empirical predictions for ecosystem function and diversity from linearized consumer–resource interactions. *Oikos* 114:71–83.

Wilson, W.G., Lundberg, P., Vázquez, D.P., Shurin, J.B., Smith, M.D., Langford, W., Gross, K.L., and Mittelbach, G.G. (2003). Biodiversity and species interactions: Extending Lotka–Volterra theory to predict community properties. *Ecology Letters* 6:944–952.

Wright, J.P., Gurney, W.S.C., and Jones, C.G. (2004). Patch dynamics in a landscape modified by ecosystem engineers. *Oikos* 105:336–348.

Wright, J.P., and Jones, C.G. (2004). Predicting effects of ecosystem engineers on patch-scale species richness from primary productivity. *Ecology* 85:2071–2081.

Wright, J.P., and Jones, C.G. (2006). The concept of organisms as ecosystem engineers. Ten years on: Progress, limitations and challenges. *Bioscience* 56:203–209.

Wright, J.P., Jones, C.G., Boeken, B., and Shachak, M. (2006). Predictability of ecosystem engineering effects on species richness across environmental variability and spatial scales. *Journal of Ecology* 94:815–824.

Wright, J.P., Jones, C.G., and Flecker, A.S. (2002). An ecosystem engineer, the beaver, increases species richness at the landscape scale. *Oecologia* 132:96–101.

12

MODEL STUDIES OF ECOSYSTEM ENGINEERING IN PLANT COMMUNITIES

Ehud Meron, Erez Gilad, Jost von Hardenberg,
Antonello Provenzale, and Moshe Shachak

12.1 ⟢ INTRODUCTION

The dynamics and spatial organization of ecological communities are strongly affected by various feedbacks between the biotic and abiotic environments. The realization that organisms can modify the abiotic environment, rather than merely being affected by it, has received much attention since the introduction of the ecosystem engineering concept by Jones et al. in 1994. Numerous case studies of ecosystem engineering have appeared since then, providing data on the engineering process and how it affects organismal, population, community, or ecosystem ecology (Wright and Jones 2006). Feedback relationships between two processes generally imply the inadequacy of studying unidirectional influences alone; the processes are coupled and affect one another at any instant of time. Studying the bidirectional relationships between biotic and abiotic processes, including their large-scale and long-time consequences, calls for the development and study of dynamic models (Ellner and Guckenheimer 2006). Such models can provide powerful complementary tools for unraveling mechanisms of ecosystem engineering at the single-patch and landscape scales, and along environmental gradients.

Despite the extensive empirical work that has been devoted to ecosystem engineering, very few mathematical models addressing engineering aspects have appeared (Cuddington and Hastings 2004, Gilad et al. 2004, Rietkerk et al. 2004, Wright et al. 2004). In this chapter we consider plant communities in water-limited systems and present model studies of ecosystem engineering along rainfall or consumer-pressure gradients, and across different levels of organization. Our work is motivated in part by field studies of plant interactions along environmental gradients (Bertness and Callaway 1994, Bertness and Ewanchuk 2002, Bertness and Hacker 1994, Brooker and Callaghan 1998, Callaway and Walker 1997, Callaway et al. 2002, Greenlee and Callaway 1996, Maestre et al. 2003, Pugnaire and Luque 2001), which report on changes from competition to facilitation as abiotic stresses or consumer pressures increase. In water-limited systems such changes have been observed in woody-herbaceous communities under conditions of increased aridity. Facilitation in this case is manifested by the growth of annuals, grasses, and other species under the canopy of woody plants (Pugnaire and Luque 2001), and might reflect a change in the ecosystem engineering strength of the woody life-form.

The biotic–abiotic feedbacks considered in this work couple biomass densities to surface-water flow and soil-water density. Three feedback processes are modeled: reduced evaporation by shading ("shading feedback"), increased infiltration at vegetation patches ("infiltration feedback"), and water uptake by plants' roots ("uptake feedback"). The first two processes concentrate the water resource at vegetation patches, thus acting as positive feedbacks, while the third process depletes the water resource and acts as a negative feedback. Water uptake, however, also induces a positive biomass-feedback loop, due to root augmentation in response to plant growth. As the plant grows, the augmented root system probes larger soil volumes, takes up more water, and further accelerates the growth of the plant. To capture this effect, it is necessary to model explicitly the non-locality of water uptake: Uptake at a given spatial point has a contribution from distant plants whose roots extend to that point. This aspect of the water-uptake process, which strongly bears on ecosystem engineering and its resilience to environmental changes and disturbances, has not been modeled in earlier works (Okayasu and Aizawa 2001; Rietkerk et al. 2002, 2004).

In a series of recent papers, we have developed and analyzed a set of single- and multispecies models of plants in water-limited systems that incorporate these three feedback processes (Gilad et al. 2004, 2006a,

2006b). We present here a synthetic review of these studies emphasizing aspects of ecosystem engineering. The outline of this review is as follows. In the second section we present a spatially explicit dynamic model for plant communities in drylands, and explain how it captures the various biomass-water feedbacks. We then apply the model to a single life-form, studying conditions under which it functions as an ecosystem engineer by concentrating the water resource (third section). Considering next the version of the model for two life-forms, we study (in the fourth section) the response of herbaceous life-forms to the engineering of woody life-forms at different levels of patch organization. We conclude in the fifth section with a few remarks on the significance of detailed modeling of biomass-water feedbacks for studying ecosystem engineering, and with a note on future directions.

12.2 ◆ A MATHEMATICAL MODEL FOR PLANT COMMUNITIES IN DRYLANDS

We consider plant communities consisting of n life-forms in water-limited environments. Depending on the particular context, a life-form can represent a single species, or a community of species whose traits fall in a narrow range of values in comparison to species of other communities. The environments to be considered represent levels of organization ranging from single-patch plots to many-patch landscapes.

The model we present here was originally proposed in Gilad et al. 2004, 2006a and extended to multiple life-forms in Gilad et al. 2006b. The extended model contains $n + 2$ dynamical variables: n biomass variables, $B_i(\mathbf{X}, T)$ ($i = 1, \ldots, n$), representing biomass densities above ground level of the n life-forms in units of [kg/m^2], a soil-water variable, $W(\mathbf{X}, T)$, describing the amount of soil water available to the plants per unit area of ground surface in units of [kg/m^2], and a surface-water variable, $H(\mathbf{X}, T)$, describing the height of a thin water layer above ground level in units of [mm]. (Since the density of water is approximately 10^3 kg/m^{-3}, water height expressed in mm is equivalent to water height in kg/m^2). The model equations are

$$\frac{\partial B_i}{\partial T} = G_B^i[B_i, W]B_i(1 - B_i / K_i) - M_i B_i + D_{B_i} \nabla^2 B_i \quad i = 1, \ldots, n$$

$$\frac{\partial W}{\partial T} = I(\{B_i\})H - N\left(1 - \sum_{i=1}^{n} R_i B_i / K_i\right)W - W\sum_{i=1}^{n} G_W^i[B_i] + D_W \nabla^2 W$$

$$\frac{\partial H}{\partial T} = P - I(\{B_i\})H + D_H \nabla^2 H^2 + 2D_H \nabla H \cdot \nabla Z + 2D_H H \nabla^2 Z, \quad (1)$$

where $\{B_i\}$ stands for all biomass densities, $\nabla^2 = \partial_X^2 + \partial_Y^2$, and $\mathbf{X}$ and T are the space and time coordinates. The quantity $G_B^i[\mathrm{yr}^{-1}]$ represents the growth rate of the ith life-form, $G_W^i[\mathrm{yr}^{-1}]$ represents its soil-water consumption rate, and $K_i[\mathrm{kg/m^2}]$ is its maximum standing biomass. The quantity $I[\mathrm{yr}^{-1}]$ represents the infiltration rate of surface water into the soil, the parameter $P[\mathrm{mm/yr}]$ stands for the precipitation rate, $N[\mathrm{yr}^{-1}]$ represents the soil-water evaporation rate, and $R_i > 0$ describes the reduction in evaporation rate due to shading by the ith life-form. The parameter $M_i[\mathrm{yr}^{-1}]$ describes the biomass loss rate of the ith life-form due to mortality and various disturbances (e.g., grazing). The terms $D_{B_i}\nabla^2 B_i$ and $D_W\nabla^2 W$ represent, respectively, local seed dispersal of the ith life-form, and soil-water transport in nonsaturated soil (Hillel 1998). Finally, the non-flat ground surface height [mm] is described by the topography function $Z(\mathbf{X})$ where the parameter $D_H[\mathrm{m^2/yr(kg/m^2)^{-1}}]$ represents the phenomenological bottom friction coefficient between the surface water and the ground surface.

The equations for the biomass densities and the soil-water density are phenomenological, while the equation for the surface-water variable is derived from shallow-water theory. The transport term $D_H\nabla^2(H^2)$ follows from the assumption of a Rayleigh-type bottom friction (linearly proportional to the flow velocity). The term $2D_H\nabla H \cdot \nabla Z$ describes changes in surface-water height due to water flow on a slope, and the term $2D_H H\nabla^2 Z$ describes the accumulation of surface water in lower areas, where $\nabla^2 Z > 0$, or the flow of surface water away from higher areas, where $\nabla^2 Z < 0$.

Equations 1 model all three biomass-water feedbacks. The infiltration feedback is modeled through the explicit form of the infiltration rate I. A monotonously increasing dependence of I on biomass density is assumed in order to capture the positive nature of this feedback; the larger the biomass density the higher the infiltration rate[1] and the more soil water available to the plants.

The explicit dependence of the infiltration rate on the biomass density is a generalization of an earlier form used in single-species models (Gilad et al. 2004, 2006a; HilleRisLambers et al. 2001; Walker et al. 1981)

$$I(\mathbf{X},T) = A \frac{\sum_i Y_i B_i(\mathbf{X},T) + Qf}{\sum_i Y_i B_i(\mathbf{X},T) + Q}, \tag{2}$$

---

[1] Various factors contribute to the higher infiltration rate of surface water into vegetated soil as compared with bare soil, including biological crusts that grow on bare soil and reduce the infiltration rate (Campbell et al. 1989, West 1990), and soil mounds, formed by litter accumulation and dust deposition, that intercept runoff (Yair and Shachak 1987).

where $A[\text{yr}^{-1}]$, $Q[\text{kg}/\text{m}^2]$, Y_i, and f are constant parameters and $Y_1 = 1$. Two distinct limits of this form for the infiltration rate are noteworthy. When $\Sigma_i Y_i B_i \rightarrow 0$, this quantity represents the infiltration rate in bare soil, $I = Af$. When $\Sigma_i Y_i B_i \gg Q$ it represents infiltration rate in fully vegetated soil, $I = A$. The parameter Q represents a reference biomass density beyond which the biomass density approaches its full capacity to increase the infiltration rate. It is normally small relative to the maximum standing biomass, implying a weak dependence of the infiltration rate on the biomass density at high density values. The infiltration contrast (between bare and vegetated soil) is quantified by the parameter f, defined to span the range $0 \leq f \leq 1$. When $f \ll 1$ the infiltration rate in bare soil is much smaller than the rate in vegetated soil. Such values can model bare soils covered by biological crusts (Campbell et al. 1989, West 1990). As f gets closer to 1, the infiltration rate becomes independent of the biomass densities B_i. The parameter f measures the strength of the positive feedback due to increased infiltration at vegetation patches. The smaller f the stronger the feedback effect.

The uptake feedback is modeled through the explicit forms of the growth rate G_B^i and of the consumption rate G_W^i. These forms capture the non-local nature of the uptake process by the root system, as well as the augmentation of the root system in response to biomass growth (Gilad et al. 2004, 2006a, 2006b). Water uptake obviously acts as a negative feedback; water availability increases biomass growth but biomass growth decreases water availability through water consumption. The uptake process, however, also acts as a positive feedback when root augmentation is taken into account; as the biomass grows the root system extends in size, probes larger soil volumes, and takes up more water.

The growth rate G_B^i at a point $\mathbf{X}$ at time T is modeled by the following non-local form:

$$G_B^i(\mathbf{X}, T) = \Lambda_i \int_\Omega G_i(\mathbf{X}, \mathbf{X}', T) W(\mathbf{X}', T) d\mathbf{X}',$$

$$G_i(\mathbf{X}, \mathbf{X}', T) = \frac{1}{2\pi S_i^2} \exp\left[-\frac{|\mathbf{X} - \mathbf{X}'|^2}{2[S_i(1 + E_i B_i(\mathbf{X}, T))]^2} \right], \tag{3}$$

where $\Lambda_i[(\text{kg}/\text{m}^2)^{-1}\text{yr}^{-1}]$ represents the plant's growth rate per unit amount of soil water, the Gaussian kernel G_i $(\mathbf{X}, \mathbf{X}', T)$ $[\text{m}^{-2}]$ represents the distribution of the root system, and the integration is over the entire physical domain Ω.[2] According to this form, the biomass growth rate depends not

---

[2] The kernel G_i is normalized such that for $B_i = 0$ the integration over an infinite domain equals unity.

only on the amount of soil water at the plant location, but also on the amount of soil water in the neighborhood spanned by the plant's roots. A measure for the root-system size is given by $S_i(1 + E_iB_i(\mathbf{X},T))$ [m], where $E_i[(\text{kg}/\text{m}^2)^{-1}]$ quantifies the root augmentation per unit biomass, beyond a minimal root-system size S_i. The parameter E_i measures the strength of the uptake feedback due to root augmentation; the larger E_i the stronger the feedback effect of the ith life-form.

The soil-water consumption rate at a point $\mathbf{X}$ at time T is similarly given by

$$G_W^i(\mathbf{X}, \text{T}) = \Gamma_i \int_\Omega G_i(\mathbf{X}', \mathbf{X}, T) B_i(\mathbf{X}', T) d\mathbf{X}', \tag{4}$$

where $\Gamma_i[(\text{kg}/\text{m}^2)^{-1}\text{yr}^{-1}]$ measures the soil-water consumption rate per unit biomass of the ith life-form. The soil-water consumption rate at a given point is due to all plants whose roots extend to this point. Note that $G_i(\mathbf{X}',\mathbf{X},T) \neq G_i(\mathbf{X},\mathbf{X}',T)$.

The shading feedback is quantified by the parameters R_i in the equation for W. It is a positive feedback, but unlike the infiltration feedback, the increased soil-water density under a vegetation patch, due to reduced evaporation, does not involve depletion of soil water in the patch neighborhood. As a consequence, the shading feedback is not expected to induce spatial instabilities leading to vegetation patterns.

It is advantageous to express the model Equations 1 in terms of nondimensional variables and parameters as defined in Table 12.1. The nondimensional form of the model equations is

$$\frac{\partial b_i}{\partial t} = G_b^i b_i(1 - b_i) - \mu_i b_i + \delta_{b_i} \nabla^2 b_i \quad i = 1, \dots, n$$

$$\frac{\partial w}{\partial t} = Ih - v\left(1 - \sum_{i=1}^n \rho_i b_i\right)w - w\sum_{i=1}^n G_w^i + \delta_w \nabla^2 w$$

$$\frac{\partial h}{\partial t} = p - Ih + \delta_h \nabla^2(h^2) + 2\delta_h \nabla h \cdot \nabla\varsigma + 2\delta_h h\nabla^2\varsigma, \tag{5}$$

where $\nabla^2 = \partial_x^2 + \partial_y^2$ and t and $\mathbf{x} = (x,y)$ are the nondimensional time and spatial coordinates. The infiltration term now reads

$$I(\mathbf{x}, t) = \alpha \frac{\sum_i \psi_i b_i(\mathbf{x}, t) + qf}{\sum_i \psi_i b_i(\mathbf{x}, t) + q}, \tag{6}$$

TABLE 12.1 Relations between non-dimensional and dimensional variables and parameters. Note that according to these relations $\lambda_1 = \mu_1 = \sigma_1 = 1$.

Quantity	Scaling	Quantity	Scaling
b_i	B_i/K_i	p	$\Lambda_1 P/(NM_1)$
w	$\Lambda_1 W/N$	γ_i	$\Gamma_i K_i/M_1$
h	$\Lambda_1 H/N$	η_i	$E_i K_i$
v	N/M_1	ρ_i	R_i
λ_i	Λ_i/Λ_1	σ_i	S_i/S_1
μ_i	M_i/M_1	δ_{b_i}	$D_{B_i}/(M_1 S_1^2)$
α	A/M_1	δ_w	$D_W/(M_1 S_1^2)$
q	Q/K_1	δ_h	$D_H N/(M_1 \Lambda_1 S_1^2)$
x	X/S_1	ψ_i	$Y_i K_i/K_1$
t	$M_1 T$	ζ	$\Lambda_1 Z/N$

the growth rate term G_b^i is

$$G_b^i(\mathbf{x}, t) = v\lambda_i \int_\Omega g_i(\mathbf{x}, \mathbf{x}', t) w(\mathbf{x}', t) d\mathbf{x}',$$

$$g_i(\mathbf{x}, \mathbf{x}', t) = \frac{1}{2\pi\sigma_i^2} \exp\left[-\frac{|\mathbf{x} - \mathbf{x}'|^2}{2[\sigma_i(1 + \eta_i b_i(\mathbf{x}, t))]^2}\right], \tag{7}$$

and the soil-water consumption rate is

$$G_w^i(\mathbf{x}, t) = \gamma_i \int_\Omega g_i(\mathbf{x}', \mathbf{x}, t) b_i(\mathbf{x}', t) d\mathbf{x}'. \tag{8}$$

Noteworthy is the form of the nondimensional precipitation rate,

$$p = \Lambda_1 P/(NM_1), \tag{9}$$

which implies the equivalence of decreasing the precipitation rate P and increasing the biomass loss rate M_1.

The studies of Equations 5 presented in the following sections are mostly numerical. Analytical studies include linear stability analysis of stationary uniform solutions, and are described in detail in Gilad et al. 2006a, 2006b. Numerical studies employ a fast algorithm for calculating the non-local growth and water-consumption rates (Equations 7 and 8).[3]

---

[3] The biomass dependence of the kernel G_i in Equation 3 rules out the use of standard convolution algorithms.

The algorithm is described in Gilad and von Hardenberg 2006. We note that the numerical solutions described here are robust and do not depend on delicate tuning of any particular parameter.

12.3 ➥ ECOSYSTEM ENGINEERING IN THE MODEL

The model equations for a single life-form ($n = 1$) have two uniform solutions describing bare soil ($b_1 = 0$, $w = p/v$, $h = p/(\alpha f)$) and uniform vegetation (b_1 is a nonzero constant). In addition, there are nonuniform solutions describing vegetation patterns (Gilad et al. 2004; von Hardenberg et al. 2001; Klausmeier 1999; Lefever and Lejeune 1997; Meron et al. 2004; Okayasu and Aizawa 2001; Rietkerk et al. 2002, 2004; Shnerb et al. 2003; Valentin et al. 1999; Yizhaq et al. 2005). These pattern solutions vary from gaps in uniform coverage at high rainfall to vegetation stripes at intermediate rainfall to vegetation spots at low rainfall. In the low-rainfall regime there is a bistability range where stable spot-pattern solutions coexist with stable bare-soil solutions. This range gives rise to single-patch solutions. We begin studying ecosystem engineering using these solutions. Throughout this work we define engineering as the capacity of a plant to concentrate soil water beyond the level pertaining to bare soil. We will occasionally use the terms *positive engineering* and *negative engineering* (Jones et al. 1997) to distinguish between soil-water concentration and soil-water depletion relative to the level of soil water in bare soil.

The actual soil-water distribution in and around a biomass patch area is determined by the relative strengths of the various biomass-water feedbacks. We will mainly be concerned here with the counter-effects of the infiltration feedback (soil-water concentration) and the uptake feedback (soil-water depletion). The strengths of these feedbacks are controlled by the parameters f and η_1, respectively.

ECOSYSTEM ENGINEERING AND RESILIENCE

We may expect the engineering capacity of a plant, in terms of soil-water concentration, to increase as the infiltration feedback becomes stronger relative to the uptake feedback. Is there a price the system has to pay for attaining high engineering? To answer this question we studied the spatial distributions of the biomass and soil-water variables at various values of f and η_1 (Gilad et al. 2004). The results are summarized in Figure 12.1 and indicate the existence of a trade off between the engineering capacity of a plant and its resilience to disturbances; conditions

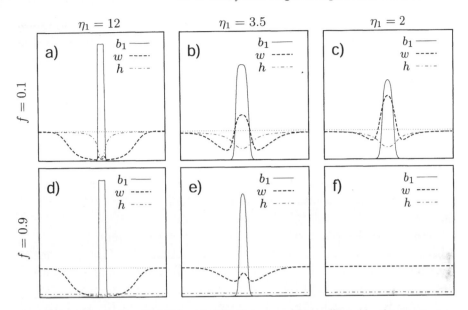

FIGURE 12.1 Spatial profiles of the variables b_1, w, and h as affected by the parameters that control the main positive biomass-water feedbacks, f (infiltration feedback) and η_1 (uptake feedback). The profiles are cross sections of two-dimensional solutions of the model equations (Equations 5–8). In all panels, the horizontal dotted lines denote the soil-water level at bare soil. Strong infiltration feedback and weak uptake feedback (panel c) lead to high soil-water concentration reflecting strong engineering. Strong uptake feedback results in soil-water depletion and no engineering, irrespective of the infiltration-feedback strength (panels a, d). While the species characterized by $\eta_1 = 2$ is the best engineer under conditions of strong infiltration contrast (panel c), it leads to low system resilience; the engineer along with the micro-habitat it forms completely disappear when the infiltration contrast is strongly reduced, e.g., by crust removal (panel f). A species with somewhat stronger uptake feedback ($\eta_1 = 3.5$) still acts as an ecosystem engineer (panel b) and also survives disturbances that reduce the infiltration contrast (panel e), thereby retaining the system's resilience. Parameter values are $\nu = \delta_w = 3.333$, $\alpha = 33.333$, $q = 0.05$, $\delta_h = 333.333$, $\eta_1 = 3.5$, $\gamma_1 = 16.667$, $p_1 = 0.95$, and $\delta_{b_1} = 0.033$, with $P = 75\,\text{mm/yr}$. Panels a and d span a horizontal range of $14\,\text{m}$ while all other panels span $3.5\,\text{m}$. The vertical range in all panels is $[0,1]\,\text{kg/m}^2$ for the biomass density, and $[0,187.5]\,\text{kg/m}^2$ for the soil-water density. Reprinted with permission from Gilad et al. (2004). Copyright 2007 by the American Physical Society.

that favor ecosystem engineering, resulting in water-enriched patches or microhabitats, imply low resilience, and conditions that favor high resilience imply weak or no engineering.

Shown in Figure 12.1 are spatial profiles of b_1, w, and h for a single patch of the ecosystem engineer at decreasing values of η_1, representing species with different root-extension properties, and for two extreme values of f. The value $f = 0.1$ models high infiltration rates under engineer's patches and low infiltration rates in bare soil, which may result from a biological crust covering the bare soil. The value $f = 0.9$ models high infiltration rates everywhere. This case may describe, for example, uncrusted sandy soil. Engineering effects resulting in soil-water concentration appear only in the case of (1) low infiltration in bare soil, and (2) engineer species with limited root-extension capabilities, $\eta_1 = 3.5$ or $\eta_1 = 2$ (panels b and c in Figure 12.1). The soil-water density under an engineer's patch in this case exceeds the soil-water density level of bare soil (shown by the dotted lines), thus creating opportunities for species that require this extra amount of soil water to colonize the water-enriched patch.

While a weak uptake feedback enhances the engineering ability, it reduces the resilience of the ecosystem engineer (and all dependent species) to disturbances. Figure 12.1f shows the response of an engineer species with the highest engineering ability to concentrate water ($\eta_1 = 2$, Figure 12.1c) to a disturbance that strongly reduces the infiltration contrast ($f = 0.9$). We continue referring to crust removal, but other disturbances that reduce the infiltration contrast, such as erosion of bare soil, will have similar effects. The engineer, and consequently the microhabitat it forms, disappear altogether for two reasons: (1) surface water infiltrates equally well everywhere and the plant patch is no longer effective in trapping water, and (2) the engineer's roots are too short to collect water from the surrounding area.

Resilient ecosystem engineers are obtained with strong infiltration feedbacks and moderate uptake feedbacks ($\eta_1 = 3.5$) as Figure 12.1e shows. Removal of the crust (by increasing f) destroys the micro-habitats (soil-water density is smaller than the bare soil's value) but the engineer persists. Once the crust recovers the ecosystem engineer resumes its capability to concentrate water and the micro-habitats recover as well. It is also of interest to comment that, when the uptake feedback is too strong, the plant persists but the engineering is negative (Figure 12.1a, d).

ECOSYSTEM ENGINEERING ALONG ENVIRONMENTAL GRADIENTS

The spatial soil-water distribution induced by a given ecosystem engineer can vary along an environmental gradient. Figure 12.2a shows solutions of the model equations along a rainfall gradient. The line $\mathcal{B}$ shows the soil-water content in bare soil while the line S shows the maximal

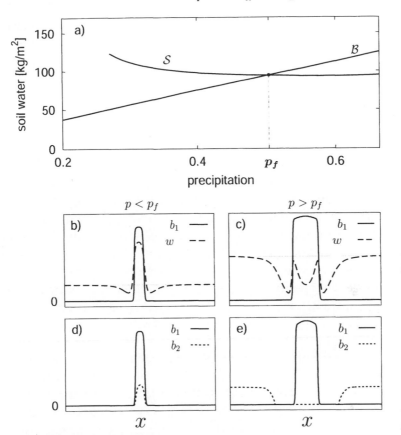

FIGURE 12.2 Solutions of the model for a single life-form (a, b, c), showing a transition from negative to positive engineering as precipitation decreases, and solutions of the model for two life-forms (d, e), showing the corresponding transition from competition to facilitation. The lines $\mathcal{B}$ and $\mathcal{S}$ in panel (a) show, respectively, the soil-water density in bare soil and under a b_1 patch as functions of precipitation. Above (below) $p = p_f$ the water content under the b_1 patch is lower (higher) than in bare soil, implying negative (positive) engineering. Panels b–e show spatial profiles of b_1, b_2, and w in the competition range $p > p_f$ (c, e) where the herbaceous life-form, b_2, is excluded by the woody life-form, b_1, and in the facilitation range $p < p_f$ (b, d) where b_2 grows under the b_1 canopy. Precipitation values are $p = 0.25$ (187.5 mm/yr) for b, d; $p = 0.6$ (450 mm/yr) for c, e, and $p_f = 0.5$ (378 mm/yr). Other parameter values are $\nu = \delta_w = 1.667$, $\alpha = 16.667$, $q = 0.05$, $f = 0.1$, $\delta_h = 416.667$, $\eta_1 = 3.5$, $\eta_2 = 0.35$, $\gamma_1 = 2.083$, $\gamma_2 = 0.208$, $\rho_1 = 0.95$, $\rho_2 = \psi_2 = 0.005$, $\delta_{b_1} = \delta_{b_2} = 0.167$, $\sigma_2 = 1$, $\lambda_2 = 10$, and $\mu_2 = 4.1$. Reprinted from Gilad et al. 2006b.

water density under an engineer patch. The two lines intersect at $p = p_f$ suggesting a crossover from negative engineering at high precipitation $(p > p_f)$, where the soil-water density under a b_1 patch is lower than in bare soil, to positive engineering at low precipitation $(p < p_f)$, where the soil-water density under a patch exceeds that of bare soil. Figures 12.2b, c show examples of spatial profiles of b_1 and w in the negative-engineering range (c) and in the positive-engineering range (b). Note that the line S terminates at some low precipitation value. Below that value the ecosystem engineer (b_1) no longer survives the dry conditions and a catastrophic shift (Scheffer et al. 2001) to bare soil occurs.

The model offers the following mechanism for this crossover. As the system becomes more arid, the engineer's patch area becomes smaller and the water uptake decreases significantly. The infiltration rate at the reduced patch area, however, decreases only slightly because of its weak biomass dependence for $b_1 \gg q$ (see Equation 6). As a result a given area of the patch in a more arid environment traps nearly the same amount of surface water, but a significantly smaller amount of soil water is consumed in that area due to fewer individuals in the surrounding region, as demonstrated in Figure 12.3, and also due to the lower biomass density. The outcome is an increased soil-water density at the patch area and stronger engineering. Two factors prevent the engineer from exhausting the soil water for its own growth: its carrying capacity (maximum standing biomass), which limits the local growth; and the depletion of soil water in the immediate vicinity of the patch, which prevents its expansion. We assume here that the carrying capacity of the woody engineer represents factors that limit its own growth but do not necessarily limit the herbaceous-species growth. Only in that case can the herbaceous species benefit from the mesic patches the woody engineer forms.

12.4 ➤ APPLYING THE MODEL TO WOODY-HERBACEOUS SYSTEMS

In the previous section we studied conditions for ecosystem engineering by plants that result in the formation of mesic patches. We now use the model equations for two life-forms ($n = 2$) to study the response of a second life-form to the mesic patches (or habitats) the engineer life-form has created. We study this response at three levels of patch organization: single patch, a few interacting patches, and patch patterns at the landscape scale.

Motivated by recent field studies of plant interactions and engineering in woody-herbaceous communities (Pugnaire and Luque 2001, Wright et al. 2006), we choose the two life-forms, b_1 and b_2, to represent, respectively, a woody engineer and an herbaceous life-form. Accordingly, we choose the maximum standing biomass of the woody life-form and its root-system size to be significantly larger than those of the herbaceous life-form while its growth and mortality rates are taken to be significantly

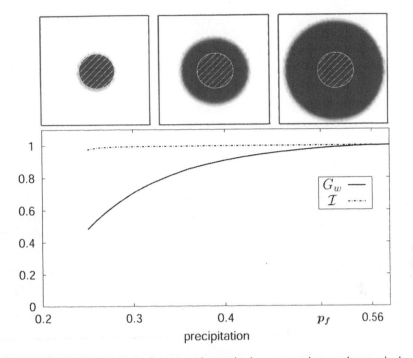

FIGURE 12.3 Water balance under a woody patch along an aridity gradient, calculated using the model equations for a single life-form. Shown are the water uptake rate G_w (solid line) and the water infiltration rate I (dotted line) per constant area size (hatched circles in upper panels) in a b_1 patch as precipitation changes. As the system becomes more arid, the b_1 patch area becomes smaller (see upper panels) and the water uptake from the same area size decreases significantly. The infiltration rate, however, decreases only slightly. Consequently, the soil-water density per unit area of a woody patch increases as the system becomes more arid. The values of G_w and I are normalized with respect to their maximal value and correspond to the model solutions denoted by the S curve in Figure 12.2. Reprinted from Gilad et al. 2006b.

lower. We confine ourselves to the case of strong infiltration feedback ($f \ll 1$) and moderate uptake feedback of the woody engineer ($\eta_1 \sim \mathcal{O}(1)$). These conditions are often realized in drylands where biological soil crusts increase the infiltration contrast and the woody vegetation consists of shrubs (Shachak et al. 1998). With this parameter choice we find that the herbaceous vegetation is strongly affected by the woody vegetation, but the woody vegetation is hardly affected by the herbaceous one.

UNIFORM AND PATTERN SOLUTIONS

Before embarking on non-uniform patch solutions of the model Equations 5 for two life-forms, we consider stationary uniform solutions and study their existence and stability. The model equations (with $n = 2$) have four stationary uniform solutions:

- $\mathcal{B}$: bare soil ($b_1 = 0$, $b_2 = 0$).
- $\mathcal{V}_1$: uniform woody vegetation ($b_1 \neq 0$, $b_2 = 0$).
- $\mathcal{V}_2$: uniform herbaceous vegetation ($b_1 = 0$, $b_2 \neq 0$).
- $\mathcal{M}$: uniform mixed woody-herbaceous vegetation ($b_1 \neq 0$, $b_2 \neq 0$).

The bare-soil solution, $\mathcal{B}$, is given by

$$b_1 = 0, \ b_2 = 0, \ w = p/\nu, \ h = p/(\alpha f). \tag{10}$$

We do not present here the explicit forms of the other solutions (the mathematical expressions are too long (Gilad et al. 2006b)) but show instead the biomass densities of the first three solutions as functions of the precipitation rate p in the bifurcation diagram displayed in Figure 12.4. At very low precipitation rates, p, the bare-soil solution $\mathcal{B}$ is the only stable solution. As p is increased, a threshold is reached at which the bare-soil solution loses stability. If $\mu_2/\lambda_2 < 1$ the threshold is given by $p_{b_2} = \mu_2/\lambda_2$, and the bare-soil solution loses stability to uniform herbaceous vegetation (the $\mathcal{V}_2$ solution branch). If $\mu_2/\lambda_2 > 1$ the threshold is given by $p_{b_1} = 1$, and the bare-soil solution loses stability to uniform woody vegetation, typically in a subcritical bifurcation (see Figure 12.4). The instability, however, leads to non-uniform pattern solutions, S, because the uniform woody vegetation solution, $\mathcal{V}_1$, that already exists at the instability point due to its subcritical nature, and is unstable to non-uniform perturbations (dotted part of $\mathcal{V}_1$). The uniform woody vegetation solution becomes stable only at relatively high p values (solid part of $\mathcal{V}_1$). The uniform mixed woody-herbaceous solution, $\mathcal{M}$, is unstable in the whole parameter regime considered here and is not shown in Figure 12.4. The reader is referred to Gilad et al. 2006b for further details about the uniform solutions and their linear stability.

Non-uniform solutions of the model equations in the low precipitation range describe spot patterns. Spot-pattern solutions consist of woody patches, coexisting with or excluding the herbaceous life-form, separated by either bare soil or uniform herbaceous vegetation. The line S in Figure 12.4 represents some of these solution branches.

ECOSYSTEM ENGINEERING AT THE SINGLE-PATCH LEVEL

Single-patch solutions are realizable in precipitation ranges where stable spot-patterns (S branch in Figure 12.4) coexist with either stable

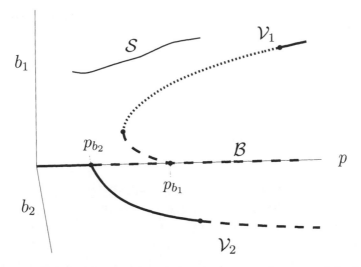

FIGURE 12.4 Bifurcation diagram showing homogeneous and pattern solution branches of the woody-herbaceous system, evaluated by numerical integration of Equations 5. The solution branches $\mathcal{B}$, $\mathcal{V}_1$, and $\mathcal{V}_2$ represent, respectively, uniform bare soil, uniform woody vegetation, and uniform herbaceous vegetation. The branch S represents the amplitudes of spots patterns. Solid lines represent stable solutions, and dashed and dotted lines represent solutions unstable to uniform and non-uniform perturbations, respectively. The thresholds $p_{b_2} = \mu_2/\lambda_2$ and $p_{b_1} = 1$ correspond to 307.5 mm/yr and 750 mm/yr, respectively. All other parameters are as in Figure 12.2.

bare soil ($\mathcal{B}$ branch) or stable uniform herbaceous vegetation ($\mathcal{V}_2$ branch). To study the structure of these solutions we resort to the results presented in the section "Ecosystem Engineering Along Environmental Gradients," about the engineering of a single life form along a rainfall gradient, applying them to the woody life-form. Figure 12.2a shows a crossover, as p is decreased below p_f, from a patch of the woody engineer, whose overall effect is water exploitation (negative engineering), to a patch where the overall effect is water concentration (positive engineering).

We now study the response of an herbaceous life-form to the negative and positive engineering of the woody life-form. Solving the model equations numerically, starting with a woody biomass patch and small randomly distributed herbaceous biomass, leads to the results displayed in Figures 12.2d, e. An herbaceous life-form that is excluded from the patch area at high precipitation rates ($p > p_f$), due to the negative engineering of the woody life-form (panel e), survives and grows at low precipitation rates ($p < p_f$) solely due to the positive engineering of the woody life-form and the mesic conditions it forms in the patch area (panel d).

Positive engineering, and consequently facilitation of herbaceous vegetation growth, develops not only upon increasing the aridity stress, but also when consumer pressure is increased. This result can readily be inferred from Equation 9. According to this equation decreasing p or the dimensional precipitation rate, P, is equivalent to increasing the loss rate of the woody biomass, M_1, which can model the enhancement of a consumer pressure such as grazing.

These results are consistent with recent field observations of woody-herbaceous interactions along rainfall gradients (Pugnaire and Luque 2001, Wright et al. 2006), and more generally, with the conceptual theory of facilitation in stressed environments (Brooker and Callaghan 1998, Bruno et al. 2003, Callaway and Walker 1997).

ECOSYSTEM ENGINEERING AT THE LEVEL OF INTERACTING PATCHES

At the level of a few interacting patches, competition over the soil-water resource among the woody patches exerts water stress on each patch. The effect of this "biotic" stress is similar to the effect discussed in the previous subsection of an abiotic aridity stress on a single, isolated patch. Figure 12.5 shows the response of an herbaceous life-form b_2 to sparse (a) and dense (b) woody patches b_1. When the patches are sufficiently sparse and effectively isolated, the woody life-form competes with the herbaceous life-form and excludes it (Figure 12.5a). However, when the patches are dense enough, coexistence of the two life-forms within the patches becomes possible (Figure 12.5b).

The mechanism of positive engineering and facilitation in the case of dense woody patches is similar to that of single patches under aridity stress (see the section "Ecosystem Engineering Along Environmental Gradients"). The competition for water reduces the b_1 patch size and consequently the soil water consumption. As a result, more soil water is left for the herbaceous life-form allowing its coexistence with the woody life-form.

ECOSYSTEM ENGINEERING AT THE LANDSCAPE LEVEL

At the landscape level, symmetry-breaking vegetation patterns can appear (Gilad et al. 2004, Rietkerk et al. 2004). At this scale environmental stresses or consumer pressures may affect interspecific interactions by

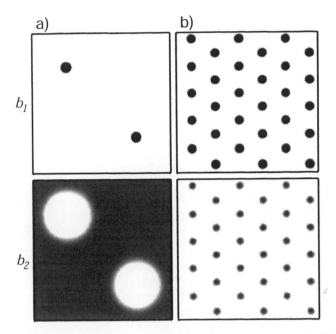

FIGURE 12.5 Model solutions showing a transition from competition to facilitation as a result of intraspecific woody-patch competition over the water resource. Shown are the spatial distributions of the woody and herbaceous biomass densities (b_1 and b_2, respectively), for sparsely scattered b_1 patches (a) and for a closely packed hexagonal pattern of b_1 patches (b). The smaller b_1 patch size in case (b) reflects stronger woody-patch competition. For the chosen environment and species traits, the b_2 life-form is excluded from b_1 patches and their close neighborhoods when the patches are sparsely scattered, but coexists with b_1 (within its patches) when the patches are closely packed. The intraspecific woody-patch competition leads to positive engineering and interspecific facilitation. The parameters used are given in Figure 12.2 except for the following: $\nu = \delta_w = 3.333$, $\alpha = 33.333$, $\delta_h = 333.333$, $\gamma_1 = 8.333$, $\gamma_2 = 0.833$, $\delta_{b_1} = \delta_{b_2} = 0.033$, $\mu_2 = 4.3$, and $p = 0.55$ (82.5 mm/yr). Reprinted from Gilad et al. 2006b.

shifting the system from one pattern state to another. Figure 12.6 (top row) shows a global transition from vegetation bands to vegetation spots of the woody life-form on a slope as a result of a local clear-cut along one of the bands. The mechanism of this transition is as follows (Gilad et al. 2004, 2006a). The clear-cut allows for more runoff to accumulate at the band segment just below it. As a result this segment grows faster, draws more water from its surroundings, and induces vegetation decay at the nearby band segments. The decay of the vegetation in the nearby segments allows for more runoff to accumulate at the next band down-

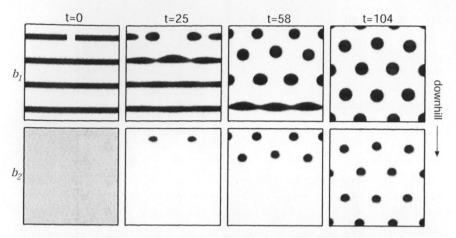

FIGURE 12.6 Facilitation induced by a pattern shift at the landscape level. Shown is a sequence of snapshots at different times (t is in years) describing a transition from vegetation bands to vegetation spots on a slope induced by a local clear cut along one of the bands (b_2 is randomly distributed at $t = 0$). In the banded pattern the b_1 life-form excludes the b_2 life-form, but in the spotted pattern they coexist due to enhanced runoff concentration. The slope angle is 15°, the precipitation is $p = 1.6$ (240 mm/yr), and all other parameters are as in Figure 12.5 except for $\gamma_1 = 16.667$, $\gamma_2 = 1.667$, and $\mu_2 = 4.8$. Reprinted from Gilad et al. 2006b.

hill. The whole process continues repeatedly until the whole pattern transforms into a spot pattern. As shown by Gilad et al. 2006a, the transition to spots is accompanied by higher soil-water densities under vegetation patches, for each spot experiences a bare area uphill twice as large as the bare area between successive bands, and therefore absorbs more runoff.

Numerical integration of the model Equations 5 indeed shows that herbaceous species (b_2) that are excluded by the woody engineer (b_1) in the banded pattern can coexist with the woody engineer in the spotted pattern. The transition from banded to spotted vegetation involves, in effect, a *facilitation front* propagating downhill; as bands gradually break into spots, patches with higher soil-water density are formed, facilitating the growth of the herbaceous life-form. This is an example of a cross-scale effect where a pattern transition at the landscape scale changes the engineering strength at the single-patch scale and induces facilitation. The fact that the transition from bands to spots takes place at *constant* environmental conditions indicates it is a pure spatial patterning effect, unlike the single-patch facilitation induced by an aridity stress as discussed earlier.

12.5 ● CONCLUDING REMARKS

We studied here a mathematical model for plant communities in water-limited systems, addressing the question of ecosystem engineering. Two ingredients of the model are particularly significant for understanding ecosystem engineering by plants: the infiltration feedback and the uptake feedback between the engineer's biomass and water. The infiltration feedback induces positive engineering by trapping surface water in vegetation patches, and creating habitats for other plant species. The uptake feedback induces negative engineering by exploiting the soil-water resource and depleting its content in and around vegetation patches. Dominance of the uptake feedback leads to competition and exclusion of other species, whereas dominance of the infiltration feedback results in facilitation and coexistence with other species.

In modeling the uptake feedback we took into account the non-local nature of this process (water uptake at a given point can be due to a distant plant whose roots extend to that point) and the augmentation of the root system in response to biomass growth. Modeling these aspects of the uptake process allowed studying the resilience of ecosystem engineers to environmental changes, and the trade-off between engineering and resilience, as the relative strength of the two feedbacks changes. This trade-off is significant for understanding the stability and functioning of water-limited ecosystems on micro (patch) and macro (watershed) spatial scales (Shachak et al. 1998). The soil-moisture accumulation at an engineer patch accelerates litter decomposition and nutrient production, and culminates in the formation of fertility islands. Watershed-scale disturbances that are incompatible with the resilience of a dominant engineer life-form can destroy the engineer patches and the fertility islands associated with them, thereby damaging the stability and functioning of the ecosystem.

Modeling the root-system augmentation in response to biomass growth also allows studying how different plant developmental strategies for coping with aridity stresses, affect ecosystem engineering and resilience. Assuming, for example, a functional dependence, $\eta_1 = \eta_1(p)$, that accounts for the capability of plants to further extend their root systems in response to aridity stresses (Bloom et al. 1985), can result in a crossover back to negative engineering as the precipitation rate p drops to levels approaching the survival limit of the engineer (Gilad et al. 2006b). This result may shed new light on recent controversial studies of interspecific interactions along environmental-stress gradients (Maestre and Cortina 2004, Maestre et al. 2005).

The explicit modeling of the infiltration and uptake feedbacks represents a "first-principle" approach whereby community and landscape properties, such as spatial structures and plant interactions, *emerge* as solutions of the model equations rather than being preset in formulating the model. This makes the model a useful platform for studying a variety of other problems (Sheffer et al. 2006, Yizhaq et al. 2005). An example of a related problem we have begun studying concerns mechanisms of species-diversity change in woody-herbaceous systems along environmental gradients. Trade-offs in herbaceous species traits, such as tolerance to grazing and shading, response to variable rainfall (Goldberg and Novoplansky 1997, Tielbörger and Kadmon 2000), and dispersal range (Nathan 2006), can easily be incorporated into the model and studied at different levels of woody-patch organization (Gilad et al. 2006b). Such studies can elucidate the roles played by single-patch engineering, patch competition, and patch patterning at the landscape level, in changing herbaceous life-form composition along a rainfall gradient or in response to a consumer pressure. Another problem of interest concerns the interspecific interactions between different woody engineers along gradients of environmental stresses, and the effects of these interactions on ecosystem engineering and herbaceous vegetation growth. In studying these problems various extensions of the model platform can be considered. Examples of such extensions include the introduction of stochastic rainfall, pulsed rainfall events, stochastic disturbances (fires), long-distance seed dispersal, and seedling dynamics.

ACKNOWLEDGMENTS

We thank Ariel Novoplansky, Efrat Sheffer, Yael Seligmann, Hezi Yizhaq, and Assaf Kletter for helpful discussions and comments. This research was supported by the James S. McDonnell Foundation, by the Israel Science Foundation, and by the Center for Complexity Science.

REFERENCES

Bertness, M.D., and Callaway, R.M. (1994). Positive interactions in communities. *Trends in Ecology & Evolution* 9:191–193.

Bertness, M.D., and Ewanchuk, P.J. (2002). Latitudinal and climate-driven variation in the strength and nature of biological interactions in New England salt marshes. *Oecologia* 132:392–401.

Bertness, M.D., and Hacker, S.D. (1994). Physical stress and positive associations among marsh plants. *American Naturalist* 144:363–372.

Bloom, A.J., Chapin, F.S., III, and Mooney, H.A. (1985). Resource limitation in plants— an economic analogy. *Annual Review of Ecology and Systematics* 16:363–392.

Brooker, R.W., and Callaghan, T.V. (1998). The balance between positive and negative plant interactions and its relationship to environmental gradients: A model. *Oikos* 81:196–207.

Bruno, J.F., Stachowicz, J.J., and Bertness, M.D. (2003). Inclusion of facilitation into ecological theory. *Trends in Ecology & Evolution* 18:119–125.

Callaway, R.M., and Walker, L.R. (1997). Competition and facilitation: A synthetic approach to interactions in plant communities. *Ecology* 78:1958–1965.

Callaway, R.M., Brooker, R.W., Choler, P., Kikvidze, Z., Lortiek, C.J., Michalet, R., Paolini, L., Pugnaire, F.I., Newingham, B., Aschehoug, E.T., Armas, C., Kikodze, D., and Cook, B.J. (2002). Positive interactions among alpine plants increase with stress. *Nature* 417:844–848.

Campbell, S.E., Seeler, J.S., and Glolubic, S. (1989). Desert crust formation and soil stabilization. *Arid Soil Research and Rehabilitation* 3:217–228.

Cuddington, K., and Hastings, A. (2004). Invasive engineers. *Ecological Modeling* 178:335–347.

Ellenr, P., and Guckenheimer, J. (2006). *Dynamic Models in Biology*. Princeton, NJ: Princeton University Press.

Gilad, E., von Hardenberg, J., Provenzale, A., Shachak, M., and Meron, E. (2004). Ecosystem engineers: From pattern formation to habitat creation. *Physical Review Letters* 93:0981051.

Gilad, E., and von Hardenberg, J. (2006). A fast algorithm for convolution integrals with space and time variant kernels. *J. Comp. Phys.* 216:326–336.

Gilad, E., von Hardenberg, J., Provenzale, A., Shachak, M., and Meron, E. (2007). A mathematical model of plants as ecosystem engineers. *Journal of Theoretical Biology* 244:680–691.

Gilad, E., Shachak, M., and Meron, E. (2006b). Dynamics and spatial organization of plant communities in water limited systems. Submitted to *Theoretical Population Biology*.

Goldberg, D., and Novoplansky, A. (1997). On the relative importance of competition in unproductive environments. *Journal of Ecology* 85:409–418.

Greenlee, J.T., and Callaway, R.M. (1996). Effects of abiotic stress on the relative importance of interference and facilitation. *American Naturalist* 148:386–396.

von Hardenberg, J., Meron, E., Shachak, M., and Zarmi, Y. (2001). Diversity of vegetation patterns and desertification. *Physical Review Letters* 87:198101.

Hillel, D. (1998). *Environmental Soil Physics*. San Diego: Academic Press.

HilleRisLambers, R., Rietkerk, M., Van den Bosch, F., Prins, H.H.T., and de Kroon, H. (2001). Vegetation pattern formation in semi-arid grazing systems. *Ecology* 82:50–61.

Jones, C.G., Lawton, J.H., and Shachak, M. (1994). Organisms as ecosystem engineers. *Oikos* 69:373–386.

Jones, C.G., Lawton, J.H., and Shachak, M. (1997). Positive and negative effects of organisms as ecosystem engineers. *Ecology* 78:1946–1957.

Klausmeier, C.A. (1999). Regular and irregular patterns in semiarid vegetation. *Science* 284:1826–1828.

Lefever, R., and Lejeune, O. (1997). On the origin of Tiger Bush. *Bulletin of Mathematical Biology* 59:263–294.

Maestre, F.T., Bautista S., and Cortina, J. (2003). Positive, negative and net effects in grass-shrub interactions in Mediterranean semiarid grasslands. *Ecology* 84: 3186–3197.

Maestre, F.T., and Cortina, J. (2004). Do positive interactions increase with abiotic stress? A test from a semiarid steppe. *Proc. R. Soc. Lond. B* (Suppl.) 271:S331–S333; Are *Pinus halepensis* afforestations useful as a restoration tool in degraded semiarid Mediterranean areas? *Forest Ecology and Management* 198:303–317.

Maestre, F.T., Valladares, F., and Reynolds, J.F. (2005). Is the change of plant–plant interactions with abiotic stress predictable? A meta-analysis of field results in arid environments. *Journal of Ecology* 93:748–757.

Meron, E., Gilad, E., von Hardenberg, J., Shachak, M., and Zarmi, Y. (2004). Vegetation patterns along a rainfall gradient. *Chaos, Solitons and Fractals* 19:367–376.

Nathan, R. (2006). Long-distance dispersal of plants. *Science* 313:786–788.

Okayasu, T., and Aizawa, Y. (2001). Systematic analysis of periodic vegetation patterns. *Progress of Theoretical Physics* 106:705–720.

Pugnaire, F.I., and Luque, M.T. (2001). Changes in plant interactions along a gradient of environmental stress. *Oikos* 93:42–49.

Rietkerk, M., Boerlijst, M.C., Van Langevelde, F., HilleRisLambers, R., Van de Koppel, J., Kumar, L., Prins, H.H.T., and De Roos, A.M. (2002). Self-organization of vegetation in arid ecosystems. *American Naturalist* 160:524–530.

Rietkerk, M., Dekker, S.C., de Ruiter, P.C., and Van de Koppel, J. (2004). Self-organized patchiness and catastrophic shifts in ecosystems. *Science* 305:1926–1929.

Scheffer, M., Carpenter, S., Foley, J.A., Folke, C., and Walkerk, B. (2001). Catastrophic shifts in ecosystems. *Nature* 413:591–596.

Sheffer, E., Yizhaq, H., Gilad, E., Shachak, M., and Meron, E. (2006). Why do plants in resource deprived environments form rings? Submitted to *Ecological Complexity*.

Shachak, M., Sachs, M., and Moshe, I. (1998). Ecosystem management of desertified shrublands in Israel. *Ecosystems* 1:475–483.

Shnerb, N.M., Sarah, P., Lavee, H., and Solomon, S. (2003). Reactive glass and vegetation patterns. *Physical Review Letters* 90:0381011.

Tielbörger, K., and Kadmon, R. (2000). Temporal environmental variation tips the balance between facilitation and interference in desert plants. *Ecology* 81:1544–1553.

Valentin, C., d'Herbès, J.M., and Poesen, J. (1999). Soil and water components of banded vegetation patterns. *Catena* 37:1–24.

Walker, B.H., Ludwig, D., Holling, C.S., and Peterman, R.M. (1981). Stability of semiarid savanna grazing systems. *Journal of Ecology* 69:473–498.

West, N.E. (1990). Structure and function in microphytic soil crusts in wildland ecosystems of arid and semi-arid regions. *Advances in Ecological Research* 20: 179–223.

Wright, J.P., Gurney, S.C., and Jones, C.G. (2004). Patch dynamics in a landscape modified by ecosystem engineers. *Oikos* 105:336–348.

Wright, J.P., and Jones, C.G. (2006). The concept of organisms as ecosystem engineers ten years on: Progress, limitations, and challenges. *BioScience* 56:203–209.

Wright, J.P., Jones, C.G., Boeken, B., and Shachak, M. (2006). Predictability of ecosystem engineering effects on species richness across environmental variability and spatial scales. *Journal of Ecology* 94:815–824.

Yair, A., and Shachak, M. (1987). Studies in watershed ecology of an arid area. In *Progress in Desert Research*, M.O. Wurtele and L. Berkofsky, Eds. Roman and Littlefield Publishers.

Yizhaq, H., Gilad, E., and Meron, E. (2005). Banded vegetation: Biological productivity and resilience. *Physica A* 356(1):139–144.

13

BALANCING THE ENGINEER–ENVIRONMENT EQUATION: THE CURRENT LEGACY

Kim Cuddington and Alan Hastings

13.1 ⚊ INTRODUCTION

As argued in Hastings et al. (2007), models of ecosystem engineers (*sensu* Jones et al. 1997) can help clarify what makes engineers distinct, and more importantly, establish when and why they should be considered separately when attempting to make predictions about natural systems. Given this argument, models that explicitly incorporate ecosystem engineers are surprisingly few in number. However, these models do range from general and heuristic approaches (e.g., Gurney and Lawton 1996) to mechanistic and system-specific studies (e.g., Gilad et al. 2004). In addition, some of these models have been solely or primarily concerned with the evolutionary consequences of engineering (e.g., Laland et al. 1999), while others have focused solely on the ecological effects (e.g., Cuddington and Hastings 2004, Wright et al. 2004).

We define ecosystem engineering models as those that include a relationship between the engineer species and the environmental state, and that subsequently link between the environmental state and some biotic characteristic of the system. By definition, then, we are identifying models of ecosystem engineering using a process-based, rather than an outcome-based, definition (see Jones and Guttiérez, Ch 1, Wilson, Ch 3, and Cuddington, Ch 4 on this issue). We consider models that include the process of an ecosystem engineer modifying the abiotic environment, regardless of the outcome of this interaction. However, we exclude models that examine only the abiotic consequences of an engineer, but that make no link to ecological or evolutionary consequences (engineer → abiotic models). For example, Weiler (2005) studies the effect of earthworm burrows on water flow in soils, but we consider this to be a hydrological model, not an ecological model, because no connection to a biotic component of the system is described.

We provide a survey of the current state of ecosystem engineer modeling with the goals of deducing general principles and paving the way for future efforts. The majority of the models that are included in this review are those at the population level (reflecting the current state of the art), where the action of engineering affects the environment, which in turn affects the engineer (engineer ↔ abiotic models). We also survey community-level models that include an engineer's effect on the environment and some relationship with another species or whole community of species. Following the definition of Jones et al. (1997), and Jones and Guttiérez (2007), we include examples without feedback to the engineer population for these community models. We suggest that the minimum requirement for this type of model is a relationship between the engineer, the environment, and some feature of the biotic community (engineer → abiotic → biotic or biotic → engineer → abiotic relationships). Of course, we do not exclude more complex models that include feedback from the abiotic environment to the engineer (engineer ↔ abiotic → biotic models), or feedback from the community to the engineer (biotic ↔ engineer → abiotic models). We also examine an ecosystem model where the role of an engineer species is considered. In general, ecosystem models include some recognition of the relationship between the general biotic community and the abiotic environment, but to be considered an engineer model, the ecosystem engineer must be described specifically, as well as the general biotic community (e.g., engineer → abiotic ↔ biotic models). Although it is not necessarily true that the same abiotic characteristic will be considered for both the engineer's effect and that of the biotic community (engineer → abiotic1 → biotic → abiotic2 models would also qualify). In general, global circulation models that add some biotic effects do not meet this criterion (e.g., Gunson et al. 2006).

Our goal is to explore some of the general conclusions to be drawn from models of ecosystem engineering, and point out some new directions. We begin with population models, starting with the simple and heuristic and moving to the detailed and specific. We note that one potentially important characteristic of engineer species, the propensity for the abiotic modifications to persist after death, has yet to be examined in more detailed models. From here we move on to cover the rather smaller literature on community and ecosystem models. We find that there is a need for more modeling efforts at this level, since engineer species have the potential to cause large-magnitude impacts on communities. Finally, most authors that have constructed ecosystem engineering models tip the balance in favor of either more detailed physical descriptions, or more detailed biotic descriptions. We predict that progress in the construction of ecosystem engineering models will require a balance in the detail of both aspects.

13.2 ⬤ POPULATION MODELS OF ECOSYSTEM ENGINEERS: THE SIMPLEST CASES

Clearly, the major difference between models of ecosystem engineers and standard models in population or community ecology is that models of ecosystem engineers must explicitly include a description of the abiotic environment. Thus, we begin with the simplest model of an ecosystem engineer, which must have, as a minimum, two variables: one representing population size, and the other representing the current environmental state.

Cuddington et al. (unpublished; see also Wilson, Ch 3) have analyzed what they consider to be the simplest version of this relationship between an engineer species and the environment. They present a model of the relationship between the relevant features of the environmental state, E, measured as a continuous variable, and the population level of the engineer species, N, also measured as a continuous variable. The environmental state, E, is some abiotic factor of interest. They suggest that this factor could be univariate, such as temperature, or multivariate, such as a combination of salinity and current flow. They divide the effect of the environment on population density into two components, density-independent effects determined by α, and density-dependent effects given by β. The effect of ecosystem engineering is to push the environment away from its normal state, $E = 0$, while the environment tries to return back to that state at a rate ρ. So the full system is

$$\frac{1}{N}\frac{dN}{dt} = ((\pm 1 + \alpha E) - (1 + \beta E))N$$
$$\frac{dE}{dt} = -\rho E + N, \tag{1}$$

where the sign of 1 in the first term of the first equation is given by that portion of density-independent growth rate that is not determined by feedback from the environmental characteristic of interest. These authors analyze this model assuming that the environment responds much faster than the engineer population. That is, they assume that E is at quasi-equilibrium to obtain

$$\frac{1}{N}\frac{dN}{dt} = \left(\left(\pm 1 + \frac{\alpha}{\rho}N\right) - \left(1 + \frac{\beta}{\rho}N\right)\right)N. \tag{2}$$

The model suggests that there are two types of ecosystem engineers: those that must engineer to survive, or obligate engineers (where the first term of Equation 2 is −1), and those that will increase when rare even in the absence of engineering, or non-obligate engineers (the first term of Equation 2 is +1). For both of these cases, the effect of engineering on the density-dependent growth rate relative to its effects on density-independent growth determines population dynamics. There may be multiple basins of attraction and, in some cases, populations can undergo what the authors call runaway growth, where the population grows until limited by a factor that is external to this simple formulation.

These authors also note that the rate at which the environment recovers determines population dynamics, and suggest that ecosystems can be classified as either robust or susceptible to engineering depending on the magnitude of ρ. In particular, increased susceptibility can shift a system from one possessing a stable equilibrium to one that has explosive population growth. For obligate engineers, slower environmental return rates can cause certain extinction to give way to systems with a positive stable state. That is, the effect of ecosystem engineering is fundamentally context dependent.

One of the limitations of this model is that a continuously increasing function is used to describe the effect of the engineering species on the environment. The authors suggest that this linear relationship is one reason why the simple model predicts runaway growth. In the very first attempt to model ecosystem engineers, Gurney and Lawton (1996) used a different approach, which sets a natural limit to the effects of the ecosystem engineer on the environment. They describe the effects of an obligate engineer in terms of altering the proportion of the landscape in three discrete habitat states, a different approach than in the previous model where the environment was described as a continuous variable. "Virgin" habitat is not occupied by the engineer, but could be colonized and modified; "habitable" landscape is occupied and modified; while "degraded" habitat can no longer be inhabited, although it recovers to the "virgin" state at a fixed rate. Since the total quantity of habitat is constant, (1 = Virgin + Habitable + Degraded), and only "habitable" area produces new engineers, population growth is limited by available habitat, and this model does not predict runaway growth.

As is expected by this verbal description, the model reduces to a system of ordinary differential equations, and consists in its simplest form of three equations. One equation describes the dynamics of the engineer species, which depends on the amount of habitat that is in the modified state. Two other equations describe the dynamics of habitat as a function of the engineer, by tracking the proportion of habitable and virgin landscape. This model obviously incorporates the basics of an ecosystem engineer model, namely an explicit inclusion of a variable describing the state of the habitat, and feedback to the engineer population. It does, however, restrict the classification of the environment.

The model predicts two potentially stable states, one with no engineers, and one with a positive engineer population. The zero engineer state is unstable for engineers that can individually modify the virgin habitat at a rate that compensates for habitat degradation. These populations will approach a positive stable state. If individual modification rates are too slow relative to degradation rates, and individuals do not cooperate, the zero-engineer state becomes stable and the population will not persist. The situation is more complicated for engineers that cooperate. Examination of particular situations suggests that if a cooperating population can overcome habitat decay rates there may be two basins of attraction, one leading to extinction, another to a positive engineer density. This positive equilibrium may become unstable and generate population oscillations where there is a high degree of cooperation and slowly recovering habitat. There is a direct analogy here to the simple model of Cuddington et al. (unpublished), which predicts similar destabilization for positive density dependence. Finally, several alternatives to the basic model, which include modifications of how the delays are incorporated, show that the results are relatively robust.

This implicit patch model approach can be extended in a variety of ways, as has been done by Wright et al. (2004). These authors incorporate a number of features that both test the robustness of the original model, and look at how additional biological aspects could change the dynamics. One important change is allowing immigration, or, in other words, the addition of individuals to the system in the absence of engineering, which means that the steady state without the engineer is not possible. Again we see a direct relationship to the very simple population model proposed by Cuddington et al. That is, the zero density engineer state is stable only where engineering *must* occur for the population to increase.

A second modification is essentially focused on the form of the time delays, by adding an additional patch type, a "partially recovered" patch, which could be recolonized by the engineer species, but which is either preferred or not preferred by them. This modification has a less dramatic effect on model predictions.

Continuing this implicitly spatial approach, but instead describing the environmental state as a continuous variable, Cuddington and Hastings (2004) developed a model that included a continuous description of habitat quality. The model analysis was focused on the dynamics of an invasive species, but the formalism could be used to look at other questions as well. The authors make a number of simplifying assumptions in order to develop a model that can be easily studied. The scale used to measure habitat quality is an implicit one, relative to the optimum value for the species under consideration. A typical measure of habitat quality could be tidal height (for an invasive salt marsh cord grass, *Spartina*), or water infiltration rate (for desert vegetation). The dynamic processes they include in the model are the modification of the habitat quality by the engineering species, and the response of the environment to modification, as well as the population dynamics of the species, including growth and spread. The engineering species does not, of course, change the total area available, but it does change the proportion of the total area found at different values of habitat quality. By assuming that the habitat quality varied smoothly over space, the further simplifying assumption can be made that all processes that affect nearby habitat also affect habitat that is of similar quality. Environmental forcing then reduces the effect of habitat alteration, by spreading the effect over a larger region. For example, as sediments accumulate at the base of a seaweed, wave action and turbulent flow simultaneously spread the accumulated particles over a larger region. The population dynamics also include both local processes (growth at a given location) and dispersal that would affect other locations.

The two variables in the system are a density function for the proportion of area with habitat quality x, $H(x)$, and a density function for proportion of the area of habitat quality x, which is occupied by the engineer $N(x)$. Thus, a general integro-difference equation format was used for the model of the form

$$N_{t+1}(x) = \int k_N(x, y) f(N_t(y), H_t(y)) dy$$
$$H_{t+1}(x) = \int k_H(x, y) g(N_t(y), H_t(y)) dy, \tag{3}$$

where $f(N_t(y),H_t(y))$ describes the growth of the invading species, k_N gives its spread across the environmental gradient due to biological processes such as dispersal of seeds or clonal growth, $g(N_t(y),H_t(y))$ characterizes the effect of the invader on its habitat, and k_H describes other forces in the environment that smooth the effect of habitat modification over a larger area. More details, and specific functions, are in Cuddington and Hastings (2004).

In general, this model predicts that standard invasion models, which do not include habitat modification, could underestimate both spatial spread rates, and population densities of engineer species invading sub-optimal habitats. The model also suggests that spread rates could be faster in previously modified environments, and thus control efforts should focus on ameliorating habitat changes as well as eliminating the engineering population.

We find some common themes in the simple models we have reviewed here. In environments that are robust to engineering, obligate engineers may be unable to persist, or may require a threshold density to do so. Cooperation of engineers can lead, however, to the destabilization of stable states, producing faster population growth than expected or oscillations. Further, environmental response time can be an important determinant of dynamics, while previously modified habitat may present an opportunity of engineers to subvert the rate of environmental response.

13.3 ● POPULATION MODELS: SPATIALLY EXPLICIT AND MECHANISTICALLY DETAILED CASES

At the other end of the spectrum, modeling of engineers can include more detail of engineering activities in specific environments. These models typically include spatially explicit descriptions of abiotic interactions, as well as equations that describe the specific mechanism by which the engineer species modifies the environment. In addition, these models are often specifically focused on the emergent properties of large-scale spatial processes.

Several models have focused on interactions between water and vegetation, illustrating the essential importance of the engineering concept for understanding the generation of spatial pattern and its subsequent dynamics in semiarid environments. These models range from the fairly general to quite detailed. Klausmeier (1999) provides a simple, yet mechanistic, model describing the effect of vegetation on soil-water infiltration rates, and the subsequent patterns of vegetation that form. He uses a partial differential equation formulation that describes the spatial dynamics of water, w, and plant biomass, n. Water is supplied uniformly and lost due to evaporation. Plants also use water at a rate related to their biomass; however, they alter the soil-water infiltration rates as their biomass increases as well. Simply linear functions are used to describe the response of plants to increased water, and the effect of plant biomass on soil-water infiltration rates. The nondimensionalized system has only three parameters:

$$\frac{dw}{dt} = a - w - wn^2 + v\frac{\partial w}{\partial x}$$
$$\frac{dn}{dt} = wn^2 - mn + \left(\frac{\partial^2}{\partial x^2} + \frac{\partial^2}{\partial y^2}\right), \tag{4}$$

where m describes plant biomass loss, a indicates water input, and v gives the speed at which water flows downhill. In a nonspatial analysis, the authors find that the model has two stable equilibria for ecologically reasonable parameters: one where there are no plants, $n = 0$, and a vegetated state. In the spatial domain, as the result of a Turing-type phenomena, these states are connected by banding patterns: strips of vegetation divided by bare ground. As water input, a, decreases, or plant mortality, m, increases, the system moves from completely vegetated to banded vegetation of increasing wavelength to no vegetation. Irregular patterns can also emerge where the nonspatial model has a limit cycle or is excitable, but the authors suggest that parameter values that cause these patterns are ecologically unrealistic.

Gilad et al. (2004) present a more detailed model in the same vein. These authors describe the positive feedbacks between vegetation and surface-water infiltration, and the soil-water uptake by plants, using a set of three coupled equations that describe aboveground biomass, soil-water density, and surface-water height (also see details in Meron et al., Ch 12). The states of bare-soil and uniform vegetation are stable states for this system, but again, the bare-soil state loses stability as precipitation exceeds a threshold value. In this region, patterned distributions of vegetation form, ranging from bands to spots depending on precipitation level. The authors also use this model to begin answering questions about the effect of the cyanobacteria-shrub engineering system on community-scale characteristics (see the section "Community and Ecosystem Models"). At the end of this contribution, the authors note that when the water-flow alterations of the engineer system are coupled to different degrees of slope, particular patterns of vegetation can be more beneficial than others. Gilad et al. (2004) suggest that, as precipitation decreases, bands will give way to regularly patterned spots of vegetation, but counterintuitively, these spots will accumulate *more* water than vegetation bands in a higher precipitation environment. Thus, the emergent spatial arrangement of the system determines the engineering outcome.

Yizhau et al. (2005) expand on this model analysis by more thoroughly examining the multistability of the patterned vegetation states. They note that the banded patterns have two responses to decreased rainfall. Either there will be an increase in the ratio of interband distance to bandwidth, with no change in wave number, or there will be a decrease in wave number. As might be expected, total vegetative biomass increases with wave number, but so does total water consumed. In fact, the ratio of water consumed to biomass decreases with wave number such that higher wave number patterns consume less water, and are, in this sense, more productive. However, the authors also note that patterns with a smaller wave number are more resilient to changes in precipitation.

In a similar coupling of a spatially explicit approach and detailed mechanism, Durán and Herrmann (2006) explore the relationship between vegetation and dune stabilization. This work shares methodological similarities with other models that focus on flow dynamics (e.g., Klausmeier 1999, Gilad et al. 2004). These authors describe the effect of established vegetation as adding a roughness factor that absorbs part of the momentum transformed to the soil by wind. That is the total shear stress acting on sand grains is given as $\tau_s = \tau \big/ \left(1 + m\beta\frac{\rho_v}{\sigma}\right)$, where τ is the total wind shear stress, m is a tuning parameter, β is the ratio of plant to surface drag coefficients, ρ_v is vegetation cover density, and σ is the ratio of plant basal to frontal area. Thus, if plants can establish in part of a dune, they will locally slow down wind, inhibiting sand erosion and enhancing accretion. Of course, sand dynamics also affect plant growth: Noncohesive sand is eroded by wind, which can denude roots and increase evaporation in general. Sand erosion or accumulation can kill plants, and this effect is included in the model.

These authors use a separation of timescales to analyze their model, since wind flow is on a much faster timescale than that of vegetation growth and surface sand erosion. Therefore they use quasi-equilibrium solutions for wind surface shear stress and sand flux, and describe this biotic–abiotic interaction using coupled differential equations to describe the sand surface and vegetation height. Plants first invade locations with low erosion or deposition rates, but stabilization of the dune depends on competition between sand transport and vegetation growth. They characterize this relationship using a dimensionless "fixation" index, which incorporates dune volume, wind stress, and vegetation growth velocity.

In a similar vein, D'Alpaos et al. (2006) describe the effect of shear stress and sedimentation alteration by *Spartina alterniflora* on the development of tidal creeks, expanding on a previous analysis by Mudd et al. (2004). In this detailed model, the elevation of the salt marsh platform determines *Spartina* biomass, while *Spartina* in turn elevates the marsh platform. *Spartina*'s abiotic effects of reducing flow speed, increasing sedimentation, and reducing turbulent flow are predicted to alter tidal creek development, through their effects on the marsh platform. The authors investigate model behavior for selected parameters and find that the channel width to depth ratio decreases as mudflats are converted to salt marsh.

13.4 ⟿ POPULATION MODELS: CASES WITH AN EVOLUTIONARY FOCUS

The role of evolutionary forces in determining the niche of an ecosystem engineer has also received considerable attention (e.g., Odling-Smee et al. 2003). The implication is that ecosystem engineering (or, equivalently, niche construction) is both affecting and affected by natural selection. These models fall into the same general class of engineer–environment feedback models as the ecological models we have described.

In this volume, Wilson (Ch 3) relates the simple population model of Cuddington et al. (unpublished) to evolutionary consequences. He illustrates how a mutant could increase in a population by responding positively to environmental changes that are made by the wild-type population activities. Given the feedback he describes, it is possible for selection to favor enhanced environmental modification.

In the first modeling contributions on this topic, Laland et al. (1996, 1999) use a set of recursion equations to examine the dynamics of traits that code for environmental alteration, "recipient" traits whose fitness depends on such alteration, and the amount of resource in the environment. A simple two-locus engineering model shows that novel evolutionary dynamics can result from ecosystem engineering, with a key role played by temporal scales. Engineering can lead to the fixation of deleterious alleles, the elimination of stable polymorphisms, and a stabilization of previously unstable polymorphisms. One characteristic of ecosystem engineering is to lengthen the ecological timescale, which thus means that the evolutionary and ecological processes are operating on more similar timescales. As a result periods of evolutionary inertia or momentum can emerge.

In a set of simulation models, Hui and coauthors extend the investigation of the evolutionary consequences of engineering to the spatial domain (Hui et al. 2004, Hui and Yue 2005, Han et al. 2006). For example, Hui et al. (2004) find that engineering can alter the genetic dynamics and diversity of metapopulations on a lattice environment, while concurrently altering environmental heterogeneity.

Several authors have examined the particular case of the evolution of flammability in plants (Bond and Midgley 1995, Kerr et al. 1999, Schwilk and Kerr 2002). Mutch (1970) speculates that higher levels of flammability could arise if plants had the ability to pass on their genes in spite of periodic fires. The question arises as to how exactly increased flammability could spread in a population, since the possessor of these alleles would be more likely to die. Kerr et al. (1999) develop a haploid model with separate loci for flammability and response to fire. They track flammable and less flammable alleles, as well as alleles that code for additional or reduced success in gaps created by fire. They describe gap frequency as a linear function of the frequency of highly flammable plants. These authors find that the presence of flammability enhancing traits can redirect the evolution of other traits, through the effect that the flammability alleles have on the environment. Stable polymorphisms of flammability are possible, as well as stable and unstable oscillations of genotypes. The magnitude of the engineering effect determines when such behavior is possible. Schwilk and Kerr (2002) extend this work to the spatial domain by tracking diploid genotype frequency on a lattice. They find that if mating, dispersal, and production of fire gaps are all strictly local processes, flammability may increase in frequency without any direct fitness benefit.

13.5 ● COMMUNITY AND ECOSYSTEM MODELS

In general, the models that describe ecosystem engineers have focused on the population dynamics or genetics of the engineer species. Yet, one of the most striking characteristics of engineers is their ability to alter the dynamics of species with which they have no trophic relationship (e.g., via engineer → abiotic → biotic relationships). That is, their effects on these species would be completely ignored in a food web or nutrient flow modeling framework. As well, some of the most urgent management problems are related to the engineering activities of invasive (Crooks 2002) or threatened species (e.g., corals).

Although there are few contributions in this area, the modeling approaches of these papers span the heuristic to spatially explicit and mechanistic approaches. We begin with models that focus on engineers that directly or indirectly interact with one or two species, move on to models that consider more general community features, such as species richness, and conclude with an ecosystem-level model.

In one of the simpler model frameworks, Wilson and Nisbet (1997) analyze one- and two-species models of sessile organisms that compete for space, but that have positive interactions in the form of the amelioration of environmental stress, or in promotion of germination success through nurse plant interactions. They do this in a spatially explicit framework in terms of patch occupancy with local transition rules along smooth environmental stress gradients. They find discontinuities of population distributions in the one-species case, and sharp boundaries of species composition in the two-species case.

Gilad et al. (2004) analyze the effects of water-flow alteration by desert shrubs with banded patterns of distribution, and extrapolate to community-level impacts. They identify the conditions under which the soil-water density levels and distribution will benefit other species. For some parameter values the soil-water level immediately adjacent to the shrub exceeds a bare-soil state, and will create new habitat for plant species that would not grow in the absence of engineers. Later, they extend their model to include the effects of consumers on the engineer species, engineering on a herbaceous species, and the effects of the herbaceous species on the woody engineer (Meron et al., Ch 12, summarized by Wilson, Ch 14).

Van de Kopel et al. (2002) examine the role of herbivores in determining the pattern of vegetation banding in semiarid systems. Herbivores consume the vegetation, which in turns alters soil-water infiltration rates. So in this model, instead of the engineer affecting another species through a modification of the abiotic environment as in Meron et al. (Ch 12) (engineer ↔ abiotic → biotic model), a member of the community alters the abiotic environment by affecting an engineer species (biotic ↔ engineer ↔ abiotic). These authors assume that rainfall is constant and describe a bounded area of grassland of size A square meters where surface water, O, soil water, W, plant biomass, P, and herbivore density, H, are given as the following:

$$\frac{\partial O}{\partial t} = R - F(O,P) - r_O O + D_O\left(\frac{\partial^2 O}{\partial x^2} + \frac{\partial^2 O}{\partial y^2}\right)$$

$$\frac{\partial W}{\partial t} = F(O,P) - r_W O - U(W,P) + D_W\left(\frac{\partial^2 W}{\partial x^2} + \frac{\partial^2 W}{\partial y^2}\right)$$

$$\frac{\partial P}{\partial t} = G(W,P) - D(P) - C(P,H) + D_P\left(\frac{\partial^2 P}{\partial x^2} + \frac{\partial^2 P}{\partial y^2}\right)$$

$$\frac{\partial W}{\partial t} = \frac{1}{A}\iint\limits_{x,y \in d} E(P(x,y)), H(x,y)dxdy - E(P)H. \tag{5}$$

Here, R is rainfall rate, $F(O,P)$ is soil-water infiltration rate, r_O and r_W are water losses through drainage and evaporation for surface and soil water respectively, D_O gives the surface-flow rates, $U(W,P)$ is plant-water uptake, D_W gives soil-water diffusion rates, $G(W,P)$ is plant-growth rate, $D(P)$ is density-dependent plant mortality, $C(P,H)$ is herbivore consumption rate, D_P describes plant spread, and $E(P)$ is the per capita rate of herbivore emigration, which is inversely proportional to plant density. The functions $F(O,P)$, $U(W,P)$, $G(W,P)$, and $C(P,H)$ are all simple, linear, and increasing.

As in other studies, the authors use a separation of timescales to simplify the analysis. They assume that surface water, soil water, and herbivores respond more quickly than plants, and use this assumption of quasi-equilibrium to reduce the problem to a single equation based on the average plant standing crop, P_{avg}, and the average herbivore density, H_{avg}:

$$\frac{\partial P}{\partial t} = gW(P, P_{avg})P - d(P) - cPH(P, P_{avg}). \tag{6}$$

Van de Koppel et al. (2002) find that herbivores with the mobility to select foraging locations at a fine scale can move a system from a vegetated to an unvegetated state, which is stable, by passing through various states of vegetation patterning.

Cordes et al. (2005) link an individual-based population model to a diffusion-advection model for sulfide in order to describe a hypothesized relationship between three different engineering groups in deep water vent systems. It is speculated that the tubeworm *Lamellibranchia luymesi* releases sulfate into hydrocarbon-rich sediments through a "rootlike" system. The sulfate is reduced to sulfide by bacteria, which are, in turn, often associated with methane-oxidizing or hydrocarbon-degrading species. This positive association between all three groups is demonstrated to be a plausible explanation for the extreme longevity of the tubeworms in the face of sulfide depletion.

We are aware of only two models that examine the effects of engineers on aggregate community properties (Wright and Jones 2004, Wilson and Wright 2007). Wright and Jones (2004) present a conceptual framework that relates ecosystem engineering to primary productivity and consequently to species richness. They note that engineering involves a change to the availability of resources in a patch, which may affect primary productivity. That is, engineered patches could have higher or lower resource availability and consequently higher or lower primary productivity than nonengineered areas. Since primary productivity is frequently related to species richness (e.g., reviewed by Grove 1999), they suggest that primary productivity be used as a metric to compare the effects of very different types of engineering, in very different environments, on species richness. They use a hump-shaped curve to describe the relationship between productivity and species richness. They conclude that, in high-productivity environments, an engineer that further increases productivity will decrease species richness, while an engineer that decreased productivity would increase richness. In low-productivity environments the opposite relationships would pertain.

In this volume, Wilson and Wright (Ch 11) also describe the environmental effects of an engineer on community species richness using a Lotka–Volterra modeling framework. They allow the environment to alter the mean and variance of intraspecific and interspecific interaction rates. They find that species richness will increase with increasing average population growth rates and will decrease with increasing average interaction rates, but is relatively insensitive to variance in these measures (Wilson and Wright, Ch 12, and the summary by Wilson, Ch 14). These authors claim their findings can be related to the predictions of Wright and Jones (2004) if species respond similarly to increasing productivity and if competition increases with increased productivity.

Byers et al. (2006) use a general conceptual framework of alternative stable states to describe the potential effect of ecosystem engineers on the management of ecosystems. They develop a conceptual model for the restoration of an ecosystem, and describe restoration as the effort to move an ecosystem from one stable state, S^*, to another, desired, stable state, D^*. The state of the ecosystem is described by a set of abiotic conditions, A, and biotic conditions, B. These variables are multivariate, where abiotic conditions might include features such as temperature, wind or current speed, and precipitation or salinity, and biotic conditions could include characteristics such as species composition, primary productivity, and the presence or absence of particular functional groups or species. These authors note that the interdependence of biotic and abiotic factors means that, in some cases, a perturbation to the system that will move it into the domain of attraction for the desired state may require a change to both abiotic and biotic characteristics. The basin boundary between the current state, S^*, and the desired state D^*, will be a surface in a multidimensional parameter space incorporating both abiotic and biotic parameters. They suggest, however, that ecosystem engineering can be thought of as having such a large impact on the system that it changes the location of the basin boundary between the current and desired state. That is, these species may make it easier or more difficult for restoration efforts to succeed. Indeed, these authors suggest that, in some situations, the only method of restoration will be to introduce or remove an ecosystem engineer. To understand and predict these effects a general model of the system could be phrased as a system of dynamic equations that include general abiotic variables, bioitic variables, and explicit dynamics for the ecosystem engineer, EE $\left(\text{e.g.,} \dfrac{dA}{dt} = f(A, B, EE), \quad \dfrac{dB}{dt} = g(A, B, EE), \quad \dfrac{dEE}{dt} = h(A, B, EE) \right).$ Depending on the exact system, these could be ordinary differential equations, partial differential equations, integro-difference equations, or some more complex form.

We are unaware of any detailed ecosystem-level models that describe ecosystem engineers (although DMS-DMSP plankton models may come close to fitting our classification scheme; e.g., Gunson et al. 2006). At the very largest scales, even up to a whole planet, plants clearly affect the atmospheric conditions, which in turn affect plant growth and population dynamics. We have chosen not to focus on approaches of this kind here because the feedbacks do not include the roles of particular species or functional groups that can be identified as ecosystem engineers (i.e., these are simply biotic ↔ abiotic models). Nonetheless, this kind of approach and question has much in common with other models we do review, and the ideas we present here may provide illumination for future efforts. In particular, we note that the idea of legacy effects and time-scales is central to these global questions.

13.6 — CONCLUSIONS

While there are relatively few models of ecosystem engineering, we note that this area seems on the verge of expansion, and many of the papers considered in this review have been published in the last 3 years. There is, however, a dearth of studies that deal with community- and ecosystem-level effects of ecosystem engineering. Although there are many models that deal with population- and landscape-level consequences of engineering, the connection to either community or ecosystem dynamics is generally extrapolated from the model results rather than being specifically included in the modeling framework. This seems an unfortunate area of investigation to have been neglected. Many of the most dramatic and costly effects of ecosystem engineers are on communities and ecosystems (Crooks 2002, Byers et al. 2006). However, this pattern of development is certainly understandable in this new area of modeling, especially given the potential complexity of the interplay between biotic and abiotic interactions in multiple-species systems.

We note that one feature of ecosystem engineering that can prevent it from being easily subsumed into models of trophic interactions is legacy effects (also called ecological inheritance in the niche-construction literature). These are alterations of the abiotic environment that persist after the engineer has left or died. Although the general concept of the role of timescales has received previous attention in the ecological literature (Ludwig et al. 1978, Rinaldi and Scheffer 2000, Hastings 2004), ecosystem engineering is one of the most important areas where this concept can be developed further. The role of time- and space scales in ecosystem engineer dynamics is emphasized in the review by Hastings et al. (2007).

We see a hint of an explicit discussion of legacy effects in the very simple model described by Cuddington et al. (unpublished). The environmental recovery rate determines how long engineering effects will persist. In their analysis, however, these authors assume that the abiotic environment responds more quickly to feedback than the engineer population. Gurney and Lawton (1996) and Wright et al. (2004) explore this feature of ecosystem engineering more clearly. There is an implicit time delay in their model expressed as the rate at which degraded habitat recovers to a virgin state, and in the extension of Wright et al. (2004), the move of a landscape patch through the degraded state to the partially recovered to the virgin state. This time delay is responsible for the oscillatory dynamics found for cooperating engineers, and is not an outcome that could be predicted without tracking the environmental state.

The role of such legacy effects is not explored in any of the more spatially explicit and detailed population models reviewed here. This omission is possibly because most of these studies are of flow-modifying engineers whose immediate effects may occur on very fast timescales.

From a methodological perspective, for many systems this difference in timescales between the different processes may simplify the daunting task of analyzing the full dynamical system (e.g., Cuddington et al. unpublished, Durán and Herrmann 2006, van de Koppel et al. 2002). This seems particularly true for species that act as flow modifiers. Some aspects of flow can be described at quasi-equilibrium since they change so much faster than surface and biotic characters.

It seems likely, however, that even in the context of the types of flow modification considered here, legacy effects will play a role. For example, although live plants reduce soil erosion in a dune environment (Durán and Herrmann 2006), such amelioration will probably continue for some period after the death of the plant. Alternately, legacy effects may alter biotic relationships, for example, *Spartina* roots are very persistent, and can act as seed traps after death, increasing local seedling recruitment (Lambrinos, Ch 16). The ecosystem engineer concept leads us to draw an analogy between these disparate systems (see Buchman et al., Ch 2, Jones and Guttiérez, Ch 1), and predict that there are similar legacy impacts in the effects of vegetation on soil-water infiltration rates in semiarid systems. Sites previously occupied by live plants will surely have greater infiltration rates for some period after the death of the vegetation, creating locations that may be easier to colonize by similar engineering species at a later date. Incorporating such legacy effects, however, can increase model complexity, and increase the difficulty of the modeler's task.

Another difficult task in the formulation of ecosystem engineer models is that of parameterization. Some authors have been able to make good use of literature data both in terms of parameterization (D'Alpaos et al. 2006, Wright et al. 2004, Wright and Jones 2004) and the elimination of particular asymptotic outcomes based on ecological and environmental feasibility (Klausmeier 1999). Although, one can imagine that such efforts may require the investigation of quite different literatures and consequently may be quite time consuming. In many cases, however, the relationships in the model formulation have not been investigated empirically and parameters are simply not available (e.g., as Gurney and Lawton 1996 conclude).

The same problem can hamper the selection of reasonable functional forms and relationships. Often the relationships between the engineer and the environment have not been investigated in the field, and the modelers must simply select those functions they judge to be reasonable. However, many theoreticians are more familiar with either physical or biological relationships. Thus models that come from authors with a

TABLE 13.1 Summary of the ecosystem engineer models reviewed, including classification by organizational level and detail.

Level	Type	Reference	Brief Description
Population or functional group	Conceptual and general	Cuddington et al. unpublished	Engineer density–environment; obligate and non-obligate engineers
		Wilson 2007	Selection for engineering traits
		Laland et al. 1996, 1999	Two-loci genetic environment
	Simple, spatially implicit	Gurney and Lawton 1996, Wright et al. 2004	Obligate, cooperating engineers and discrete habitat quality states
		Cuddington and Hastings 2004	Continuous variable of habitat quality; invasive spread
		Kerr et al. 1999	Percentage of fire gap patches and frequency of flammability
	Detailed, spatially explicit	Klausmier 1999	Vegetation-water model
		Gilad et al. 2004, Yiazhaq et al. 2005	Vegetation-water model; tracks surface flow
		Durán and Herrmann 2006	Dune stability of plants
		D'Alpaos et al. 2006	*Spartina* effects on tidal creeks
		Hui et al. 2004	Genetic-environment metapopulation dynamics
		Schwilk and Kerr 2002	2D frequency of flammability and fire gaps on a lattice
Community	Conceptual and general	Wright and Jones 2004	Engineer–species richness
		Wilson and Wright 2007	Lotka–Volterra species richness
	Simple, spatially explicit	Wilson and Nisbet 1997	one- and two-species lattice–environmental stress
	Detailed, spatially explicit	Gilad et al. 2004, Meron et al. 2007	Vegetation–water–nonengineer plants
		van de Koppel et al. 2002	Herbivores–vegetation–water
		Cordes et al. 2005	Tubeworms–sulfide–sulfate reducing bacteria; methane–hydrocarbon reducing bacteria
Ecosystem	Conceptual and general	Byers et al. 2006	Engineer–ecosystem

physical environment orientation tend to have detailed descriptions of abiotic relationships, but very simple assumptions regarding ecological interactions, while models built by those with an ecological focus have the opposite bias. Clearly striking the optimal level of detail for both sides of the engineer–environment interaction is an ongoing challenge.

REFERENCES

Bond, W., and Midgley, J. (1995). Kill they neighbour: An individualistic argument for the evolution of flammability. *Oikos* 73:79–85.

Cordes, E., Arthur, M., Shea, K., Arvidson, R., and Fisher, C. (2005). Modeling the mutualistic interactions between tubeworms and microbial consortia. *Public Library of Science Biology* 3:(e77).

Crooks, J. (2002). Characterizing ecosystem-level consequences of biological invasions: The role of ecosystem engineers. *Oikos* 97:153–166.

Cuddington, K., and Hastings, A. (2004). Invasive engineers. *Ecological Modelling* 178:335–347.

Cuddington, K., Wilson, W., and Hastings, A. (unpublished). A simple model of ecosystem engineers.

D'Alpaos, A., Lanzoni, S., Mudd, S., and Fagherazzi, S. (2006). Modeling the influence of hydroperiod and vegetation on the cross-sectional formation of tidal channels. *Estuarine Coastal and Shelf Science* 69:311–324.

Durán, O., and Herrmann, H. (2006). Vegetation against dune mobility. *Physical Review Letters* 97:(188001).

Gilad, E., von Hardenberg, J., Provenzale, A., Shachak, M., and Meron, E. (2004). Ecosystem engineers: From pattern formation to habitat creation. *Physical Review Letters* 93:(098105).

Grace, J. (1999). The factors controlling species density in herbaceous plant communities: An assessment. *Evolution and Systematics* 2:1–28.

Gunson, J., Spall, S., Anderson, T., Jones, A., Totterdell, I., and Woodage, M. (2006). Climate sensitivity to ocean dimethylsulphide emissions. *Geophysical Research Letters* 33:(L00701).

Gurney, W., and Lawton, J. (1996). The population dynamics of ecosystem engineers. *Oikos* 76:273–283.

Han, X., Li, Z., Hui, C., and Zhang, F. (2006). Polymorphism maintenance in a spatially structured population: A two-locus genetic model of niche construction. *Ecological Modelling* 192:160–174.

Hastings, A. (2004). Transients: The key to long-term ecological understanding? *Trends in Ecology and Evolution* 19:39–45.

Hastings, A., Byers, J., Crooks, J., Cuddington, K., Jones, C., Lambrinos, J., Talley, R., and Wilson, W. (2007). Ecosystem engineering in space and time. *Ecology Letters* 10:153–164.

Hui, C., and Yue, D. (2005). Niche construction and polymorphism maintenance in metapopulations. *Ecological Research* 20:115–119.

Hui, C., Li, Z., and Yue, D. (2004). Metapopulation dynamics of distribution, and environmental heterogeneity induced by niche construction. *Ecological Modelling* 177:107–118.

Jones, C.G., Lawton, J.H., and Shachak, M. (1997). Positive and negative effects of organisms as physical ecosystem engineers. *Ecology* 78:1946–1957.

Kerr, B., Schwilk, D., Bergmann, A., and Feldmann, M. (1999). Rekindling an old flame: A haploid model of the evolution and impact of flammability in resprouting plants. *Evolutionary Ecology Research* 1:807–833.

Klausmeier, C. (1999). Regular and irregular patterns in semiarid vegetation. *Science* 284:1826–1828.

Laland, K., Odling-Smee, F., and Feldman, M. (1996). On the evolutionary consequences of niche construction. *Journal of Evolutionary Biology* 9:293–316.

——. (1999). Evolutionary consequences of niche construction and their implications for ecology. *Proceedings of the National Academy of Science USA* 96:10242–10247.

Ludwig, D., Jones, D.D., and Holling, C. (1978). Qualitative analysis of insect outbreak systems: The spruce budworm and forest. *Journal of Animal Ecology* 47:315–332.

Mudd, S., Fagherazzi, S., Morris, J., and Furbish, D. (2004). Flow, sedimentation, and biomass production on a vegetated salt marsh in South Carolina: Toward a predictive model of marsh morphological and ecologic evolution. In *The Ecogeomorphology of Salt Marshes*, Estuarine and Coastal Studies Series, S. Fagherassi, M. Marani, and L. Blum, Eds. American Geophysical Union, Washington, D.C. pp. 165–188.

Mutch, R. (1970). Wildland fires and ecosystems—a hypothesis. *Ecology* 51:1046–1051.

Odling-Smee, F., Laland, K., and Feldman, M. (2003). *Niche Construction: The Neglected Process in Evolution*. Princeton, NJ: Princeton University Press.

Rinaldi, S., and Scheffer, M. (2000). Geometric analysis of ecological models with slow and fast processes. *Ecosystems* 3:507–521.

Schwilk, D., and Kerr, B. (2002). Genetic niche-hiking: An alternative explanation for the evolution of flammability. *Oikos* 99:43–442.

Van de Koppel, J.M., Rietkere, M., van Langevelde, F., Kumar, L., Klausmeier, C., Fryxell, J., Hearne, J., van Andel, J., de Ridder, N., Skidmore, A., Stroosnijder, L., and Prins, H. (2002). Spatial heterogeneity and irreversible vegetation change in semi-arid grazing systems. *American Naturalist* 159:209–218.

Weiler, M. (2005). An infiltration model based on flow variability in macropores: Development, sensitivity analysis and applications. *Journal of Hydrology* 310:294–315.

Wilson, W., and Nisbet, R. (1997). Cooperation and competition along smooth environmental gradients. *Ecology* 78:2004–2017.

Wright, J., Gurney, S., and Jones, C. (2004). Patch dynamics in a landscape modified by ecosystem engineers. *Oikos* 105:336–348.

Yizhaqm, H., Gilad, E., and Meron, E. (2005). Banded vegetation: Biological productivity and resilience. *Physica* A 356:139–144.

14

SYNTHESIS OF ECOSYSTEM ENGINEERING THEORY

William G. Wilson

The work by Meron et al. (Ch 12) demonstrates beautifully the outcomes that can arise from the feedback between an organism and a component of its environment, water. They consider the situation of a plant that affects its environment in three different ways. First, the plant takes up soil moisture to grow, and in water-stressed environments, plant growth is limited by soil moisture. Second, the plant alters the soil surface around itself in such a way as to enhance soil water infiltration. Third, the plant's leaves shade the soil, reducing soil water loss, thereby also enhancing water concentration. Enhancing water infiltration and reducing evaporation are positive feedbacks to plant growth: Growth results in greater infiltration and shading, hence more available water, and thus more growth.

The uptake of water represents the straightforward use of a resource, but it extends outward from the plant and takes place throughout the entire root structure. This uptake profile is in contrast to water infiltration, which takes place close to the plant and diffuses outward. Mathematically, uptake is represented by a kernel G(plant,soil,time), and its value indicates the relative amount of water taken by the plant at its location from a particular soil location. Taking the product of the soil water concentration and the kernel, and then integrating it over all soil locations gives the amount of water taken up by the plant. Similarly, integrating, over all plant locations, the product of the soil water concentration and the kernel gives the amount of water lost from a particular place in the soil. This soil water depletion at a particular location represents an inhibition to plant growth, but water infiltration where a plant sits represents an activation of plant growth. Their situation is far more complicated, incorporating soil topography and species competition, but the general feature of short-range activation and long-range inhibition can lead to interesting spatial results. In particular, stable pattern formation can erupt under a variety of scenarios.

In their chapter of this book, Wilson and Wright (Ch 11) examine the consequences of the environment modification performed by ecosystem engineers on various community measures, thereby considering the passive effects of ecosystem engineering, not the feedbacks involved between the engineer and the environment. The Lotka–Volterra model they use considers a large community of species, each species described by a set of linearized interactions and a species-specific maximum growth rate. Extending previous theoretical work involving such models, they outline general predictions for the implications at the community level due to changes in the "environment" that determine species-level growth rates and interaction strengths. These species-level parameters are taken from distributions parameterized by a mean and standard deviation, and the fundamental assumption of their work is that environmental modifications lead to changes in the means and standard deviations. Community consequences of environmental change then become a multistep question. How does environmental change alter the distributions of species-level growth rates and interaction strengths, as understood by their mean and variance? Then, how do these distributional changes affect important measures of the community?

Their examination of community-level impacts focuses primarily on the realized species richness given a larger pool of potential species. In terms of the species-level distributions of interaction rates, they consider four governing parameters: the mean and variance of the density-independent growth rates, and the mean and variance of the net (intraspecific plus interspecific) interaction strengths a species experiences. What they find is that, given an environmental change that increases these four parameters, only the increase in mean growth rate increases the community's species richness. Increases in the other three distributional parameters lead to a decrease in species richness. One measure of interest to community ecology is how species richness varies with productivity. They outline the assumptions and connections with previous work in their description of the oft-mentioned hump-shaped species-richness curve with increasing fertility.

Cuddington and Hastings (Ch 13) demonstrate the breadth of ecosystem engineering models represented in the literature. Mentioned prominently many times within this book, and deservedly so, the work of Gurney and Lawton (1996) represents a prime example of what can result when an organism both uses and abuses its environment. In essence, their model considers an environment with three states, degraded, virgin, and usable, with a species that turns "virgin" habitat into "usable" habitat, but eventually degrades it, at which point it must then recover back to the virgin state. The feedbacks inherent to the model, coupled with particular functional forms for the various processes, result in very interesting dynamics and useful insights (Wright et al. 2004). In another important example, Cuddington and Hastings (2004) provide an excellent, and in some ways more general, model in which they identify "habitat quality" as a continuous environmental state. A species that can modify this distribution of resource quality alters its own invasive properties.

The models that Cuddington and Hastings (Ch 13) review are important examples and have interesting results. Many of these models emphasize feedbacks between an environmental state variable and the population density of an organism, and though often interesting, such feedbacks are not new to ecologists, neither empiricists nor theorists. These ecosystem engineering models have much in common, conceptually, with other resource–consumer models, the primary difference between any two models likely being the structure of the resource variables or the specific functions assigned to different processes. Few theorists modeling resource–consumer interactions really care whether the resource is biotic or abiotic, and maybe all have, at one time or another, even asserted that their model applies universally, but they do care deeply about the resulting dynamics and ecological conclusions. Even though logistic growth is often immediately taken off the shelf for resource growth, implying a self-activating resource, consideration of an "open system" (allowing resource to appear from nowhere) probably makes the biotic–abiotic distinction meaningless. But the essence of the preceding ecosystem engineering models, and what makes them really theoretically interesting, is the feedback between the two state variables.

When it comes down to theory, models with two state variables representing either predator–prey, resource–consumer, or ecosystem engineer–environment will have many commonalities and similar methods of solution. Indeed, they might all, under some limited conditions, have the identical equations. To a theorist, the words assigned to the variables and parameters become irrelevant, and the common dynamics are the fascination. This argument is not meant to imply that there is nothing new to learn when considering aspects unique to abiotic resources and the coupling between biotic and abiotic parts of the ecosystem, but certainly there is much to be gained by considering the similarities with biotic resources.

Are the empirical descriptions and concepts of ecosystem engineering relevant to theoretical questions? Certainly I've laid out my biases regarding the definition of ecosystem engineering in Chapter 3, doubting the distinctiveness of the concept from plain old environment modification. On the theory side of ecosystem engineering, progress will be made not by parsing out whether or not the engineering is being performed through a physical state change, but from a deeper understanding of the underlying mechanistic processes and the feedbacks between the many possible biotic and abiotic state variables. This is what the science of ecology is all about. If this argument holds, then the theory of ecosystem engineers will continue along ecological theory's traditional path, formulating models that range from those based on specific systems, which add important details to connect with and be falsified by empirical observations, to those based on general ideas, which strip away unnecessary details to address mechanisms that operate across many ecological situations.

REFERENCES

Cuddington, K., and Hastings, A. (2004). Invasive engineers. *Ecological Modelling* 178:335–347.

Gurney, W.S.C., and Lawton, J.H. (1996). The population dynamics of ecosystem engineers. *Oikos* 76:273–283.

Wright, J.P., and Jones, C.G. (2004). Predicting effects of ecosystem engineers on patch-scale species richness from primary productivity. *Ecology* 85:2071–2081.

IV

SOCIO-ECONOMIC ISSUES AND MANAGEMENT SOLUTIONS

The final section of the book focuses on applications and management issues. The world is becoming much more dominated by human activities, and the concept of ecosystem engineers has great potential to aid in what will be increasingly more active management of ecosystems. The chapters in this section take a variety of approaches, ranging from case studies focused on particular species, or particular systems, to broader overviews. What all the chapters have in common is an attempt to apply the concepts covered earlier in this book to questions where humans are truly involved.

15

RESTORING OYSTER REEFS TO RECOVER ECOSYSTEM SERVICES

Jonathan H. Grabowski and Charles H. Peterson

15.1 ⬦ INTRODUCTION

Overharvesting of wild oysters and environmental mismanagement in estuaries around the world have resulted in the loss of fisheries income and collapse of an ecologically important ecosystem engineer and its associated ecosystem goods and services. The decline of the eastern oyster, *Crassostrea virginica* (Gmelin 1791), throughout the mid-Atlantic and southeastern U.S., has reduced landings to 1–2% of the historic peaks approximately a century ago in many estuaries such as the Chesapeake Bay and eastern North Carolina (Frankenberg 1995, Heral et al. 1990, Newell 1988, Rothschild et al. 1994). Declines in the abundance of oysters are a consequence of degradation of oyster reefs via destructive harvesting practices as well as overfishing, oyster disease, sedimentation, and water quality degradation, which collectively have greatly reduced the quantity and quality of intact reef habitat (Frankenberg 1995, Rothschild et al. 1994). Although restoration efforts have proceeded for several decades in estuaries throughout the eastern U.S., these efforts have traditionally focused on reversing the trend of declining landings rather than rebuilding sustainable oyster reefs that create habitat and other ecosystem services (Peterson et al. 2003).

Ecosystem Engineers

TABLE 15.1 Ecosystem services that are provided by oyster reef habitat.

Ecosystem Service	Benefit/Value
1. Production of oysters	(↑ market & recreational value)
2. Water filtration & concentration of pseudofeces	(↓ suspended solids, turbidity, phytoplankton biomass, & microbial production; & ↑ denitrification, submerged aquatic vegetation [SAV], & recreational use)
3. Provision of habitat for epibenthic inverts	(↑ biodiversity & productivity)
4. Carbon sequestration	(↓ greenhouse gas concentrations)
5. Augmented fish production	(↑ market & recreational value)
6. Stabilization of adjacent habitats and shoreline	(↑ SAV & salt marsh habitat; ↓ effects of sea-level rise [SLR])
7. Diversification of the landscape & ecosystem	(↑ synergies among habitats)

Oyster reefs are valued for the wide diversity of ecosystem goods and services that they provide (Table 15.1). Oyster reefs are the only hard substrate in a predominately soft-sediment environment (Lenihan 1999, Lenihan and Peterson 1998). The biogenic structure formed by vertically upright oyster aggregations creates habitat for dense assemblages of mollusks other than oysters, polychaetes, crustaceans, and other resident invertebrates (Bahr and Lanier 1981, Lenihan et al. 2001, Rothschild et al. 1994, Wells 1961). Juvenile fish and mobile crustaceans also recruit to and utilize oyster reefs as refuge and foraging grounds, so that oyster reefs augment the tertiary productivity of estuaries (Breitburg et al. 2000; Coen and Luckenbach 2000; Coen et al. 1999; Grabowski et al. 2005; Harding and Mann 2001, 2003; Luckenbach et al. 2005; Peterson et al. 2003; Rodney and Paynter 2006; Soniat et al. 2004; Tolley and Volety 2005). Removal of filter-feeding oysters from estuaries such as in the Chesapeake Bay and the Pamlico Sound has resulted in trophic restructuring that promotes planktonic and microbial organisms over demersal and benthic flora and fauna (Baird et al. 2004, Dame et al. 1984, Jackson et al. 2001, Newell 1988, Paerl et al. 1998, Ulanowicz and Tuttle 1992). Oysters also promote pelagic fauna by preventing primary production from entering microbial loops and thus allowing it to pass up the food chain to bottom-feeding fishes, crabs, and higher-order predators like red drum, tarpon, and dolphins (Coen et al. 1999, Peterson and Lipcius 2003). By filtering nutrients, sediments, and phytoplankton from the water column, oyster reefs also structured estuarine communities historically and promoted the health of other estuarine habitats such as submerged aquatic vegetation (SAV) by increasing light penetration and minimizing negative effects of eutrophication.

In addition to filtering water and providing critical habitat for transient and resident fish and invertebrates, oyster reefs perform several other important ecosystem services. Oyster reefs attenuate wave energy and reduce erosion of other valuable habitats such as salt marshes and SAV (Henderson and O'Neil 2003, Meyer et al. 1997). Oysters sequester carbon from the water column as they form calcium carbonate shells. As a carbon sink, oyster reefs potentially reduce concentrations of greenhouse gases (Peterson and Lipcius 2003). Oyster reefs also promote denitrification by concentrating deposition of feces and pseudofeces, which potentially enhances watershed management activities aimed at reducing anthropogenic N and promotes greater benthic plant production (Newell 2004, Newell et al. 2002). Finally, oyster reefs are an important component of the estuarine landscape. The location of an oyster reef could influence landscape-scale processes, such as providing a corridor between shelter and foraging grounds (Micheli and Peterson 1999, Peterson et al. 2003).

Previous attempts to assess the monetary value of ecosystem services provided by oyster reefs are limited, which inhibits the ability of managers to evaluate alternative habitat restoration options and make informed choices about how to manage restored oyster reefs. A more holistic understanding of the value of these services will also guide future restoration efforts. Here we discuss how to quantify the economic value of each of the ecosystem services provided by oyster reefs. We also provide quantitative estimates of the value of some specific functions (i.e., oyster harvests, water quality improvements, and recreational and commercial fishery benefits) where data are available in order to compare the value of harvesting oysters in a traditional fishery to the monetary value of providing other oyster reef services.

15.2 — EVALUATING ECOSYSTEM SERVICES PROVIDED BY OYSTER REEFS

OYSTER REEFS AS A FISHED COMMODITY

Consumer demand for oysters continues to promote wild oyster fisheries in the U.S. where populations are still viable. Increased oyster landings in the Gulf of Mexico have partly compensated for lost productivity throughout many historically productive regions such as Delaware Bay, the Chesapeake Bay, Pamlico Sound, and the south-Atlantic coast of the

U.S. Even though the precipitous decline in landings in the Chesapeake during the 1980s was partly buffered by increased oyster prices, the loss in dockside value after adjusting for inflation from 1980 to 2001 is estimated at 93% (National Research Council 2004). Given that catch per unit effort (CPUE) also decreased by 39% during this time period (National Research Council 2004) and fuel costs have increased, the erosion of profits experienced by the industry during the past couple of decades is even greater than 93%. This decline in CPUE largely reflects increased regulations restricting harvesting practices and decreased oyster biomass in the oyster habitat. For instance, Rothschild et al. (1994) divided the total harvest value by the estimated amount of total oyster bottom, which they estimated declined by 50% between 1890 and 1991, and determined that a century of overharvesting reduced the annual oyster yield in Maryland per unit oyster bottom from $550 \, g/m^2$ in 1890 to $22 \, g/m^2$ in 1991.

We calculated the commercial value of oysters per unit of reef area in two ways. First, we multiplied the oyster yields reported in Rothschild et al. (1994) by the dockside market price ($3.01/lb of oysters in 1991) in coastal Maryland. Overharvesting reduced the value of oyster yields from $36.45 per $10 \, m^2$ of oyster bottom in 1890 to $1.46 per $10 \, m^2$ in 1991 (both values are in 1991 dollars). This reduction in value may be slightly underestimated if it is partly a consequence of decreased harvesting effort rather than a decrease in the density of harvestable oysters per unit of reef. However, it is unlikely that this pattern is largely due to reduced effort given that increases in oyster prices over the past two decades would likely motivate greater harvesting effort, and fishery-independent sampling efforts have determined that the density of living oysters in the Chesapeake Bay is two to three orders of magnitude below historic levels (Rothschild et al. 1994).

Second, we calculated the harvest potential using data from oyster reef projects in which oyster reefs were restored in coastal North Carolina. Using data on oyster densities of legally harvestable sizes in the Neuse River Estuary (Lenihan and Grabowski 1998, Lenihan and Peterson 2004), we estimated that subtidal reefs in this region contain 0.6–1.6 bushels of oysters per $10 \, m^2$ worth $12.80–$32.00. The value of oysters on these reefs is roughly comparable to historic yields in Maryland a century ago and approximately an order of magnitude greater than the present oyster fishery production from reefs in Maryland. This difference is largely a consequence of experimental reefs in North Carolina having not been subjected to continual harvest because they were designated as reef sanctuaries. Furthermore, these results suggest that harvesting by traditional methods such as dredging and hand-tonging would likely result in rapidly decreasing oyster yields in subsequent years if harvesting were initiated on these reefs in North Carolina (Lenihan and Peterson 2004).

BEYOND OYSTERS: VALUING ADDITIONAL ECOSYSTEM GOODS AND SERVICES PROVIDED BY OYSTER REEFS

Although the value of oyster landings throughout the eastern U.S. has been recorded since the nineteenth century, economic evaluations of the additional ecosystem services provided by oyster reefs are limited. This paucity of economic data is partly a reflection of viewing oysters narrowly as a fishery resource to exploit rather than holistically as an ecosystem engineer that should be managed as a provider of a multitude of goods and services. Of further concern, a century of overharvesting has left most if not all ecosystems across the coastal U.S. with two to three orders of magnitude fewer oysters (Frankenberg 1995, Heral et al. 1990, Rothschild et al. 1994), and existing oyster reefs may be so degraded that they do not perform the same services as intact historic reefs (Dame et al. 2002, Jackson et al. 2001, Lenihan and Peterson 1998, Newell 1988). However, scientists have utilized oyster reef restoration over the past two to three decades to investigate how oyster reefs function (Coen et al. 1999, Dame et al. 1984, Grabowski et al. 2005, Harding and Mann 1999, Lenihan et al. 2001, Meyer et al. 1996, Newell et al. 2002, Peterson et al. 2003, Zimmerman et al. 1989). This information has assisted managers not only in shifting from an exploitable-resource to a valued-habitat view of oyster reefs but also in enhancing the ability to recover goods and services through oyster reef restoration. Incorporation of ecological data into economic models that integrate the value of each of these services will further managers' capacity to decide among management options and alternative restoration designs in order to maximize the value created by restored reefs.

Oysters as a biofilter

Oysters are filter feeders that feed upon suspended particles in the water column, pumping such a high rate of water flow that they are considered an important biofilter that helps maintain system functioning (Baird and Ulanowicz 1989, Grizzle et al. 2006, Newell 1988). The decline of oyster populations in estuaries along the eastern U.S. has coincided with increased external nutrient loading into these coastal systems (Paerl et al. 1998). Collectively these ecosystem perturbations have increased bottom-water hypoxia and resulted in restructured food webs dominated by phytoplankton, microbes, and pelagic consumers that include many nuisance species rather than benthic communities supporting higher-level consumer species of commercial and recreational value (Breitburg 1992, Jackson et al. 2001, Lenihan and Peterson 1998, Paerl et al. 1998, Ulanowicz and Tuttle 1992).

Perhaps one of the most compelling examples of the consequences of loss of filtration capacity is Newell's (1988) estimate that oyster populations in the Chesapeake Bay in the late 1800s were large enough to filter a volume of water equal to that of the entire Bay every 3.3 days, whereas reduced populations currently in the Bay would take 325 days. Two other examples of the filtration capacity of bivalves include the introductions of the clam (*Potamocorbula amurensis*) in San Francisco Bay and zebra mussels (*Dreissena polymorpha*) in the Great Lakes, which have demonstrated how dramatically suspension feeding by bivalves can remove suspended solids and nutrients from the water column (Alpine and Cloern 1992, Carlton 1999, Klerks et al. 1996, MacIsaac 1996). Although the decline in oyster populations undoubtedly contributed to the decline in water quality in the Chesapeake over the past century, the application of quantified changes in water quality as a consequence of small-scale restoration studies to larger-scale, estuarine-wide management of water quality presents some significant challenges.

Experimental manipulation of oyster populations has demonstrated that oysters can influence water quality by reducing phytoplankton biomass, microbial biomass, nutrient loading, and suspended solids in the water column. Other potential water quality benefits could result by concentrating these materials as pseudofeces in the sediments, stimulating sediment denitrification, and producing microphytobenthos (Dame et al. 1989). For example, Porter et al. (2004) manipulated the presence of oysters in 1000-l tanks and found that oysters increased light penetration through the water column by shifting algal production from phytoplankton to microphytobenthos-dominated communities. Microphytobenthos biomass subsequently reduced nutrient regeneration from the sediments to the water column. Cressman et al. (2003) determined that oysters in North Carolina decreased chlorophyll a levels in the water column by 10–25% and fecal coliform levels by as much as 45% during the summer. Grizzle et al. (2006) developed a method to measure seston *in situ* and subsequently demonstrated that this method more precisely identifies differences in seston than traditional techniques conducted in a laboratory, suggesting that studies relying upon laboratory analyses may underestimate the effects of oysters on seston. Nelson et al. (2004) transplanted oyster beds in small tributaries in coastal North Carolina and noted that some small reefs reduced total suspended solids and chlorophyll a levels. Laboratory studies also have found that bivalves influence local plankton dynamics and reduce turbidity levels (Prins et al. 1995, 1998). On the other hand, Dame et al. (2002) removed oysters from four of eight creeks in South Carolina and noted that the presence of oyster reefs explained little of the variability in chlorophyll a, nitrate, nitrite, ammonium, and phosphorous. In general, bivalve control of phytoplankton biomass is thought to be most effective when bivalve biomass is high and water depth is shallow (Officer et al. 1982), so that small-scale restoration efforts or restorations in deeper water may not necessarily achieve detectable gains in water quality.

Attributing a single value per unit of oyster reef restored may be inappropriate if the relationship between the spatial extent of oyster reef habitat and water quality is nonlinear (Dame et al. 2002). For instance, in some estuaries large-scale restoration efforts may be necessary before water quality is measurably improved because nutrient and suspended solid loading rates currently far surpass the filtration capacity of the oyster populations present in these bays. Future research efforts that provide empirical data on this functional relationship will greatly benefit attempts to model the economic services provided by oyster reefs. Because these inherent difficulties exist in generalizing the economic benefits associated with oyster filtration, one alternative approach would be to quantify the cost of providing a substitute for this service. As a natural biofilter that removes suspended solids and lowers turbidity, oyster reefs are analogous to wastewater treatment facilities. Thus the filtration rate of an individual unit of oyster reef can be quantified and compared to the cost of processing a similar amount of suspended solids and nutrients with a waste treatment facility.

Larger-scale restoration efforts within shallow coastal embayments designed to achieve improvements in water quality could have substantial indirect benefits of great economic value. First, by decreasing water turbidity (i.e., by filtering suspended solids) and suppressing nutrient runoff, oyster reefs can promote the recovery of SAV in polluted estuaries (Peterson and Lipcius 2003). Newell and Koch (2004) modeled the effects of oyster populations on turbidity levels and found that oysters, even at relatively low biomass levels (i.e., 25 g dry tissue weight m^{-2}), were capable of reducing suspended sediment concentrations locally by nearly an order of magnitude. This reduction would result in increased water clarity that would potentially have profound effects on the extent of SAV in estuaries such as the Chesapeake Bay.

Recognized as extremely important nursery grounds for many coastal fish species (Thayer et al. 1978), vegetated habitats such as SAV have been reduced in estuaries such as the Chesapeake Bay by agricultural runoff, soil erosion, metropolitan sewage effluent, and resultant N loading from all of these sources as well as atmospheric deposition. Nitrogen loading at levels of 30 kg N ha^{-1} yr^{-1} within Waquoit Bay, Massachusetts, resulted in the loss of 80 to 96% of the total extent of seagrass beds, and seagrass beds were completely absent in embayments with loading rates that doubled this amount (Hauxwell et al. 2003). They also found that nitrogen loading increased growth rates and standing stocks of phytoplankton, which likely caused severe light limitation to SAV by reducing light penetration through the water. Kahn and Kemp (1985) created a bioeconomic model to estimate damage functions for commercial and recreational fisheries associated with the loss of SAV in the Chesapeake Bay and determined that a 20% reduction in total SAV in the Bay results in a loss of 1–4 million dollars annually in fishery value. If improvements in water quality from oyster reef habitat increase the amount of SAV in the estuary, then the value of augmented fishery resources created by this additional SAV should be attributed to oyster reefs.

Second, improvements in water quality in general are valued by the general public who use estuarine habitats for activities such as swimming, boating, and sportsfishing. For instance, Bockstael et al. (1988, 1989) surveyed residents in the Baltimore–Washington area in 1984 and determined that their annual aggregate willingness to pay in increased taxes for moderate (i.e., ~20%) improvements in water quality (i.e., decreased nitrogen and phosphorous loading and increased sportsfishing catches) was over $100 million. The National Research Council (2004) used the consumer price index to adjust estimates reported in the preceding studies to 2002 price levels and reported that a 20% improvement in water quality along the western shore of Maryland relative to conditions in 1980 is worth $188 million for shore beach users, $26 million for recreational boaters, and $8 million for striped bass sportsfishermen. Although there are several potential sources of error in these estimates, they may be underestimated given that improvements in the Chesapeake Bay water quality will likely result in increased recreation in the Bay and these analyses did not include the value that U.S. residents outside of the Baltimore–Washington area place on Bay resources despite the nation-wide recognition and utilization of the Chesapeake Bay (Bockstael et al. 1988). Evaluation of ecosystem services provided by oyster reefs should include assessment of this suite of benefits if larger-scale oyster restoration efforts achieve measurable improvements in water quality in estuaries along the Atlantic and Gulf coasts of the U.S.

Oysters as habitat for fish

Several studies have used restoration efforts to assess the role of oyster reefs as critical habitat for commercially and recreationally important fish species (Coen et al. 1999, Grabowski et al. 2005, Harding and Mann 1999, Lenihan et al. 2001, Meyer et al. 1996, Peterson et al. 2003, Zimmerman et al. 1989). Our ability to quantify the value of fish provided per unit area of oyster reef will be dependent upon several factors, such as (1) whether oysters successfully and regularly recruit to the reef and create vertical relief that provides habitat for important prey species; (2) the amount of existing oyster reef habitat already available locally; (3) whether other habitats are functionally redundant to oyster reef habitat and consequently compensate for oyster reef degradation; (4) an oyster reef's location (or landscape setting) within the network of oyster reefs and other important estuarine habitats that already exist; and (5) the biogeographic region (or ecosystem) where it is located.

Given the context dependency of oyster reef community processes, assessments of economic benefits for commercial and recreational fisheries must incorporate knowledge of the life history and ecology of local fish species. For instance, while water quality improvements in the Chesapeake Bay would generate value for striped bass fishermen if catches increased, improvements in water quality in the estuaries of the southeast would not provide this particular value because striped bass do not extend south of Cape Lookout, North Carolina. Peterson et al. (2003) reviewed existing data on oyster reef restoration efforts from the southeast U.S. and determined that each $10 \, m^2$ plot of restored oyster reef habitat produces an additional $2.6 \, Kg \, yr^{-1}$ of production of fish and large mobile crustaceans for the functional lifetime of the reef. Because these efforts were focused on the southeast U.S., species that utilize oyster reef habitat located in other estuaries in the U.S. such as striped bass but are not indigenous to this region were excluded from the analyses. On the other hand, those species utilizing an oyster reef in the southeast U.S. clearly include ecological equivalents for species restricted to other biogeographic regions, so the degree of enhancement of fish production may be similar.

Using 2001–2004 dockside landing values from the southeastern U.S. and Gulf of Mexico (National Marine Fisheries Service 2006), we converted the amount of augmented production per each of the 13 species groups that were augmented by oyster reef habitat in Peterson et al. (2003) to a commercial fish landing value (Table 15.2). We then calculated the streamline of cumulative benefits provided by a $10 \, m^2$ oyster reef for the functional lifetime of the reef (Figure 15.1). Future landings values were discounted at a rate of 3% to adjust for the opportunity cost of capital adjusted for inflation. Our estimates suggest that a $10 \, m^2$ reef that lasts 50 years would produce finfish valued at $98.06 in 2004 dollars, whereas harvesting this same reef for oysters destructively after 5 years would reduce this finfish value to $17.45 in 2004 dollars. Although a

TABLE 15.2 The commercial fisheries value of augmented fish created by oyster reef restoration in the southeast U.S.

	Augmented Fish Production[a] (Kg/10 m²)	Commercial Fish Price[b] ($/Kg)	Augmented Fish Value[c] ($/yr/10 m²)
Sheepshead Minnow	0.000	$ —	$ —
Bay Anchovy	0.019	$ —	$ —
Silversides (3 spp.)	0.002	$ —	$ —
Gobies	0.644	$ —	$ —
Blennies	0.050	$ —	$ —
Sheepshead	0.586	$ 1.17	$ 0.69
Stone Crab[d]	0.653	$ 6.75	$ 0.88
Gray Snapper	0.114	$ 3.43	$ 0.39
Toadfish	0.022	$ 4.95	$ 0.11
Gag Grouper	0.293	$ 4.82	$ 1.41
Black Sea Bass	0.046	$ 2.90	$ 0.13
Spottail Pinfish	0.005	$ 1.14	$ 0.01
Pigfish	0.135	$ 0.60	$ 0.08
		Total ($/yr/10 m²):	**$3.70**

[a] Estimates of annual augmented fish and crustacean biomass produced per 10 m² of restored oyster reef are from Peterson et al. (2003).
[b] Individual fish landing prices were derived from the National Marine Fisheries Service commercial landings online database (National Marine Fisheries Service 2006).
[c] Augmented fish values were calculated by multiplying augmented fish production values by the commercial fish price for each species group.
[d] Because Peterson et al. (2003) estimated the total biomass of stone crabs but commercial landings price is derived from only the weight of the claws, we divided our fish value estimate by 5 (i.e., we estimated that the claws account for 20% of the total weight of stone crabs).

50-year life span may seem extremely long given that oyster diseases have hampered many recent restoration efforts, reefs historically persisted for centuries prior to mechanical harvesting began and some recent reef sanctuaries in North Carolina that are currently intact were constructed over 2 decades ago (Powers et al. unpublished data). This difference illustrates that how a restored reef is managed (i.e., whether destructive harvesting of oysters is permitted or if the reef is protected) will largely influence the streamline of ecosystem goods and services that it provides.

Comparing the value of augmented fish production with oyster harvests revealed that consideration of ecosystem services provided by oyster reefs more broadly could enhance the value derived from oyster reef habitat. Specifically, the value of oyster harvests from 10 m² of reef

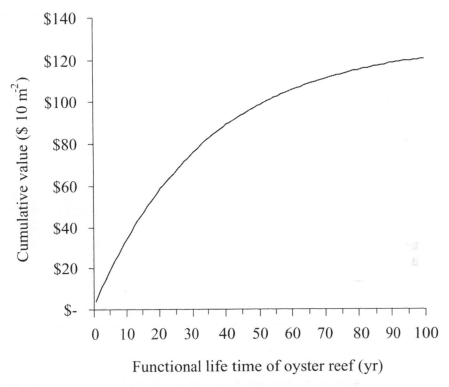

FIGURE 15.1 The long-term projection of cumulative value of enhanced fish and mobile crab production per 10 m² of restored oyster reef habitat for various hypothetical alternative lifetimes of functionality of a restored reef. Values were discounted at an annual rate of 3% to adjust for the opportunity cost of capital. The functional lifetime of the reef is influenced by management of the reef (i.e., whether it is set aside as a sanctuary or destructively harvested), oyster recruitment levels, and the incidence of oyster diseases that can decimate living oyster populations prior to establishing the vertical relief and structural complexity associated with intact reef habitat.

habitat using our preceding estimates (in the section "Oyster Reefs as a Fished Commodity") of $36.45 in year 5 and $1.45 in each subsequent year discounted at a rate of 3% totals to $63.97 for a reef with a 50-year life span. This estimate is 34.8% less than the value of fish produced by a similar amount of reef habitat during this time span. Scaling the estimate of finfish value up to a 1-acre reef sanctuary that lasts 50 years would result in ~$40,000 in additional value from commercial finfish and crustacean fisheries. If this value can be extrapolated to entire estuaries, the total value of augmented fish production from oyster reef habitat far surpasses the economic value of oyster landings over the past decade in many estuaries throughout the eastern U.S. This estimate undoubtedly is subject to potential sources of error due to the unpredictability of future fishery resources. For instance, the provision of ecosystem goods and services by oyster reefs may be dependent upon the amount of oyster reef habitat already in the system, such that the marginal value of each unit of restored oyster reef may vary as more and more reefs are restored in the system. In particular, restoration of extensive amounts of reef habitat within some estuaries may result in reef-related species that are limited by factors other than habitat availability. However, this initial estimate may be conservative because the abundances of many fish and crustacean species that utilize oyster reef habitat also have been dramatically reduced from decades of overfishing, so that recent studies investigating fish use of restored reef habitat likely underestimated the potential abundance of these species. Enhancing our understanding of these processes to regional scales will be especially important if restoration efforts ever begin to approach historical levels of intact oyster bottom given that the amount of shell bottom in 1884 in just the Maryland portion of the Chesapeake Bay was estimated at 279,000 acres (Rothschild et al. 1994).

Other ecosystem services provided by oysters

Oysters create biogenic structure by growing in vertically upright clusters that provide habitat for a wide diversity of densely aggregated invertebrates (Bahr and Lanier 1981, Lenihan et al. 2001, Rothschild et al. 1994, Wells 1961). Although few of these species (i.e., mollusks other than oysters, polychaetes, crustaceans, and other resident invertebrates) are of commercial or recreational value to fishermen, they are consumed by many valuable finfish and crustacean species and thus indirectly benefit fisheries (Grabowski et al. 2005, Peterson et al. 2003). Given that we have already calculated the benefit of oyster reef habitat to fish in the previous section, we did not ascribe additional value to these benthic invertebrates that reside on oyster reefs. However, evaluations of estuarine biodiversity and its maintenance should include consideration of oyster reef habitats given that they can contain one to two orders of magnitude more macro-invertebrates than adjacent mud bottom (Grabowski et al. 2005).

Oyster reefs attenuate wave energy and stabilize other estuarine habitats such as salt marshes (Meyer et al. 1997). Oyster reefs also promote sedimentation, which potentially benefits the establishment of SAV (Henderson and O'Neil 2003). Oyster reefs are a living breakwater that can and will rise at rates far in excess of any predicted sea-level rise rate and thus help stabilize shoreline erosion and habitat loss, which are otherwise predicted to be dramatic in many coastal estuaries if left unprotected by natural buffers (Reed 1995, 2002; Zedler 2004). Although there are currently insufficient data with which to quantify the generality of these processes and assess their economic value, these services are indicative of the integrated mosaic of estuarine habitats. Thus a more complete evaluation of the ecosystem services provided by oyster reef habitat will require not only a greater understanding of its role in influencing coastal geology within the estuarine landscape but also evaluations of the services provided by the habitats oyster reefs promote.

Investigation of how the landscape setting of restored oyster reefs influences these ecological processes will be pivotal to future assessments. For instance, an oyster reef located in between a salt marsh and SAV may be an important corridor for predators moving among habitats (Micheli and Peterson 1999, Peterson et al. 2003). Conversely, some ecosystem services provided by oyster reefs in these vegetated landscapes may be redundant. For instance, Grabowski et al. (2005) found that restored oyster reefs in vegetated landscapes do not affect juvenile fish abundances, whereas oyster reefs restored on mudflats isolated from SAV and salt marshes augment juvenile fish abundances and potentially increase fish productivity within estuaries. The landscape setting of an oyster reef will also influence other processes such as oyster recruitment and survivorship (Grabowski et al. 2005), which in turn could affect filtration rates and subsequent removal of seston from the water column. Ecological studies that quantify landscape and ecosystem-scale variation in these processes will enhance our ability to model spatial variability in the value of services provided by oyster reefs.

15.3 ❧ CHALLENGES AND CONCLUSIONS

Evaluation of the ecosystem services provided by oyster reefs could assist coastal managers in readjusting management schemes to maximize the benefits of restoration efforts and consequently shift to an ecosystem approach to fisheries management. For instance, comparison of oyster harvest values with other services reveals the importance of evaluating ecosystem services rather than continuing to exploit oyster reefs for the oyster harvest value. Although the value of oyster harvests may initially measure up to other benefits such as the value of augmented fish production to the commercial fishery, the consequences of destructive oyster sampling either require continual restoration efforts or would result in oyster harvest levels similar to severely degraded estuaries along the eastern U.S. Whether oyster reefs can sustain less destructive oyster harvesting techniques (i.e., how quickly oysters grow to replace losses by harvest) such as diver collection of oysters by hand is unclear and merits further investigation.

Given that the value of augmented commercial fish landings surpasses oyster harvest values, the entire suite of ecosystem services that are sustained by intact reefs probably greatly exceeds the value currently derived from oyster harvests. Oyster restoration efforts at larger scales that enhance water quality potentially result in even larger benefits such as increased recreational use, heightened willingness to consume seafood, and reduced need for construction of wastewater purification systems. Water quality improvements from oyster restoration efforts and their economic values are more difficult to quantify; however, current estimates of the willingness of boaters, beach users, and recreational fishermen from the Chesapeake Bay to pay for local improvements in the Bay's water quality suggest that oyster restoration efforts capable of achieving significant gains in water quality will result in economic returns derived from these changes that far exceed the value of current oyster landings.

Lack of quantitative information on several of the other ecosystem services hinders a more complete evaluation of the suite of benefits provided by oyster reefs and, subsequently, hampers the ability of regulators to implement a more ecosystem-based approach to managing coastal resources. Insufficient data currently exist to fully evaluate local (landscape-scale) and regional variability in ecosystem services. This information would allow coastal managers to determine which reefs provide disproportionately valuable service and should be conserved as sanctuaries. It would also help managers identify those that are less valuable and could be harvested without as much concern for the ecosystem consequences. Placing oyster reefs in the greater context of the estuary requires landscape-scale data with simultaneous evaluation of each habitat across multiple trophic levels, which is difficult to obtain. However, larger-scale restoration efforts to assess the recovery of ecosystem services are currently being conducted in the Gulf of Mexico and in several estuaries along the East Coast of the United States. These studies will greatly enhance our ability to develop more holistic economic models that account for spatial variability in the provision of ecosystem goods and services by oyster reefs.

REFERENCES

Alpine, A.E., and Cloern, J.E. (1992). Trophic interactions and direct physical effects control phytoplankton biomass and production in an estuary. *Limnology and Oceanography* 37:946–955.

Bahr, L.M., and Lanier, W.P. (1981). *The Ecology of Intertidal Oyster Reefs of the South Atlantic Coast: A Community Profile.* Washington, D.C.: U.S. Fish and Wildlife Service.

Baird, D., Christian, R.R., Peterson, C.H., and Johnson, G.A. (2004). Consequences of hypoxia on estuarine ecosystem function: Energy diversion from consumers to microbes. *Ecological Applications* 14:805–822.

Baird, D., and Ulanowicz, R.E. (1989). The seasonal dynamics of the Chesapeake Bay ecosystem. *Ecological Monographs* 59:329–364.

Bockstael, N.E., McConnell, K.E., and Strand, I.E. (1988). Benefits from improvement in Chesapeake Bay water quality. In *Benefit Analysis Using Indirect or Imputed Market Methods,* Vol. III, U.S. EPA-811043-01-0. Washington, D.C.: U.S. Environmental Protection Agency.

——. (1989) Measuring the benefits of improvements in water quality: The Chesapeake Bay. *Marine Resource Economics* 6:1–18.

Breitburg, D.L. (1992). Episodic hypoxia in Chesapeake Bay: Interacting effects of recruitment, behavior, and physical disturbance. *Ecological Monographs* 62:525–546.

Breitburg, D.L., Coen, L.D., Luckenbach, M.W., Mann, R., Posey, M., and Wesson, J.A. (2000). Oyster reef restoration: Convergence of harvest and conservation strategies. *Journal of Shellfish Research* 19:371–377.

Carlton, J.T. (1999). Molluscan invasions in marine and estuarine communities. *Malacologia* 41:439–454.

Coen, L.D., and Luckenbach, M.W. (2000). Developing success criteria and goals for evaluating oyster reef restoration: Ecological function or resource exploitation? *Ecological Engineering* 15:323–343.

Coen, L.D., Luckenbach, M.W., and Breitburg, D.L. (1999). The role of oyster reefs as essential fish habitat: A review of current knowledge and some new perspectives. In *Fish Habitat: Essential Fish Habitat and Rehabilitation*, L.R. Benaka, Ed. Bethesda, MD: American Fisheries Society Symposium 22.

Cressman, K.A., Posey, M.H., Mallin, M.A., Leonard, L.A., and Alphin, T.D. (2003). Effects of oyster reefs on water quality in a tidal creek estuary. *Journal of Shellfish Research* 22:753–762.

Dame, R., Bushek, D., Allen, D., Lewitus, A., Edwards, D., Koepfler, E., and Gregory, L. (2002). Ecosystem response to bivalve density reduction: Management implications. *Aquatic Ecology* 36:51–65.

Dame, R.F., Spurrier, J.D., and Wolaver, T.G. (1989). Carbon, nitrogen and phosphorous processing by an oyster reef. *Marine Ecology Progress Series* 54:249–256.

Dame, R.F., Zingmark, R.G., and Haskins, E. (1984). Oyster reefs as processors of estuarine materials. *Journal of Experimental Marine Biology and Ecology* 83:239–247.

Frankenberg, D. (1995). *Report of North Carolina Blue Ribbon Advisory Council on Oysters.* Raleigh, NC: North Carolina Department of Environment, Health, and Natural Resources.

Grabowski, J.H., Hughes, A.R., Kimbro, D.L., and Dolan, M.A. (2005). How habitat setting influences restored oyster reef communities. *Ecology* 86:1926–1935.

Grizzle, R.E., Greene, J.K., Luckenbach, M.W., and Coen, L.D. (2006). A new *in situ* method for measuring seston uptake by suspension-feeding bivalve molluscs. *Journal of Shellfish Research* 25:643–649.

Harding, J.M., and Mann, R. (1999). Fish species richness in relation to restored oyster reefs, Piankatank River, Virginia. *Bulletin of Marine Science* 65:289–300.

——. (2001). Oyster reefs as fish habitat: Opportunistic use of restored reefs by transient fishes. *Journal of Shellfish Research* 20:951–959.

——. (2003). Influence of habitat on diet and distribution of striped bass (*Morone saxatilis*) in a temperate estuary. *Bulletin of Marine Science* 72:841–851.

Hauxwell, J., Cebrian, J., and Valiela, I. (2003). Eelgrass *Zostera marina* loss in temperate estuaries: Relationship to land-derived nitrogen loads and effect of light limitation imposed by algae. *Marine Ecology Progress Series* 247:59–73.

Henderson, J., and O'Neil, L.J. (2003). *Economic Values Associated with Construction of Oyster Reefs by the Corps of Engineers*, EMRRP Technical Notes Collection (ERDC TN-EMRRP-ER-01). Vicksburg, MS: U.S. Army Engineer Research and Development Center. Accessed at http://www.wes.army.mil/el/emrrp.

Heral, M., Rothschild, B.J., and Goulletquer, P. (1990). *Decline of Oyster Production in the Maryland Portion of the Chesapeake Bay: Causes and Perspectives*, CM-1990-K20. Copenhagen, Denmark: International Council for the Exploration of the Sea, Shellfish Committee.

Jackson, J.B.C., Kirby, M.X., Berger, W.H., Bjorndal, K.A., Botsford, L.W., Bourque, B.J., Bradbury, R.H., Cooke, R., Erlandson, J., Estes, J.A., Hughes, T.P., Kidwell, S., Lange, C.B., Lenihan, H.S., Pandolfi, J.M., Peterson, C.H., Steneck, R.S., Tegner, M.J., and Warner, R.R. (2001). Historical overfishing and the recent collapse of coastal ecosystems. *Science* 293:629–638.

Kahn, J.R., and Kemp, W.M. (1985). Economic losses associated with the degradation of an ecosystem: The case of submerged aquatic vegetation in Chesapeake Bay. *Journal of Environmental Economics and Management* 12:246–263.

Klerks, P.L., Fraleigh, P.C., and Lawniczak, J.E. (1996). Effects of zebra mussels (*Dreissena polymorpha*) on seston levels and sediment deposition in western Lake Erie. *Canadian Journal of Fisheries and Aquatic Sciences* 53:2284–2291.

Lenihan, H.S. (1999). Physical-biological coupling on oyster reefs: How habitat structure influences individual performance. *Ecological Monographs* 69:251–275.

Lenihan, H.S., and Grabowski, J.H., and Thayer, G.W. (1998). *Recruitment to and Utilization of Restored Oyster Reef Habitat by Economically Valuable Fishes and Crabs in North Carolina: An Experimental Approach with Economic Analyses*. Beaufort, NC: National Research Council, Final Report No. 1–97. National Marine Fisheries Service.

Lenihan, H.S., and Peterson, C.H. (1998). How habitat degradation through fishery disturbance enhances impacts of hypoxia on oyster reefs. *Ecological Applications* 8:128–140.

——. (2004). Conserving oyster reef habitat by switching from dredging and tonging to diver-harvesting. *Fishery Bulletin* 102:298–305.

Lenihan, H.S., Peterson, C.H., Byers, J.E., Grabowski, J.H., Thayer, G.W., and Colby, D.R. (2001). Cascading of habitat degradation: Oyster reefs invaded by refugee fishes escaping stress. *Ecological Applications* 11:764–782.

Luckenbach, M.W., Coen, L.D., Ross, P.G., Jr., and Stephen, J.A. (2005). Oyster reef habitat restoration: Relationships between oyster abundance and community development based on two studies in Virginia and South Carolina. *Journal of Coastal Research* 40:64–78.

MacIsaac, H.J. (1996). Potential abiotic and biotic impacts of zebra mussels on the inland waters of North America. *American Zoologist* 36:287–299.

Meyer, D.L., Townsend, E.C., and Murphy, P.L. (1996). *The Evaluation of Restored Wetlands and Enhancement Methods for Existing Restorations*. Final Report. Silver Spring, MD: NOAA.

Meyer, D.L., Townsend, E.C., and Thayer, G.W. (1997). Stabilization and erosion control value of oyster clutch for intertidal marsh. *Restoration Ecology* 5:93–99.

Micheli, F., and Peterson, C.H. (1999). Estuarine vegetated habitats as corridors for predator movements. *Conservation Biology* 13:869–881.

National Marine Fisheries Service. (2006). Commercial fisheries landings. Accessed at http://www.st.nmfs.gov/st1/commercial/landings/annual_landings.html.

National Research Council. (2004). *Nonnative Oysters in the Chesapeake Bay: Committee on Nonnative Oysters in the Chesapeake Bay, Ocean Studies Board, Division on Earth and Life Studies*. Washington, D.C.: The National Academies Press.

Nelson, K.A., Leonard, L.A., Posey, M.H., Alphin, T.D., and Mallin, M.A. (2004). Using transplanted oyster (*Crassostrea virginica*) beds to improve water quality in small tidal creeks: A pilot study. *Journal of Experimental Marine Biology and Ecology* 298:347–368.

Newell, R.I.E. (2004). Ecosystem influences of natural and cultivated populations of suspension-feeding bivalve molluscs: A review. *Journal of Shellfish Research* 23:51–61.

Newell, R.I.E. (1988). Ecological changes in Chesapeake Bay: Are they the result of overharvesting the American oyster, *Crassostrea virginica*? In *Understanding the Estuary: Advances in Chesapeake Bay Research*, Publication 129, M.P. Lynch and E.C. Krome, Eds. Baltimore, MD: Chesapeake Research Consortium.

Newell, R.I.E., Cornwell, J.C., and Owens, M.S. (2002). Influence of simulated bivalve biodeposition and microphytobenthos on sediment nitrogen dynamics: A laboratory study. *Limnology and Oceanography* 47:1367–1379.

Newell, R.I.E., and Koch, E.W. (2004). Modeling seagrass density and distribution in response to changes in turbidity stemming from bivalve filtration and seagrass sediment stabilization. *Estuaries* 27:793–806.

Officer, C.B., Smayda, T.J., and Mann, R. (1982). Benthic filter feeding: A natural eutrophication control. *Marine Ecology Progress Series* 9:203–210.

Paerl, H.W., Pinckney, J.L., Fear, J.M., and Peierls, B.L. (1998). Ecosystem responses to internal and watershed organic matter loading: Consequences for hypoxia in the eutrophying Neuse River Estuary, North Carolina, USA. *Marine Ecology Progress Series* 166:17–25.

Peterson, C.H., Grabowski, J.H., and Powers, S.P. (2003). Estimated enhancement of fish production resulting from restoring oyster reef habitat: Quantitative valuation. *Marine Ecology Progress Series* 264:249–264.

Peterson, C.H., and Lipcius, R.N. (2003). Conceptual progress towards predicting quantitative ecosystem benefits of ecological restorations. *Marine Ecology Progress Series* 264:297–307.

Porter, E.T., Cornwell, J.C., and Sanford, L.P. (2004). Effect of oysters *Crassostrea virginica* and bottom shear velocity on benthic-pelagic coupling and estuarine water quality. *Marine Ecology Progress Series* 271:61–75.

Prins, T.C., Escaravage, V., Smaal, A.C., and Peeters, J.C.H. (1995). Functional and structural changes in the pelagic system induced by bivalve grazing in marine mesocosms. *Water Science and Technology* 32:183–185.

Prins, T.C., Smaal, A.C., and Dame, R.F. (1998). A review of the feedbacks between bivalve grazing and ecosystem processes. *Aquatic Ecology* 31:349–359.

Reed, D.J. (1995). The response of coastal marshes to sea-level rise: Survival or submergence? *Earth Surface Processes and Landforms* 20:39–48.

———. (2002). Sea-level rise and coastal marsh sustainability: Geological and ecological factors in the Mississippi delta plain. *Geomorphology* 48:233–243.

Rodney, W.S., and Paynter, K.T. (2006). Comparisons of macrofaunal assemblages on restored and non-restored oyster reefs in mesohaline regions of Chesapeake Bay in Maryland. *Journal of Experimental Marine Biology and Ecology* 335:39–51.

Rothschild, B.J., Ault, J.S., Goulletquer, P., and Heral, M. (1994). Decline of the Chesapeake Bay oyster population: A century of habitat destruction and overfishing. *Marine Ecology Progress Series* 111:29–39.

Soniat, T.M., Finelli, C.M., and Ruiz, J.T. (2004). Vertical structure and predator refuge mediate oyster reef development and community dynamics. *Journal of Experimental Marine Biology and Ecology* 310:163–182.

Thayer, G.W., Stuart, H.F., Kenworthy, W.J., Ustach, J.F., and Hall, A.B. (1978). *Habitat Values of Salt Marshes, Mangroves, and Seagrasses for Aquatic Organisms.* In *Wetland Functions and Values: The State of Our Understanding,* P.E. Greeson, J.R. Clark, and J.E. Clark, Eds. Minneapolis, MN: American Water Resource Association.

Tolley, S.G., and Volety, A.K. (2005). The role of oysters in habitat use of oyster reefs by resident fishes and decapod crustaceans. *Journal of Shellfish Research* 24:1007–1012.

Ulanowicz, R.E., and Tuttle, J.H. (1992). The trophic consequences of oyster stock rehabilitation in Chesapeake Bay. *Estuaries* 15:298–306.

Wells, H.W. (1961). The fauna of oyster beds, with special reference to the salinity factor. *Ecological Monographs* 31:239–266.

Zedler, J.B. (2004). Compensating for wetland losses in the United States. *Ibis* 146(S1):92–100.

Zimmerman, R.J., Minello, T.J., Baumer, T., and Castiglione, M. (1989). *Oyster Reef as Habitat for Estuarine Macrofauna.* NOAA Tech. Memm. NMFS-SEFC-249, Galveston, TX: NOAA.

16

MANAGING INVASIVE ECOSYSTEM ENGINEERS: THE CASE OF *SPARTINA* IN PACIFIC ESTUARIES

John G. Lambrinos

16.1 ● INVASIVE ENGINEERS CAUSE UNIQUE PROBLEMS

Growing concern over the ecological and economic impact of invasive plants has spawned costly efforts to control or eradicate them. In the U. S. alone at least 9.6 billion dollars are spent annually on invasive plant management (Pimentel et al. 2005). A basic assumption of these programs is that reducing invader density will result in proportional reductions in invader impact that will in turn trigger recovery to pre-invasion conditions. Control programs, however, often fail to achieve this goal (Holmes and Richardson 1999, Cummings et al. 2005). One reason is that control can spawn complex community and ecosystem dynamics that are not the simple reversal of invasion processes (Blossey 1999, Zavaleta et al. 2001). Predicting these complex dynamics is made even more difficult by the fact that we usually have a poor understanding of the mechanisms that drive invader impact in the first place (Levine et al. 2003).

Ecosystem engineering is one important mechanism of invader impact that can create unique difficulties for management. Many of the most pernicious invaders are strong ecosystem engineers that alter the abundance and distribution of physical habitat in invaded landscapes (Crooks 2002, D'Antonio and Meyerson 2002). These physical state changes can persist in the invaded landscape long after the invasive engineer that created them has been eliminated. System recovery in these cases depends not only on the biotic mechanisms governing community interactions and assembly, but also on feedbacks with the engineered habitat (Suding et al. 2004). This is a fundamentally different scenario than recovery from invaders that exert their impact only through direct biotic interactions such as predation or competition.

In this chapter I explore the particular challenges that ecosystem engineers pose for invasive plant management. I focus on the case history of the invasion of Willapa Bay, Washington, by Atlantic smooth cordgrass (*Spartina alterniflora*). This is an exemplary case of a strong invasive ecosystem engineer that is currently the focus of intensive management efforts. Throughout the chapter I also draw on the wider literature where appropriate to illustrate the generality of the issues surrounding invasive ecosystem engineers.

16.2 ➡ *SPARTINA* INVASION IN WILLAPA BAY

Several large perennial grass species in the genus *Spartina* are currently invading estuaries throughout the Pacific. These include *S. alterniflora*, *S. anglica*, *S. patens*, *S. densiflora*, and hybrids between *S. alterniflora* and the California endemic *S. foliosa* (Daehler and Strong 1999, Ayres et al. 2004). The *Spartina* invasion process is typified by the invasion history of Willapa Bay, Washington (46°40'N, 124°02'W). Willapa Bay is a large tidal estuary formed by the alluvial deposits of the Columbia River. Like other northern Pacific estuaries, its mid and low tidal elevations have historically lacked native vascular plants. In the late nineteenth century, however, a nascent commercial oyster industry probably accidentally introduced *S. alterniflora* in shipments of seed oysters from the East Coast (Civille et al. 2005). Since the 1800s *S. alterniflora* populations have spread throughout the bay's tideflats primarily as tidally borne seed. Once established, individual clones grow vegetatively producing large monocultural stands that can reach heights of 1.5 m (Davis et al. 2004). At the peak of the infestation in 2002 populations occupied about 75 km² of the bay's 230 km² of tidal flats (Murphy 2003). This caused considerable concern among local residents and land management officials. The invaded tideflat is critical habitat for a number of ecologically and economically important species including migrating shorebirds, juvenile salmonids, and commercial oysters.

16.3 ⬤ DIFFICULTIES PREDICTING SPREAD

Accurately predicting the establishment and spread of invasive populations can greatly improve the efficacy of screening programs and allow for the optimal allocation of scarce control resources (Higgins and Richardson 1996, Higgins et al. 2000). Habitat modification by ecosystem engineers complicates predictions about spread, however. Climate matching and ecological niche models are important tools for evaluating the potential of species to establish and spread in a new range (Krticos et al. 2003, Peterson 2003). These models assume that environmental conditions set the constraints on species distributions, and that these vary independently from population dynamics. The hallmark of ecosystem engineers, however, is that their population dynamics change the abiotic environment. The engineering process can itself feed back on population dynamics or patterns of spread. One way this can happen is if engineering ameliorates conditions in unfavorable habitat patches. In a spatially implicit model Cuddington and Hastings (2004) showed that habitat-modifying invaders can have significantly faster population growth rates and ultimately higher population density in suboptimal habitats than invaders that do not modify their environment. Although no modeling studies have demonstrated this, it seems plausible that engineers could also slow their own expansion rates if they created unfavorable conditions. More generally, ecosystem engineering can alter habitat heterogeneity across a range of spatial and temporal scales, and this can greatly influence invasive spread (Hastings et al. 2007, Melbourne et al. 2007). Feedbacks such as this could lower the predictive ability of static climate matching or niche models parameterized with data from the native range. However, no studies have yet evaluated whether these models perform more poorly when predicting invasion patterns for engineers compared to non-engineers.

To be useful for specific management questions, predictive models that incorporate engineering feedbacks will need to be spatially explicit. This is because although engineering processes are often easily generalized, the consequences of engineering for recruitment or spread dynamics can be highly contingent on spatially varying traits such as environmental gradients (see Crain and Bertness 2006) or patterns of propagule supply. The *S. alterniflora* invasion in Willapa Bay provides a good example. In its native range, *S. alterniflora* stands reduce flow-related physical stress, facilitating the establishment of a suite of plant species in the cobble beach habitat immediately behind stands (Bruno and Kennedy 2000). In the invasion context of Willapa Bay, however, the majority of conspecific recruitment occurs on the open tideflat in front of established beds where there is little or no engineering influence. This recruitment pattern partly reflects the fact that seedling recruitment is strongly inhibited within and immediately adjacent to high-density stands by intense light competition in these areas (Lambrinos and Bando in press).

In low-density situations, where isolated individuals have established by long-distance dispersal, the reduced intraspecific competition could allow facilitation to operate. Here the leeward side of clones could provide a favorable recruitment environment similar to that seen in New England cobble beaches. Isolated *Spartina* individuals in Willapa Bay, however, are strongly pollen limited (Davis et al. 2004a, 2004b). This pollen limitation is overcome only as clones grow vegetatively and reach an adult density that also exerts a strong competitive inhibition on seedling survival. Moreover, although seeds can potentially disperse long distances on the tide, most seeds are retained within established beds (Figure 16.1). The same physical mechanisms that reduce hydrological flow and increase sedimentation within beds likely also increase seed deposition and retention. As a consequence, established *Spartina* beds may actually be slowing expansion rates in Willapa Bay by trapping potential propagules in an unfavorable recruitment environment.

Interestingly, in San Francisco Bay some hybrid genotypes exhibit high rates of self-fertilization. In this case, isolated individuals of these genotypes can produce abundant local seed shadows as well as localized habitat modification that could facilitate seedling establishment. Indeed, the lee sides of established clones appear to provide favorable microsites for recruitment in San Francisco Bay (Sloop and Ayres unpublished data). This suggests that species-specific life history traits can interact with the spatial patterning of engineering to drive spread patterns.

16.4 ⬤ INVASION IMPACT MECHANISMS

The high spatial and temporal variability of natural systems often makes it difficult to identify appropriate restoration targets for sites degraded by invasive species or other perturbations (Landres et al. 1999, Allen et al. 2002). As a consequence, there has been a growing emphasis placed on returning function to restoration sites instead of specific species compositions or abundances (Palmer et al. 1997, Hobbs and Harris 2001). It is often not a trivial task, however, to identify precisely how perturbations such as invasion alter functional processes. Invaders can exert their impact through a number of direct and indirect mechanisms that operate over varying spatial and temporal scales (Carlton 2002,

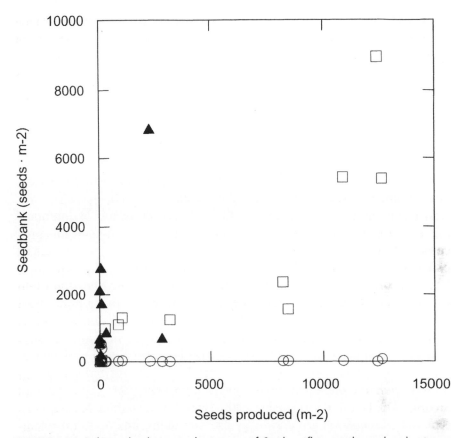

FIGURE 16.1 Relationship between the amount of *S. alterniflora* seeds produced at intact meadows (squares) and eradication sites (triangles) in fall 2003 and the amount of viable seed recovered in sediment cores at those sites and adjacent tideflats (circles) the following spring. There is a strong relationship between local seed production and the local seedbank within intact meadows ($R^2 = 0.72$, $F_{1,9} = 26.40$, $P < 0.001$), but not eradication sites ($R^2 = 0.11$, $F_{1,8} = 2.14$, $P = 0.182$) or adjacent tideflats ($R^2 = 0.00$, $F_{1,19} = 0.61$, $P = 0.445$). Many eradication sites had no seed production but had measurable seedbanks, presumably because of long-distance dispersal into the sites. Seed viability in the field is typically <1 year and there is no long-term seedbank. (Lambrinos, unpublished data).

White et al. 2006). In addition, there is a general lack of data describing invasion impact mechanisms (Levine 2003).

A clear understanding of impact mechanisms can be very helpful, however, in designing effective restoration strategies (Byers et al. 2002). The type of impact mechanism can influence the degree of system hysteresis and the timescale over which systems recover following invasive species removal. Removal of invasive populations will immediately change the direct trophic relationships in the system. For instance, eradication of invasive consumers such as rodents and goats often leads to the quick recovery of prey populations (Klinger et al. 2002, Sinclair et al. 2005). In contrast, engineered structures can continue to influence ecosystems even if invaders are eradicated. The clam *Mya arenaria* invaded Gray's Harbor, Washington, in the mid 1870s, but populations suffered a catastrophic decline in the 1890s that nearly extirpated clams from the estuary. Over 100 years later, however, large beds of *Mya* shells left from the original invasion still persist. These beds exert a significant influence on the system by providing important nursery habitat for Dungeness crabs (*Cancer magister*) (Palacios et al. 2000).

Most invaders exert their impact through a complex mix of engineering and non-engineering mechanisms. In addition, engineering processes influence ecosystems both by directly altering physical states or processes as well as by the consequence these altered physical states have on community dynamics and interactions. Pacific *Spartina* invasions are good examples, not the least reason being that unlike most invasions there has been a considerable amount of work detailing invasion impact mechanisms. *Spartina alterniflora*, *S. anglica*, and *S. alterniflora* x *S. foliosa* hybrids typically invade unvegetated tideflats at relatively low elevations and then convert these habitats to densely vegetated *Spartina* meadows at relatively high elevations (Daehler and Strong 1996, Dethier and Hacker 2005, Hacker and Dethier 2006). These types of conversions are driven by an interacting mix of engineering-mediated physical habitat changes and trophically mediated food web changes. The prodigious aboveground structure of *Spartina* causes decreased tidal flows, increased deposition of fine sediment and organic matter, and decreased light penetration to the sediment surface. The increased deposition and retention of fine sediments combine with a massive buildup of roots belowground to raise the relative elevation of invaded patches. Belowground *Spartina* biomass also contributes to increased sediment carbon and nitrogen storage, increased benthic respiration rates, and increased porewater sulfide concentrations and sediment anoxia (Neira et al. 2005, 2006). These habitat changes adversely affect the survivorship of a range of surface-feeding invertebrates such as bivalves, amphipods, and polychaetes (Neira et al. 2006). The dense belowground *Spartina* biomass also preempts the space available to infauna (Neira et al. 2005). Overall, invaded tideflats have a lower abundance and diversity of macrofauna than unvegetated tideflats (O'connell 2002, Neira 2005).

Differences in the structural form of *Spartina* beds, however, can have a markedly different effect. In San Francisco Bay, macrofauna abundance is significantly greater within areas vegetated by the native *S. foliosa* than in unvegetated or hybrid *Spartina*-invaded tideflats (Brusati and Grosholz 2006). *Spartina foliosa* is shorter, has a less dense canopy and root mass, and produces considerably less overall biomass than hybrid *Spartina* and other invasive members of the genus. This moderate amount of structure appears to facilitate benthic invertebrates perhaps by ameliorating sediment temperatures or providing a refuge from bird predation (Brusati and Grosholz 2006).

In addition to their engineering impact invasive *Spartina* populations also exert trophic impacts, although in some cases these are also mediated by engineering processes. For instance, the physical structure provided by invading hybrid *Spartina* in San Francisco Bay provides a refuge for invasive green crabs (*Carcinus maenus*), and this significantly increases the predation pressure on infauna within *Spartina* beds (Neira et al. 2006). More directly, *Spartina* influences trophic structure through its prodigious primary production. When *Spartina* invades tideflats the dominant source of primary production shifts from relatively high quality (low C:N) microalgae to relatively low quality (high C:N) *Spartina* (Tyler et al. in preparation). Stable isotope studies indicate that surface-feeding invertebrates feed mainly on microalgae, while subsurface feeders such as capitellid polychaetes and turbificid oligochaetes can incorporate significant amounts of *Spartina* detritus. In contrast to surface feeders, subsurface taxa are tolerant of the sediment conditions found within *Spartina* beds. The result is a broad shift from a trophic structure dependent on primary production to one dependent on detritus (Neira et al. 2006, Levin et al. 2006).

Time lags in population dynamics resulting from demography and life history can create large discrepancies between the time an invader becomes abundant and the time when impacts appear, even when the impact mechanisms are direct species interactions such as resource competition (Byers and Goldwasser 2001). Engineering processes that generate accumulating changes in abiotic conditions can also contribute to time lags in invasive impact. For example, fine sediment and belowground biomass slowly accumulate over decades in established *Spartina* beds (Neira et al. 2006, Tyler et al. unpublished data). The consequences of these slow drifts in abiotic conditions for community and population dynamics are likely complex. In addition, accumulating abiotic changes could also create an impact penalty for delaying removal activities at a site by making restoration cost or feasibility dependent on infestation age. These spatial and temporal complexities can make designing effective management strategies for invasive engineers difficult.

16.5 ⬤ CHOICE OF CONTROL STRATEGIES

Like the climate matching and niche models for predicting establishment and spread, many current models of invader impact are static and provide managers with little information with which to predict the spatial or temporal dynamics of impact (e.g., Parker et al. 1999). A more mechanistic understanding of impact is an important step in designing adaptive management strategies that can account for these dynamics (Byers et al. 2002).

The strategies and methods used to control invasive ecosystem engineers can influence the extent and persistence of invader impact. Such considerations are rarely incorporated into the design of control programs, however. Control methods are usually evaluated for their cost and efficacy in reducing infested area or invasive population growth. Data on removal costs are often combined with population and life history data to parameterize optimal control models (e.g., Moody and Mack 1988, Higgins et al. 2000). For the case of *S. alterniflora* in Willapa Bay, such models conclude that when control budgets are limited or uncertain the optimal strategy is to target the removal of isolated clones over established meadows, because these have the highest short- and long-term contribution to population growth per unit of removal cost (Taylor and Hastings 2004, Grevstad 2005). For ecosystem engineers that produce recalcitrant impact, however, management costs will also include the cost of restoring habitat after invasive populations have been removed. Hall and Hastings (in review) demonstrate using a linear programming model that the optimal control strategy can shift from prioritizing invader removal to prioritizing restoration of damaged habitat even when the invader has not been completely eradicated. Several conditions favor a combined removal and restoration strategy. If the invader grows slowly then diverting resources away from removal does not cause significant increases in invasive population growth or spread. In these cases investing in restoration early can significantly reduce impacts over the course of the invasion, particularly if invasion damage persists for a long time. This minimizes the total management costs (removal and restoration expenditures plus ecological cost of the invasion). Inexpensive restoration costs relative to removal also favor a combined strategy. A combined strategy is only optimal, however, if management budgets support enough removal to ensure eventual eradication.

Restoration costs will undoubtedly vary with factors such as the size and age of the invaded patch. For instance, the removal of *Spartina* seedlings or small clones may require little or no active restoration. At high-energy sites in Willapa Bay, natural processes can restore ambient sediment conditions to small invaded patches within 6 years (Tyler et al. unpublished data). In contrast, large and long-established meadows are likely to be much more recalcitrant and require significant restoration investment. Incorporating these considerations into optimal control models for invasive ecosystem engineers is urgently needed.

A number of ambitious and costly *Spartina* control programs are under way throughout the Pacific region (Shaw and Gosling 1995, Kriwoken and Hedge 2000, Hedge et al. 2003, Ayres et al. 2004, Buffet 2005, Wang et al. 2006). Like most invasive species control programs, the immediate focus of *Spartina* management has been on designing efficient and cost-effective removal strategies. This is understandable; killing *Spartina* is no trivial task. Plants have considerable belowground reserves that allow them to readily re-sprout following removal of their aboveground biomass. Even systemic herbicides such as glyphosate require multiple applications to completely kill clones. This is partly because tidal regimes severely restrict the time available for foliar absorption, while at the same time covering leaves with a muddy film that further hampers absorption (Hedge et al. 2003). The estuarine environment poses other severe logistical constraints on control activities. Only specialized equipment can effectively operate on soft tidal substrates, and personnel must be cautious to avoid becoming trapped in often quicksand-like sediments.

In Willapa Bay, the explicit goal of the *Spartina* management plan is the eradication of invasive populations throughout the bay by 2010 (Murphy 2003). Although specific restoration targets have not been defined, the implicit assumption is that *Spartina* eradication will result in the restoration of the pre-invasion tideflat habitat. The eradication program employs several distinct removal strategies. Hand-pulling is used to remove seedlings. At larger scales, heavy equipment is used to crush or mow aboveground *Spartina* biomass. Sites are then rototilled or disced to disrupt the formidable belowground root mass. A biocontrol program has been successful in establishing viable populations of the specialist planthopper *Prokelisia marginata*, but so far has not caused appreciable reductions in *Spartina* infestation (Grevstad et al. 2003). By far the most effective and cost-efficient control method to date is the application of the systemic herbicide Imazapyr by large mechanized applicators and aerial spraying (Patten and Stenvall 2003).

Besides varying in their efficacy at killing *Spartina*, the different control methods also vary in the degree to which they alter the engineered habitat (Table 16.1). Although herbicide applications kill aboveground plant parts, they leave nearly all the aboveground and belowground engineered structure intact. Mechanical methods, on the other hand,

TABLE 16.1 The effect of *Spartina* control activities on physical habitat characteristics. Control methods vary in the degree to which they alter habitat traits, and habitat traits vary in their persistence following *Spartina* death.

Engineered Habitat Trait	Effect of Control	Persistence Following Eradication	Reference
Aboveground biomass	Mechanical methods remove most structure; herbicide or biocontrol leaves structure intact.	None for mechanical methods; up to 2 years for herbicide or biocontrol.	Patten (2005)
Belowground biomass	Structure persistent with most methods. Discing may hasten decomposition.	<6 years for individual clones at high-energy sites; 20 years or longer for large beds at low-energy sites.	Tyler et al. unpublished data, McGrorty and Goss-Custard (1987)
Raised sediment elevation	All current methods leave raised elevation intact.	<6 years for individual clones at high-energy sites; up to 20 years (or longer) for large beds at low-energy sites.	Tyler et al. unpublished data, Ball (2004) McGrorty and Goss-Custard (1987)
Elevated porewater NH_4^+	All methods first lead to immediate spikes in porewater concentrations followed by gradual decreases toward uninvaded levels.	At least 3 years in large beds.	Tyler et al. unpublished data
Elevated porewater S^{2-}	All methods first lead to immediate spikes in porewater concentrations followed by gradual decreases toward uninvaded levels.	At least 3 years in large beds.	Tyler et al. unpublished data

destroy much of the aboveground structure, but still leave the root mass and accumulated sediment intact. This difference could have a profound influence on the legacy of *Spartina* impact and on the future course of restoration. Management programs, however, usually treat control and restoration as distinct sequential components. Developing optimal management strategies for invasive ecosystem engineers requires more integrated approaches because the cost or success of restoration can be influenced by the type of engineering legacy and the way in which eradication is achieved. In Willapa Bay, Imazapyr and increased management budgets have had dramatic effects in the past few years. Large and long established *Spartina* meadows have been eradicated in a few locations, and bay-wide *Spartina* acreage has been significantly reduced (Murphy 2005). There is, however, a growing recognition that eradication alone might not be enough to restore the lost tideflat habitat.

16.6 ● ALTERNATIVE RESTORATION TRAJECTORIES

The possibility that the physical legacy left by eradicated invasive engineers could retard recovery or lead to unexpected restoration outcomes is a serious concern. Byers et al. (2006) developed a conceptual model to illustrate how the introduction or removal of an ecosystem engineer can influence the likelihood of successful restoration. The model describes ecological systems in terms of the dynamics linking biotic and abiotic conditions. These dynamics define boundaries separating basins of attraction of alternative system states. There is growing appreciation that positive feedbacks between species and the abiotic environment can maintain systems in locally stable states, many of which may be undesirable from a management perspective (Hobbs and Norton 1996, Rietkerk 2004, Suding et al. 2004).

In Willapa Bay, local *Spartina* eradication removes the competitive inhibition on seedling recruitment, but leaves much of the rest of the engineered structure intact including a recalcitrant root mass and raised tidal elevation (Table 16.1). This structure still facilitates seed entrapment. Sites that had no local seed production because of eradication efforts still had viable seedbanks the following spring (Figure 16.1). *Spartina* seeds have low viability in the field and there is no persistent seedbank (Mooring et al. 1971, Woodhouse 1979, Sayce 1985). This suggests that eradication sites had acquired seed through dispersal from neighboring seed sources. Eradication sites subsequently support high densities of emerging seedlings in the spring relative to adjacent tideflats (Figure 16.2). In experimental seed additions there was no consistent difference in *Spartina* germination success between tideflats and

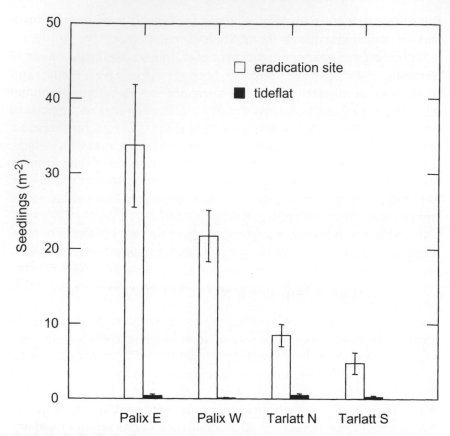

FIGURE 16.2 *Spartina alterniflora* seedlings at four eradication sites and adjacent tide-flats in May 2005. Eradication sites had previously been high-density (>300 stems/m²) *Spartina* meadows that were subjected to mechanical removal and herbicide treatments for at least the previous 2 years. Greater recruitment at eradication sites likely reflects greater seed entrapment relative to adjacent tideflats (see text). Seedling density assessed using 1 m x 1 m quadrats (n = 50). Values are mean ± SE. (Lambrinos unpublished, data).

Spartina meadows, and there is broad concordance between the size of the seedbank at a site and the density of emerging seedlings (Lambrinos and Bando in press). This suggests that the observed differences in seedling recruitment reflect the large differences in propagule supply between *Spartina* meadows or eradication sites and tideflats. Control consequently creates habitat patches that are potentially more susceptible to re-invasion than the original tideflat. A similar example comes from a terrestrial system where invasive bush lupines (*Lupinus arboreus*) are nitrogen fixers that enrich the soil underneath their canopy but also suppress recruitment under the canopy by severely reducing light availability. Herbivore outbreaks occasionally cause mass lupine die-offs, and this creates patches of nutrient-rich bare soil that are susceptible to invasion by non-native grasses (Maron and Jefferies 1999).

Although control activities create sites that favor seed deposition and early seedling emergence, they also induce short-term changes in sediment conditions that inhibit seedling establishment. Porewater ammonium and sulfide spike to phytotoxic levels immediately after removal of aboveground *Spartina* biomass (Tyler et al. unpublished data). Seedlings sown in experimental plots where aboveground *Spartina* had been clipped grew poorly relative to seedlings on unmanipulated tideflats, and did not survive more than 1 year (Lambrinos and Bando in press). This inhibition on recruitment in the short term appears to suppress the immediate recolonization of eradication sites.

Although the spikes in porewater ammonium and sulfide are transient, raised sediment elevation has a much longer legacy. Consequently eradication sites eventually become susceptible not only to *Spartina* recolonization but also to colonization by other plant species typical of higher-elevation salt marsh (Table 16.2). At these sites *Spartina* eradication appears to be facilitating the development of a salt marsh community, not the desired tideflat community (Reeder and Hacker 2004, Lambrinos, unpublished data). It is difficult to predict recovery trajectories based on short-term observations of the early post-control community, however. The plant community now at eradication sites in Willapa Bay is a unique assemblage composed of a high abundance of macroalgae and the non-native annual *Cotula coronopifolia* (Patten 2004; Table 16.2). Future recovery trajectories will depend on how these early colonists influence community assembly through their direct biotic interactions as well as how they contribute to habitat change through their own engineering feedbacks.

I apply the Byers et al. (2006) model to the Willapa Bay system (Figure 16.3). There are currently three long-term stable states in the intertidal habitat of Willapa Bay that can be defined by their abiotic and biotic properties (see Figure 16.3 and Table 16.2): a higher-elevation *Spartina* marsh (diamond), a higher-elevation salt marsh dominated by other halophytes (square), and a lower-elevation tideflat (triangle). The basins of attraction for these ecosystem states are the set of biotic and abiotic conditions that will move asymptotically to the locally stable ecosystem state. In the absence of specific mechanistic models, the boundaries between basins of attraction are defined by arbitrary, nonlinear functions of both abiotic and biotic variables (solid curved lines in Figure 16.3).

By removing some of the strong engineering feedbacks on community and ecosystem processes provided by *Spartina*, eradication

TABLE 16.2 Plant community composition of four system states in Willapa Bay, WA. Data are from neighboring sites within the same local area. Relative elevation at each site given in parentheses (data from NOAA Coastal Services Center). Values are mean ± SE of absolute percent cover measured in $0.25\,m^2$ quadrats (n = 40). The eradication site is 2 years post-control. (Lambrinos, unpublished data).

	Salt Marsh (2.5–3 m)	Tideflat (0.0 m)	*Spartina* Bed (2–2.5 m)	Recent Eradication (2–2.5 m)
Algal mat	0	6.1 ± 2.4	0	50.3 ± 7.5
Atriplex patula	3.6 ± 0.6	0	0	0
Cotula coronopifolia	0.1 ± 0.1	0	0	9.6 ± 2.9
Cuscuta salina	0.1 ± 0.1	0	0	0
Deschampsia caespitosa	6.5 ± 1.1	0	0	0
Distichlis spicata	23.4 ± 3.4	0	0	0
Glaux maritima	1.3 ± 0.9	0	0	0
Grindelia integrifolia	1.0 ± 0.3	0	0	0
Jaumea carnosa	24.7 ± 1.7	0	0	0
Plantago maritima	0.4 ± 0.2	0	0	0
Salicornia virginica	37.1 ± 3.1	0	0	5.2 ± 1.4
Spartina alterniflora	0	0	86.3 ± 3.3	0.2 ± 1.4
Triglochin maritimum	1.0 ± 0.4	0	0	0.3 ± 0.2
Zostera japonica	0	33.7 ± 5.7	0	0
Zostera marina	0	3.0 ± 1.3	0	0
Bare mud	1.5 ± 0.5	57.2 ± 5.4	13.8 ± 3.3	30.7 ± 6.1

fundamentally changes the likelihood that the system will remain in its current biotic and abiotic state or change states toward the desired end point. In the model framework this is represented by the shifting position of basin boundaries (Figure 16.3A–C). How this shift influences the restoration outcome depends on the position of the new boundaries relative to the transient system changes caused by the removal activities. Mechanical and herbicide control could remove enough structure so that natural erosion and decomposition processes lead to the restoration of the original tideflat (Figure 16.3A). However, eradication may not be enough to push the system into the basin of attraction of the desired state. In this case eradication sites could be recolonized by *Spartina* or colonized by salt marsh species (Figure 16.3B). Successful restoration would require significant amounts of additional work such as actively removing sediment to lower elevation or continued discing to promote decomposition of the *Spartina* root mass. It is possible that the initial

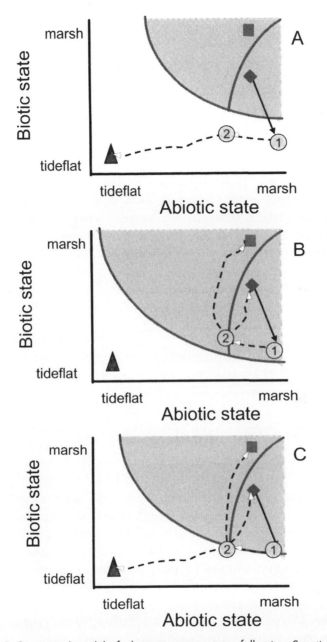

FIGURE 16.3 Conceptual model of alternate system states following *Spartina* removal. The invaded *Spartina* state (diamond), the desired tideflat state (triangle), and the alternate marsh state (square) are locally stable. Solid lines represent the human modification of the system by eradication activities; dotted lines represent autonomous system trajectories. Circles represent transitory system states following *Spartina* removal: (1) initial changes caused by control activities; (2) initial decomposition and erosion. Three restoration scenarios are possible: (A) recovery, (B) hysteresis, (C) contingency (adapted from Byers et al. 2006).

system changes following control lie very close to the boundaries of alternative system states (Figure 16.3C). In this case restoration trajectories would be strongly influenced by variation in environmental conditions that influence community assembly or the persistence of physical structure. Didham et al. (2005) hypothesize that physiologically stressful systems such as wetlands are prone to catastrophic shifts in community states caused by propagule limitation, stochastic priority effects, and changes in the regional species pool. There is evidence that recovery trajectories following *Spartina* eradication do indeed depend on the landscape context of the eradication site (Reeder and Hacker 2004).

The conceptual model suggests that, beyond removing *Spartina*, manipulating biotic conditions will do little to help achieve the desired restoration goals as long as abiotic conditions remain the same. Facilitating change in physical state is far more important for achieving the restoration goal. Such efforts are likely to be extremely costly, however. The Byers et al. model provides a good conceptual framework for designing studies that assess whether restoration efforts are required or even feasible in a particular situation. These, combined with optimization models such as the Hall and Hastings model, which account for combined removal and restoration costs, will greatly improve our ability to predict the true costs and likely success of what are increasingly more substantial and long-term investments in invasive management.

16.7 ⬤ COLLATERAL IMPACTS OF CONTROL

The degree to which ecosystems recover to pre-invasion conditions following eradication is not the only concern facing managers attempting to eradicate invasive engineers. Removing invaders can inflict collateral impacts on the invaded ecosystem. Managing the secondary effects of invasive species removal is an important component of holistic ecosystem-based management (Westman 1990, Zavaleta et al. 2001). Many of the traditional tools for predicting community dynamics such as food web analysis and competition models can be used to predict the secondary effects of removal that result from trophic interactions (Zavaleta et al. 2001). Few tools exist, however, to help predict the secondary impacts that arise from nontrophic processes.

Some engineering effects of invaders may be at least partially beneficial to native species. In many situations the invader is the primary contributor to these functions because native providers have been displaced by the invader itself or by other stressors such as urbanization. One important engineering function that invaders sometimes provide is physical habitat in the form of such things as nest sites or cover from predators. Saltcedar (*Tamarix* spp.) has replaced native riparian trees in many parts of the southwestern U.S., and now provides nesting habitat for the endangered Willow Flycatcher (*Empidonax trailii extimus*). This has complicated saltcedar management (Dudley et al. 2000). Other negative aspects of saltcedar engineering exacerbate the management difficulties. Saltcedar lowers the water table and increases soil salinity, inhibiting native recovery at eradication sites. Successfully limiting invasion impacts while maintaining Willow Flycatcher populations will require sophisticated integration of removal and restoration activities at a regional scale (Zavaleta et al. 2001).

It is often difficult to evaluate the true costs and benefits of invasive engineering to native populations. In some cases the physical habitat provided by the invader is not equivalent to that provided by native vegetation creating an ecological trap. Ecological traps occur when the behavioral cues (such as vegetation structure) used by organisms for habitat selection do not correlate well with adaptive outcomes (Schlaepfer et al. 2002). In San Francisco Bay several bird species including the endangered California Clapper Rail (*Rallus longirostris obsoletus*) and the threatened Alameda Song Sparrow (*Melospiza melodia pusillula*) nest in hybrid *Spartina*. *Spartina*, however, grows at lower tidal elevations than native vegetation. Both Clapper Rail and Song Sparrow nests placed in *Spartina* are consequently more subject to flooding and this reduces nesting success (Nordby et al 2004). Invader-engineered habitat can also differ from native habitat in its non-engineering-related properties such as food resources, or in the way the engineering influences species interactions. Aggressive Marsh Wrens (*Cistothorus palustris*) are attracted to the tall structure provided by *Spartina* where they compete with Song Sparrows by actively destroying Song Sparrow nests and eggs. This competition significantly reduces Song Sparrow nesting success in *Spartina* marshes (Nordby et al. 2004).

Sometimes invasive engineers contribute to important ecosystem services that could be disrupted by control activities. Indeed many invasive species are introduced specifically for the engineering functions they provide such as erosion control (Reichard and White 2001). In other cases invasions have produced completely unexpected services. For example, prodigious filtering of the water column by *zebra* mussels (*Dreissena polymorpha*) has dramatically improved the water quality of the heavily eutrophied Great Lakes (Fahnenstiel et al. 1995).

The degradation of engineered structures following the death of the invader can itself cause problems. Mechanical control of *Spartina* deposits large amounts of biomass into the water column over a short period of time. In highly mixed tidal estuaries such as Willapa Bay this is unlikely to cause serious problems for water quality, but in more enclosed systems such as lakes the decomposition of dead invader biomass can be a serious concern (James et al. 2002). Floating *Spartina* wrack does become entangled in the fishing gear of commercial and sports fishermen during the late summer salmon season. This has caused some antagonism from these important stakeholders to the control program (Brian Couch personal communication). *Spartina* eradication could also lead to rapid erosion of the accumulated sediment in *Spartina* beds. If this occurred it would catastrophically affect the bay ecosystem with dire repercussions for the economically important oyster and fishing industries. Available data indicate that erosion of large eradication sites is gradual and slow (Table 16.1; Ball 2004). Indeed, there is some evidence that strong episodic erosion events, which are often observed at the edge of vegetated salt marshes, are a consequence of engineering feedbacks from the vegetation. Van de Koppel et al. (2005) show that the feedbacks between salt marsh vegetation and sediment deposition create both an accumulation of sediment and increased physical stress at the seaward edge of a tidal bench. As a result, salt marsh edges in the Netherlands become increasingly vulnerable to episodic events like storms that trigger cascades of heavy erosion. Removal of vegetation may lead to more gradual and spatially homogenous patterns of erosion. The long-term erosion dynamics of this system are unknown, however.

16.8 ➤ RECOMMENDATIONS

Invasive ecosystem engineers cause conspicuous damage to native ecosystems. Even more disconcertingly, this damage can long outlive the invaders that caused it. Targeting invasive ecosystem engineers for eradication and mitigating their impacts should be a high priority. This is not an easy task. This chapter illustrates some of the unique challenges that invasive engineers such as *Spartina* pose. It also highlights some important areas where new strategies and research are needed:

1. Because ecosystem engineering can be one contributor to time lags in the transition between invasion stages or in spatial spread, nascent introductions of known engineers should be priority targets for monitoring. In addition, static risk models may not be appropriate for ecosystem engineers that dynamically modify habitat characteristics.
2. More work is needed to identify how engineering processes contribute to invader impact. Too often, control and restoration activities are

initiated without a mechanistic understanding of how the invader is causing impact or how control activities will curtail these impact mechanisms. A mechanistic understanding of impact will allow better predictions about the timescale of recovery and can be used to design better restoration strategies.

3. Control and restoration should be better integrated within comprehensive adaptive management. The success of restoration can depend on how control methods affect the persistence of engineered structure in the landscape. Optimality models that evaluate the total costs associated with an invasion, including both the cost of removal and the cost of restoration, can help prioritize eradication and restoration effort.

4. Invasive engineers increase the possibility that control will result in unexpected alternative ecosystem states. Post-control monitoring should identify these potential alternatives and quantify ecosystem changes with respect to them. If enough information is known about the mechanistic drivers of change, strategies can be designed to guide systems to the preferred state.

5. In some cases invasive engineers provide important ecosystem services whose loss must be mitigated for following eradication. In addition, removing an invasive engineer may not eliminate its physical legacy or may create novel impacts from the degradation of the engineered habitat. Assessing the longevity and spatial distribution of these impacts will depend more on understanding physical and biophysical processes than on population dynamics.

REFERENCES

Ayres, D.R., Smith, D.L., Zaremba, K., Klohr, S., and Strong, D.R. (2004). Spread of exotic cordgrasses and hybrids (*Spartina* sp.) in the tidal marshes of San Francisco Bay, California, U.S.A. *Biological Invasions* 6:221–231.

Ball, D. (2004). Monitoring the efffects of *Spartina alterniflora* eradiction on sediment dynamics in two Pacific Northwest estuaries. M.S. Thesis. Western Washington University.

Blossey, B. (1999). Before, during and after: The need for long-term monitoring in invasive plant species management. *Biological Invasions* 1:301–311.

Bruno, J.F., and Kennedy, C.W. (2000). Patch size dependent habitat modification and facilitation on New England cobble beaches by *Spartina alterniflora. Oecologia* 122:98–108.

Brusati, E.D., and Grosholz, E.D. (2006). Native and introduced ecosystem engineers produce contrasting effects on estuarine infaunal communities. *Biological Invasions* 8:683–695.

Buffet, D. (2005). *Invasive Plant Management Proposal: Spartina Anglica-English Cordgrass Management in Boundary Bay Wildlife Management Area and Surrounds, Spartina Project (#15)*. Surrey British Columbia, Canada: Ducks Unlimited Canada.

Byers, J.E., and Goldwasser, L. (2001). Exposing the mechanism and timing of impact of nonindigenous species on native species. *Ecology* 82:1330–1343.

Byers, J.E., Jones, C.G., Cuddington, K., Talley, T.S., Hastings, A., Lambrinos, J.G., Crooks, J.A., and Wilson, W.G. (2006). Using ecosystem engineers to restore ecological systems. *Trends in Ecology and Evolution* 21:493–500.

Byers, J.E., Reichard, S., Randall, J., Parker, I., Smith, C.S., Lonsdale, W.M., Atkinson, I.A.E., Seastedt, T.R., Williamson, M., Chornesky, E., and Hayes, D. (2002). Directing research to reduce the impacts of nonindigenous species. *Conservation Biology* 16(3):630–640.

Civille, J.C., Sayce, K., Smith, S.D., and Strong, D.R. (2005). Reconstructing a century of *Spartina alterniflora* invasion with historical records and contemporary remote sensing. *Ecoscience* 12:330–338.

Crain, C.M., and Bertness, M.D. (2006). Ecosystem engineering across environmental gradients: Implications for conservation and management. *Bioscience* 56:211–218.

Crooks, J.A. (2002). Characterizing ecosystem-level consequences of biological invasions: The role of ecosystem engineers. *Oikos* 97:153–166.

Cuddington, K., and Hastings, A. (2004). Invasive engineers. *Ecological Modelling* 178:335–347.

Cummings, J., Reid, N., Davies, I., and Grant, C. (2005). Adaptive restoration of sand-mined areas for biological conservation. *Journal of Applied Ecology* 42:160–170.

Daehler, C.C., and Strong, D.R. (1996). Status, prediction, and prevention of introduced cordgrass *Spartina* spp. Invasions in Pacific Estuaries, USA. *Biological Conservation* 78:51–58.

D'Antonio, C.M., and Meyerson, L.A. (2002). Exotic plant species as problems and solutions in ecological restoration: A synthesis. *Restoration Ecology* 10:703–713.

Davis, H.G., Taylor, C.M., Civille, J.C., and Strong, D.R. (2004a). An Allee effect at the front of a plant invasion: *Spartina* in a Pacific estuary. *Journal of Ecology* 92:321–327.

Davis, H.G., Taylor, C.M., Lambrinos, J.G., and Strong D.R. (2004b). Pollen limitation causes an Allee effect in a wind pollinated invasive grass (*Spartina alterniflora*). *Proceedings of the National Academy of Sciences USA* 101:13804–13807.

Dethier, M.N. and Hacker, S. D. (2005). Physical factors vs. biotic resistance in controlling the invasion of an estuarine marsh grass. *Ecological Applications* 15:1273–1283.

Didham, R.K., Watts, C.H., and Norton, D.A. (2005). Are systems with strong underlying abiotic regimes more likely to exhibit alternative stable states? *Oikos* 110:409–416.

Dudley, T.L., DeLoach, C.J., Lovich, J.E., and Carruthers, R.I. (2000). Saltcedar invasion of western riparian areas: Impacts and new prospects for control. In *Transactions of 65th North American Wildlife and Natural Resources Conference*, R.E. McCabe and S.E. Loos, Eds. Washington, D.C.: Wildlife Management Institute, pp. 345–381.

Grevstad, F.M., Strong, D.R., Garcia-Rossi, D., Switzer, R.W., and Wecker, M. (2003). Biological control of *Spartina alterniflora* in Willapa Bay, Washington using the planthopper *Prokelisia marginata*: Agent specificity and early results. *Biological Control* 27:32–42.

Hacker, S.D., and Deither, M.N. (2006). Community modification by a grass invader has differing impacts for marine habitats. *Oikos* 113:279–286.

Hacker, S.D., Heimer, D., Hellquist, S.E., Reeder, T.G., Reeves, B., Riordan, T.J., and Deither, M.N. (2001). A marina plant (*Spartina anglica*) invades widely varying habitats: Potential mechanisms of invasion and control. *Biological Invasions* 3:211–217.

Hall, R.J., and Hastings, A. (in review). Minimizing invader impacts: Striking the right balance between removal and restoration.

Hastings, A., Byers, J.E., Crooks, J.A., Cuddington, K., Jones, C.G., Lambrinos, J.G., Talley, T.S., and Wilson, W.G. (2007). Ecosystem engineering in space and time. *Ecology Letters* 10:153–164.

Hedge, P., Kriwoken, L. K., and Patten, K. (2003). A review of *Spartina* management in Washington State, U.S. *Journal of Aquatic Plant Management* 41:82–90.

Higgins, S.I., and Richardson, D.M. (1996). A review of models of alien plant spread. *Ecological Modelling* 87:249–265.

Higgins, S.I., Richardson, D.M., and Cowling, R.M. (2000). Using a dynamic landscape model for planning the management of alien plant invasions. *Ecological Applications* 10:1833–1848.

Hobbs, R.J., and Norton, D.A. (1996). Towards a conceptual framework for restoration ecology. *Restoration Ecology* 4:93–110.

Holmes, P.M., and Richardson, D.M. (1999). Protocols for restoration based on recruitment dynamics, community structure, and ecosystem function: Perspectives from South African fynbos. *Restoration Ecology* 7:215–230.

James, W.F., Barko, J.W., and Eakin, H.L. (2002). Water quality impacts of mechanical shredding of aquatic macrophytes. *Journal of Aquatic Plant Management* 40:36–42.

Klinger, R.C., Schuyler, P., and Sterner, J.D. (2002). The response of herbaceous vegetation and endemic plant species to the removal of feral sheep from Santa Cruz Island, California. In *Turning the Tide: The Eradication of Invasive Species,* IUCN SSC Invasive Species Specialist Group, C.R. Veitch and M.N. Clout, Eds. Gland, Switzerland: IUCN, pp. 141–154.

Kriticos, D.J., Sutherst, R.W., Brown, J.R., Adkins, S.W., and Maywald, G.F. (2003). Climate change and the potential distribution of an invasive alien plant: *Acacia nilotica* ssp. Indica in Australia. *Journal of Applied Ecology* 40:111–124.

Kriwoken, L.K., and Hedge, P.T. (2000). Exotic species and estuaries: Managing *Spartina anglica* in Tasmania, Australia. *Ocean and Coastal Management* 43:573–584.

Lambrinos, J.G., and Bando, K. (in press). Habitat modification inhibits seedling recruitment in populations of an invasive ecosystem engineer. *Biological Invasions.*

Levin, L.A., Neira, C., and Grosholz, E.D. (2006). Ecosystem modification by invasive cordgrass through changes in trophic function. *Ecology* 87:419–432.

Levine, J.M., Vilà, M., D'Antonio, C.M., Dukes, J.S., Grigulis, K., and Lavorel, S. (2003). Mechanisms underlying the impacts of exotic plant invasions, Proceedings of the Royal Society B. *Biological Sciences* 270:775–781.

Maron, J.L., and Jefferies, R.L. (1999). Bush lupine mortality, altered resource availability, and alternative vegetation states. *Ecology* 80:443–454.

McGrorty, S., and Goss-Custard, J.D. (1987). *A Review of the Rehabilitation of Areas Cleared of* Spartina. Dorset, England: Institute of Terrestrial Ecology (Natural Environment Research Council).

Melbourne, B., Cornell, H., Davies, K., Dugaw, C., Elmendorf, S., Freestone, A., Hall, R., S., Harrison, Hastings, A., Holland, M., Holyoak, M., Lambrinos, J., Moore, K., and Yokomizo, H. (2007). Invasion in a heterogeneous world: Resistance, coexistence or hostile takeover? *Ecology Letters* 10:77–94.

Moody, M.E., and Mack, R.N. (1988). Controlling the spread of plant invasions: The importance of nascent foci. *Journal of Applied Ecology* 25:1009–1021.

Mooring, M.T., Cooper, A.W., and Seneca, E.D. (1971). Seed germination response and evidence for height ecophenes in *Spartina alterniflora* from North Carolina. *American Journal of Botany* 58:48–55.

Murphy, K.C. (2003). *Report to the Legislature: Progress of the 2003* Spartina *Eradication Program.* Report no. PUB 805–110 (N/1/04). Olympia, WA: Washington State Department of Agriculture.

Murphy, K.C. (2005). *Report to the Legislature: Progress of the 2003* Spartina *Eradication Program.* Report no. PUB 850–151(N/1/06). Olympia, WA: Washington State Department of Agriculture.

Neira, C., Grosholz, E.D., Levin, L.A., and Blake, R. (2006). Mechanisms generating modification of benthos following tidal flat invasion by a Spartina hybrid. *Ecological Applications* 16:1391–1401.

Neira, C., Levin, L.A., and Grosholz, E.D. (2005). Benthic macrofaunal communities of three sites in San Francisco Bay invaded by hybrid *Spartina*, with comparison to uninvaded habitats. *Marine Ecology Progress Series* 292:111–126.

Nordby, J.C., Cohen, A.N., and Bessinger, S.R. (2004). *The Impact of Invasive* Spartina *Alterniflora on Song Sparrow Populations in San Francisco Bay Salt Marshes.* San Francisco, CA: Third International Conference on Invasive *Spartina*, November 8–10 2004, p. 23.

O'connell, K.A. (2002). *Effects of invasive Atlantic smooth cordgrass* (Spartina alterniflora) *on infaunal macro invertebrate communities in southern Willapa Bay, Washington.* Thesis, Western Washington University, Bellingham, WA.

Palacios, R.L., Armstrong, D.A., and Orensanz, J. (2000). Fate and legacy of an invasion: Extinct and extant populations of the soft-shell clam (*Mya arenaria*) in Grays Harbor (Washington). *Aquatic Conservation: Marine and Freshwater Ecosystems* 10:279–303.

Parker, I.M., Simberloff, D., Lonsdale, W.M., Goodell, K., Wonham, M., Kareiva, P.M., Williamson, M.H., Von Holle, B., Moyle, P.B., Byers, J.E., and Goldwasser, L. (1999). Impact: Toward a framework for understanding the ecological effects of invaders. *Biological Invasions* 1:3–19.

Patten, K., and Stenvall, C. (2003). *Control of Smooth Cordgrass* (Spartina alterniflora): *A Comparison Between Various Mechanical and Chemical Methods for Efficacy, Cost and Aquatic Toxicity.* Proceedings of the 11th International Aquatic Invasive Species Conference, pp. 306–316, Pembroke, Ontario, Canada.

Patten, K. (2004). *The Efficacy of Chemical and Mechanical Treatment Efforts in 2002 on the Control of* Spartina *in Willapa Bay in 2003.* Progress report submitted to the Willapa National Wildlife Refuge.

Patten, K. (2005). *Shorebird, Waterfowl and Birds of Prey Usage in Willapa Bay in Response to* Spartina *Control Efforts.* Report to Willapa Wildlife Refuge, Migratory Birds and Habitat Programs, Pacific Region of U.S. Fish and Wildlife Service and Pacific Joint Ventures.

Peterson, A.T. (2003). Predicting the geography of species' invasions via ecological niche modeling. *The Quarterly Review of Biology* 78:419–433.

Pimentel, D., Zuniga, R., and Morrison, D. (2005). Update on the environmental and economic costs associated with alien-invasive species in the United States. *Ecological Economics* 52(3):273–288.

Reeder, T.G., and Hacker, S.D. (2004). Factors contributing to the removal of a marine grass invader (*Spartina anglica*) and subsequent potential for habitat restoration. *Estuaries* 27:244–252.

Reichard, S.H., and White, P. (2001). Horticulture as a pathway of invasive plant introductions in the United States. *Bioscience* 51:103–113.

Rietkerk, M., Dekker, S.C., de Ruiter, P.C., and van de Koppel, J. (2004). Self-organized patchiness and catastrophic shifts in ecosystems. *Science* 305:1926–1929.

Sayce, K. (1988). *Introduced Cordgrass,* Spartina alterniflora *(Loisel.) in Salt Marshes and Tidelands of Willapa Bay, Washington,* FWSI-87058(TS). Ilwaco, WA: U.S. Fish and Wildlife Service, Willapa Bay National Wildlife Refuge.

Schlaepfer, M.A., Runge, M.C., and Sherman, P.W. (2002). Ecological and evolutionary traps. *Trends in Ecology and Evolution* 17:474–480.

Shaw, W.B., and Gosling, D.S. (1995). *Spartina* control in New Zealand—an overview. In *Proceedings of the Australasian Conference on Spartina Control,* J.E. Rash, R.C. Williamson, and S.J. Taylor, Eds. Melbourne, Australia: Victorian Government Publication, pp. 43–60.

Sinclair, L., McCartney, J., Godfrey, J., Pledger, S., Wakelin, M., and Sherley, G. (2005). How did invertebrates respond to eradication of rats from Kapiti Island, New Zealand? *New Zealand Journal of Zoology* 32:293–315.

Suding, K.N., Gross, K.L., and Houseman, G.R. (2004). Alternative states and positive feedbacks in restoration ecology. *Trends in Ecology and Evolution* 19:46–53.

Taylor, C.M., and Hastings, A. (2004). Finding optimal control strategies for invasive species: A density-structures model for *Spartina alterniflora*. *Journal of Applied Ecology* 41:1049–1057.

Van de Koppel, J., van der Wal, D., Bakker, J.P., and Herman, P.M.J. (2005). Self-organization and vegetation collapse in salt marsh ecosystems. *The American Naturalist* 165:E1–E12.

Wang, Q., An, S., Ma, Z., Zhao, B., Chen, J., and Li, B. (2006). Invasive *Spartina alterniflora*: Biology, ecology and management. *Acta Phytotaxonomica Sinica* 44:588–599.

Westman, W.E. (1990). Park management of exotic plant species: Problems and issues. *Conservation Biology* 4:251–260.

Woodhouse, W.W. (1979). *Building Saltmarshes Along the Coasts of the Continental United States,* Special report no. 4. Belvoir, VA: U.S. Army Corps of Engineers, Coastal Engineering Research Center.

Zavaleta, E.S., Hobbs, R.J., and Mooney, H.A. (2001). Viewing invasive species removal in a whole-ecosystem context. *Trends in Ecology and Evolution* 16:454–459.

17

LIVESTOCK AND ENGINEERING NETWORK IN THE ISRAELI NEGEV: IMPLICATIONS FOR ECOSYSTEM MANAGEMENT

Yarden Oren, Avi Perevolotsky, Sol Brand, and Moshe Shachak

The landscapes of the Negev are extensively modulated by natural ecosystem engineers (EE) (Shachak et al. 1995). The main modulators are *cyanobacteria* that glue together soil particles forming a biogenic soil crust (Zaady and Shachak 1994), *shrubs* that create a soil mound underneath (Shachak and Lovett 1998), *porcupines* that dig pits (Boeken et al. 1995), and *isopods* and *ants* that induce significant physical alterations in the soil (Shachak and Yair 1984, Wilby et al. 2001). All these engineers can be considered hydro-engineers since they modify water flow and storage, thus affecting other organisms in this water-limited system (Noy-Meir 1973).

In addition to natural EEs, domestic grazing and browsing animals (DGA), such as goats and sheep, introduced by humans to the rangelands of the Negev thousands of years ago (Noy-Meir and Seligman 1979, Finkelstein and Perevolotsky 1990), also have become important EEs. As components of the ecosystem, livestock have become important in the biotic engineering network of the Negev. This chapter is an attempt to integrate the domestic and natural EEs into one framework and show its effect on the structure and function of the ecosystem.

We introduce the concept of engineering network. This concept encompasses the web of interactions among processes at the landscape, ecosystem, and community levels, emerging from the activity of an assemblage of EEs. The development of the concept is based on our accumulative knowledge of functioning of natural, multiple EEs in the Negev. We further develop the EE network concept by constructing an integrated model that combines natural and domestic organisms. We conclude the chapter by discussing the utility of the model for ecosystem management issues related to pastoralism and recreation.

17.1 ⬥ ENGINEERING NETWORKS

Most EE studies have been devoted to the effect of a single species on landscape, ecosystem, and/or community processes (Wright and Jones 2006). The study of a single species that functions as an EE focuses on identification of its primary environmental modulation and secondary effects that follow the modulation. For example, in the Negev, porcupines modulate the landscape by creation of pits in the crusted soil. The pits trap and accumulate runoff, litter, and seeds. The overall effect of pit formation is a creation of water-, nutrient-, and seed-enriched patches characterized by higher productivity and diversity of plants than the unmodified background (Alkon 1999). The engineering effect of the porcupines can be depicted as a network of interactions among landscape (patch formation in the form of pits), ecosystem (water and litter flows), and community (seeds and species flows) processes. A simplified presentation of the engineering network induced by the porcupines is shown in Figure 17.1 (blue boxes and arrows). The porcupine example depicts the essence of single-species engineering, namely by creating a network of interactions among various levels of organization as a consequence of landscape modulation. This network is a result of the interactions between the traits of the engineering organism (pit digging, P1 in Figure 17.1), the function of the ecological system (conversion of rainfall into surface runoff, P2 in Figure 17.1), and the effect of ecosystem processes on species assemblage (filtering of herbaceous species, P3 in Figure 17.1).

In most ecosystems there are more than one EE; usually there is an assemblage of EEs. A principal issue concerning engineering, which has not yet been addressed, is the integrated function of a community of engineers. Addressing the effects of a group of EEs generates a new set of questions: Is the integrated effect of a network of engineers indifferent, interfering, or synergetic with the effect of each of them individually? Do the EEs control together the same state changes as they do separately? Does the EE assemblage increase the complexity of the network by increasing the number of processes and links among them?

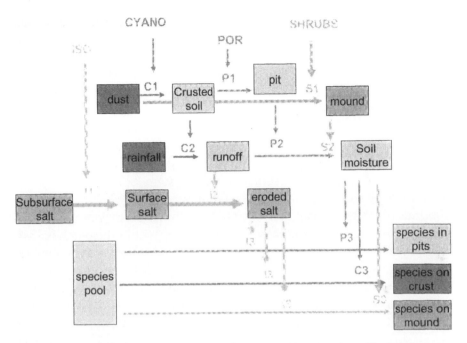

FIGURE 17.1 The engineering network in the Negev Desert and its effects on species richness. Three ecosystem engineers—cyanobacteria (CYANO, C), porcupines (POR, P), and shrubs (S)—modulate the landscape by creating three patch types: crusted soil, pits, and mounds. Landscape-level engineering controls state changes in an ecosystem process of water flow (C2, P2, S2). Water flow (P3, C3, S3) affects species assemblages on the three patch types. An additional engineer-isopod (ISO, I) modulates another ecosystem process-desalinization, by transferring salty soil from subsurface to the surface (I1). The salty soil is eroded from the system by runoff water (I2), a desalinization process that affects plant assemblages (I3). Notice that the engineering network is a network that links processes at various levels of organization (horizontal boxes and arrows) induced by the engineers (vertical arrows). (See color plate.)

Answers to these questions can provide the foundation for developing a framework for studying the "community ecology of engineers."

To explore the essence of the communal impact of engineers we analyze the multi-engineering system of the Negev. Our approach commences with an analysis of the impact generated by a single engineer and then explores changes due to the activities of an additional engineer. We then add more engineers, one at a time, and examine changes in the whole network and its effects on the ecology of the system (Figure 17.1).

To demonstrate the approach we analyze two EE systems in the Negev by adding to the porcupine-engineered network, cyanobacteria as a second EE (Figure 17.1, pink). Cyanobacteria's primary function as

EE is the formation of soil crust induced by secretion of poly-saccharides that glue together the soil particles (Eldridge and Greene 1994, Zaady and Shachak 1994) (C1 in Figure 17.1). The source of the soil particles is dust deposition (Shachak and Lovett 1998). The formation of crust significantly changes ecosystem and community processes, mostly by decreasing the rate of water infiltration into the soil (Eldridge et al. 2002). Lower water infiltration rate results in low soil moisture and generation of surface runoff (C2 in Figure 17.1). If there is no local barrier runoff water leaks rapidly out of the system and further decreases water available to the local biota within the system (Shachak et al. 1998). Runoff water can also be accumulated in sink patches such as pits or mounds (P2, S2 in Figure 17.1). Runoff accumulation in water-enriched patches increases the productivity and diversity within the patches and, thus, in the whole system (Shachak et al. 1995). In the case of the paired EEs—porcupine and cyanobacteria—their system function is complementary. Cyanobacteria stimulate runoff generation (C2 in Figure 17.1) thus creating a source of water movement in the system while the porcupine pits function as sinks (P2 in Figure 17.1). The water-enriched patch is a product of the network of the two EEs (Figure 17.1, blue and pink flows). The study of a paired EE system suggests that adding a second EE can modify the network in several ways. Firstly, by adding state variables that extend the components involved in the engineering network (dust, rainfall, and species on crust in Figure 17.1). Secondly, adding controllers over the change of states (dust to crust, rainfall to runoff, and filtering of species that can grow on crust) also modifies the network.

We can further increase the number of EEs in the engineering network by adding the activity of isopods (Figure 17.1, brown). Isopods control desalinization processes (Shachak and Yair 1984). In arid areas, rainfall with dissolved salts infiltrates into the soil. The water rapidly evaporates and salts accumulate at the soil depth to which average water infiltration reaches. This process of salinization reduces plant production and diversity (Yair and Shachak 1987). Isopods counteract salinization by transporting soil from the salty layer and depositing the high-saline soil on the soil crust (I1 in Figure 17.1). This saline soil is eroded out of the system by runoff, thus decreasing salt accumulation in the ecosystem (I2 in Figure 17.1). Adding isopods to the engineering network of porcupines and cyanobacteria demonstrates a new way of extending functionally the engineering network by adding into the complex ecological picture the process of desalinization (Figure 17.1, brown flow). This process is integrated into the engineering network in two ways: through runoffs that control the transition from deposits of saline soil to eroded soil and via the effects of the eroded saline on local species assemblages on the crust and in the porcupine pits (I3 in Figure 17.1).

We conclude our demonstration of the functioning of a multiple-engineer network by adding shrubs as yet another important EE in the Negev ecosystem (Figure 17.1, green flow). Shrubs modulate the landscape through the formation of soil mounds under their canopy (S1 in Figure 17.1). The mound is a product of long-term dust accumulation (Shachak and Lovett 1998). The mound, which is composed of uncompacted soil with little crust cover, has high water infiltration capacity and hence intercepts the flow of surface runoff and creates a water-enriched patch (S2 in Figure 17.1). In addition to water accumulation, the mound accumulates litter and traps seeds thus creating an "island of fertility" with a species-rich community (Boeken and Shachak 1998). The shrub engineering extends the EE network by adding the soil mound and plant species that characterize the mound as state variables (Figure 17.1, green flow). The shrub also controls the soil moisture regime by slowing down the aforementioned leakage of water from the system. These changes in soil moisture affect the nature of the species assemblage on the soil mound.

Some generalizations emerge from analyzing the network of interactions created by the principal EEs of the Negev.

1. An engineering network is composed of several ecological processes (horizontal arrows in Figure 17.1: soil accumulation and erosion, water flow, and species distribution) controlling each other (vertical arrows in Figure 17.1: crust formation, soil desalinization, soil moisture dynamics, and species establishment within patches).
2. EEs control state changes at one or more of the processes.
3. An engineering network links several levels of organization and creates an integrated effect.

We suggest that the concept of engineering network can help in studying the effect of human-introduced EE such as livestock or planted trees. Linking human-introduced EE to the network should follow the procedure of integrating its specific activities into the network in terms of additional state variables and controllers to already existing processes or by adding a new process.

In the next section we apply this procedure in order to elucidate the effect of livestock introduction to the Negev in the context of the engineering network. Linking the natural EEs and livestock into an integrated network will enable us to connect engineering networks and management in the last section.

17.2 ● LIVESTOCK AND ENGINEERING NETWORK

Livestock husbandry in an arid environment is strongly affected by the distribution of herbaceous vegetation biomass, which is controlled by the water regime. Since the most significant EEs in determining the water regime are the crust-forming cyanobacteria and the shrubs we chose to link livestock activities with these two EEs. The two EEs in the Negev engineering network, cyanobacteria and shrubs, create the characteristic landscape structure of arid shrublands, which consist of shrub-dominated patches set within a relatively open matrix (Shmida et al. 1986, West 1989). The matrix is dominated by a microphytic crust-forming community (West 1990, Zaady and Shachak 1994). These two contrasting patch types differ in microclimate (Burke 1989, Moro et al. 1997), soil moisture dynamics (Noy-Meir 1973, Mott and McComb 1974, Cornet et al. 1992), soil texture and nutrient composition (Blackburn 1975, West 1981, Ludwig and Tongway 1995), and microbial and animal activity (Elkins et al. 1986, Smith et al. 1994, Zaady et al. 1996), thus, offering disparate conditions for the growth of annual plants. The crusted matrix is characterized by a sparse assemblage of annuals poor in species. These annual community properties are attributed to the lower soil-water content under the crust, the physical resistance of the crust to seed germination, and heavier seed predation (Lange and Belnap 2003). In the shrub patches, however, a dense and rich annual community develops. This is mostly due to accumulation of water and nutrients under the shrub (Noy-Meir 1973, Halvorson and Patten 1975, Gutiérrez et al. 1993, Boeken and Shachak 1998a, Pugnaire et al. 1996, Shachak et al. 1998). We selected from the engineering network presented in Figure 17.1, the two-phase landscape mosaic created by the natural EEs, and its effect on productivity and diversity of annual plants as a background for integrating livestock grazing activities into the natural engineering network (Figure 17.2).

Domestic grazing and browsing animals (DGA) affect ecosystems in two general ways: removing plant biomass and disturbing soil through trampling and compaction. The two activities modulate the features of the natural environment thus influencing the existing engineering network (e.g., Creda and Lavee 1999, Graetz and Tongway 1986, Greenwood and McKenzie 2001, Mwendera and Mohamed Saleem 1997, Proffitt et al. 1993, Weltz et al. 1989). Studying the effects of grazing animals on the ecosystem is complicated by the dual and simultaneous effects of the organisms in biomass removal and soil disturbance activities that are closely coupled. Hence, understanding and predicting the outcome of animal grazing activities on the engineering network depends on the ability of decoupling the two activities.

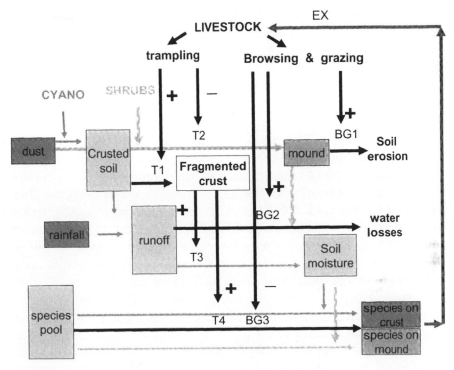

FIGURE 17.2 The joint network of natural ecosystem engineers and livestock. Livestock introduces a new state variable (fragmented crust) to the natural engineering network by trampling on the crust (T1). Crust trampling increases (+) soil moisture (T3) and species richness on crust (T4). Trampling on shrubs decreases (−) soil mound (T2). Browsing and grazing increase soil erosion (BG1) and water losses (BG2) and decrease species diversity (BG3). EX refers to plant species exploitation by grazing. (See color plate.)

Plant consumption by livestock modifies vegetation species richness and composition (Milchunas et al. 1988, Perevolotsky and Seligman 1998, Olff and Ritchie 1988) and affects the natural food web (Berger 2006). In addition, browsing modulates the landscape structure by changing individual shrub biomass and the patchy distribution of shrubs in the landscape, both affecting ecosystem processes such as water flow and soil erosion (Fuls 1992). The biomass consumption and soil disturbance effects were addressed by experimental studies that demonstrate that both activities control ecological processes at various levels of organization (e.g., Turner et al. 1993; Cole 1990; Boeken et al. 1995, 1998; Abdel-Magid et al. 1987).

In the Negev, trampling is a livestock-induced engineering that counteracts the integrated cyanobacteria–shrub engineering by fragmenting the uniform crust (T1 in Figure 17.2) and flattening the soil mound (T2 in Figure 17.2). Hence, livestock grazing in the Negev introduces engineering activities that modify the structures—soil crust and soil mound—created by the two natural EEs as well as their combined effect on water regime (T3 in Figure 17.2).

Based on field studies that include livestock grazing and browsing (Zaady et al. 2001) and simulated shrub biomass removal and crust trampling (Oren 2001) we propose a model that integrates domestic grazers and browsers (Figure 17.2, black box and arrows) into the cyanobacteria–shrub engineering network (Figure 17.2, green, blue, and red boxes and arrows). The model incorporates the basic properties of the cyanobacteria–shrub network and the engineering effects of browsing and trampling by livestock on the network, obtained from studies in the Negev (Boeken and Shachak 1998, Shachak et al. 1998, Shachak and Lovett 1998, Zaady et al. 2001, Oren 2001, Eldrige et al. 2002, Wright et al. 2006). The basic properties of the cyanobacteria–shrub–livestock network are as follows:

1. A landscape created by two engineers (shrubs and cyanobacteria) decreases soil erosion in comparison to uniform crusted landscape formed by one engineer, cyanobacteria. This effect is due to soil accumulation under the shrubs and the creation of characteristic soil mound underneath most shrubs.
2. Simulated shrub browsing disturbs the engineering function of the shrubs as a sink for soil. Removing the shrub exposes the soil mound to erosion, thus increasing soil output from the system.
3. A landscape created by two engineers (shrubs and cyanobacteria) decreases runoff water and increases soil moisture in the shrub patch in comparison with uniform crusted landscape. In this case the crust patch functions as a source and the shrub patch functions as a sink for runoff water. Water conservation is higher in landscapes modulated by shrubs and cyanobacteria since less runoff water outflows from a two-phase mosaic compared with a landscape covered only by crust.
4. Shrub browsing disturbs the engineering function of the shrubs as a sink for runoff water. Disturbance to the shrub reduces the size of the soil mound and decreases its efficiency as a sink for water.
5. Density and species richness of annuals in the two-phase mosaic is higher than in a crusted landscape. This is mainly due to the facilitation effect of the shrub, which creates a water-enriched patch and safe sites for seed accumulation and establishment.
6. Density and richness of annual plant assemblage positively respond to the engineering effect of trampling on the crust. This positive

response is attributed to increase in water infiltration into the soil underneath the crust and the consequent increase of soil moisture as well as seed trapping in the fragment crust and lower seed predation risk.

7. Simulated browsing by shrub clipping lowers annual plant density while the effect on species richness is indiscernible. This negative effect could be ascribed to the destructive effects of shrub browsing on soil mound that result in a decrease in soil moisture.

The experimental results were integrated into an EE network conceptual model that is composed of three processes (Figure 17.2, horizontal arrows): (1) dust transformation into either crusted soil controlled by cyanobacteria or into soil mounds controlled by shrubs; (2) rainfall conversion by crust into runoff and runoff conversion into soil moisture by soil mound; and (3) formation of specific species assemblages of annual plants on the crust and the soil mound. In this model, livestock modify the engineering network by crumbling of the soil crust by trampling and flattening the soil mound by shrub browsing and trampling (T1 and T2 in Figure 17.2). They modify the network in several ways (Figure 17.2, black arrows): (1) adding the fragmented crust as a state variable (Figure 17.2, black box); (2) controlling water leaking (BG2 in Figure 17.2) and soil erosion (BG1 in Figure 17.2); and (3) controlling the distribution and abundance of annual plant species (T3 and BG3 in Figure 17.2).

Some generalizations emerge from the integration of livestock into the natural engineering network:

1. Livestock adds a new state variable fragmented crust to the network. In this way livestock counteracts the cyanobacteria engineering.
2. Livestock grazing is involved in already existing processes and does not add new ones. However, it significantly intervenes in all processes in the network.

The effects shown in our model (Figure 17.2) of domestic grazing and browsing animals (DGA) on ecosystem structure and function have also been intensively studied in other ecosystems (Mwendera and Mohamed Saleem 1997, Bari et al. 1993, Cerda and Lavee 1999, Jones 2000). It is well recognized that DGA affect all levels of organization—landscape (Oksanen and Oksanen 1989, Senft et al. 1987), ecosystem (Fuls 1992, Skarpe 1991), community (Milchunas et al. 1998, Osem et al. 2002), and population levels (Crawley 1989, Bastrenta et al. 1995).

Integrating livestock into the cyanobacteria–shrub engineering network model can shed new light on the effects of domestic grazing and browsing animals via ecological exploitation and modulation (Mwendera and Mohamed Saleem 1997, Bari et al.1993, Cerda and Lavee 1999, Jones 2000).

Based on the model we suggest distinguishing between processes of exploitation vs. modulation of the environment by livestock. By exploitation we refer to the consumption or removal of plant biomass, mostly of herbaceous vegetation, by the grazing animals. By modulation we refer to changes in physical attributes of the landscape, such as those induced by trampling or by browsing on shrubs. When DGA exploit resources they function as herbivores and they are a part of the food web of the system (EX in Figure 17.2). When DGA modulate the environment they function as ecosystem engineers (EEs) and are part of the engineering network (T1, T2, T3, BG1, BG2, BG3 in Figure 17.2). The model shows that exploitation (grazing) and modulation (trampling and browsing) by livestock are coupled and form a feedback loop that operates on different levels of organization (Figure 17.2). The model depicts that livestock grazing includes interweaving exploitation and modulation pathways cascading through three levels of organization. The exploitation activities—browsing and grazing—generate two ecosystem-level processes: soil erosion and runoff production (Figure 17.2). The exploitation–modulation relationship is mediated by landscape-level modulation by creating the two-phase mosaic. Increases in soil erosion and runoff generation are consequences of changes in the size of vegetation patches due to grazing and browsing. The exploitation–modulation trajectory operates as follows: DGA induce first the complete or partial removal of shrub and herbaceous biomass. Consequently, the shrub patch structure is altered, which in turn modulates resource distribution. The modulated resource distribution affects herbaceous plant community structure, which feeds back to livestock exploitation.

The landscape modulation activity—trampling—modifies the rather homogeneous soil crust by crumbling it. This modulation increases water infiltration and stimulates annual plant species richness and density, and feeds back to exploitation by providing more plant biomass for DGA.

The two pathways demonstrate the interrelationship and feedback between exploitation and modulation. Both exploitation and modulation modify landscape structure, which, in turn, affects the plant community (Figure 17.2). Consequently, the modified plant community feeds back to the livestock foraging activities. Therefore, livestock modulation and exploitation is cyclic and is, in fact, mediated by landscape, ecosystem, and community responses to modulation and exploitation. The engineering network model with DGA implies that addressing DGA–

plant relationships merely by analyzing the herbivory impact may be limited and misleading. A comprehensive analysis of DGA–plant relationships should also incorporate the network of interactions among other levels of organization induced by modulation.

The utility of a comprehensive network model as a prerequisite for livestock management in an engineered landscape is delineated in the next section.

17.3 ☙ NEGEV DESERT MANAGEMENT: EXPLOITATION AND MODULATION

PASTORALISM

In this section we examine traditional (livestock husbandry) and modern (afforestation for recreational activity) land uses practiced in the Negev in light of the EE network model.

The traditional subsistence activity that has been practiced in the Negev Desert over millennia is nomadic pastoralism (Noy-Meir and Seligman 1979, Finkelstein and Perevolotsky 1990). *Pastoralism* relates to economy based on livestock husbandry; *nomadic* refers to continuous movement in search of essential resources, mostly water and forage, by humans and herds. First testimonies for pastoral activity in the Negev are from the Pre-Pottery Neolithic B Period, almost 8000 years ago (Avner et al. 1994). Avner and his colleagues located 1400 ancient sites, mostly related to pastoralism, within a 1200 km² area, demonstrating the long-term existence of livestock in the Negev. The Bedouin, who are the current dwellers of the Negev, proved to be prudent pastoralists and have exploited the desert resources for a long time (Abu-Rabia 1994). In the Sinai, a neighboring desert, Bedouin pastoralists demonstrated high skills and ecological understanding in exploiting the complex mountainous environment around Mt. Sinai (Perevolotsky et al. 1989, Perevolotsky et al. 2005).

In the Negev, livestock feed mainly on green annual plants that occur, following rainfall events, either in the crusted, inter-shrub area or on the soil mound underneath shrubs. During spring, when the vegetation is green and lush, the animals graze mainly in the inter-shrub area. This is the productive season in the desert and the peak of secondary productivity (milk and offspring) (Perevolotsky et al. 1989, Abu-Rabia 1994). By the early summer, annual vegetation in the crusty area withers and most of it has already been consumed. This is the migration time when Bedouin search for forage resources far from their traditional rangelands. In the fall, when the herds return from their annual migration to other ecological regions or to the stubble fields in the cultivated area, the dry forage underneath the shrubs plays a crucial role in the nutrition of the livestock. In this period, when the desert is very poor in terms of forage resources, the under-shrub dry herbaceous vegetation is the only resort the Bedouin can use before starting to supplement the livestock with costly feed such as imported grains. One may consider the soil mounds under the shrubs as small "hay barns" of the desert.

Bedouin livestock husbandry, like other pastoral systems in arid environments, is closely related to the dynamics of primary production (herbaceous forage) that is correlated with water regime and patch type. When grazing is introduced to the paired EE network, primary production on the crust increases over time due to trampling, while annual vegetation under shrubs decreases due to less runoff and direct trampling on the mound (Figure 17.3). Once grazing is excluded, the crust recovers, soil moisture decreases, and the primary production on the crust also decreases. On the shrub mound the process is reversed: More resources (water and nutrients) flow from the crust to the mound (shrub patch) and stimulate higher primary production (Figure 17.3a). In the meantime, shrub patches are also affected by livestock trampling: Primary production under shrubs decreases due to less runoff. When grazing is excluded the crust recovers and stimulates higher primary production under the shrub. Along the same line we can analyze the effect of livestock browsing on the shrub patches. Browsing reduces the size of the shrub and therefore the primary production of annuals underneath it decreases (Figure 17.3b). The exact impact of browsing is dependent on the shrub palatability (different colors in Figure 17.3b). The size of very palatable shrubs decreases rapidly while its recovery period is extended (red line in Figure 17.3b); the more adapted, less palatable shrubs demonstrate relatively small spatial changes due to browsing, therefore primary production underneath it is relatively stable and the recovery process is shorter.

As a consequence, sustainable management of such paired EE ecosystems requires a spatially shifting management, which combines grazing periods with resting (non-grazing) periods. The length and timing of these periods depend on the recovery rate of the crust and the shrubs, which is a function of local environmental conditions, shrub palatability, and pastoral systems (Figure 17.3).

Traditional pastoral management is composed of livestock husbandry practice including grazing intensity, livestock type, grazing timing and duration, and shrub removal by browsing animals such as goats and camels and by firewood collection. Sustainable range management of this environment should, therefore, incorporate our ecological knowledge

a. Crust trampling

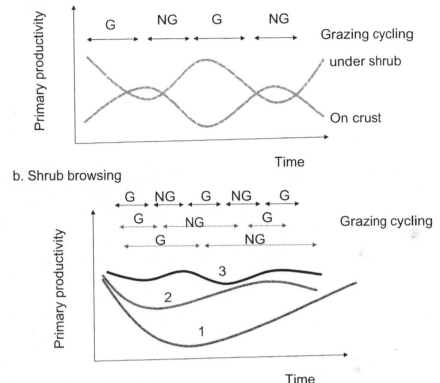

FIGURE 17.3 A conceptual model of the effects of crust trampling (a) and shrub browsing (b) on primary productivity related to shrub and crust patch types under shifting of grazing (G) and non-grazing (NG) periods. 1, 2, and 3 refer to shrub traits: 1 = high palatability and low recovery; 2 = intermediate palatability and recovery; 3 = low palatability and high recovery. (See color plate.)

about the function of the EE network. Practically, the management should aim toward establishing the correct balance among the following:

1. Undisturbed crusted surface that produces runoff, which, in turn, feeds the herbaceous vegetation underneath the shrubs.
2. The area covered by shrubs, which should not exceed a certain space since most shrubs are unpalatable and compete for space with herbaceous vegetation on the crust.
3. Certain trampling impact that improves soil moisture in the crust and supports production of herbaceous vegetation in this habitat.
4. Limited damage to the remaining shrubs by browsing or firewood collection in order to minimize the damage to soil mounds and the herbaceous vegetation that develops on them.

We infer that the traditional Bedouin livestock management, including the territoriality rules (Perevolotsky 1997) and migration cycles, comply with the four management principals, thus preventing drastic deterioration and degradation of the system.

RECREATION

Deserts do not provide many natural opportunities for recreation. The factor most limiting for this activity is the scarce sources of shade. Not many trees grow naturally in the desert and modern management seeks to artificially create centers where the long stay of visitors is facilitated. This is the ultimate goal of the Savannization project, initiated by the KKL (the Israeli Forest Authority) in the 1990s, in Sayeret Shaked Park, located in the northern Negev (~200 mm). The management concept on which this park had been established is the collection of runoff water from large enough crust surface. The runoff is accumulated along man-made contour banks, thus creating water-enriched catchments in which trees were planted (Shachak et al. 1998). The water-enriched catchments support rapid development of the trees, which later become a basis of small shade groves ("savannization"). The creation of such recreational areas is again based on the understanding and directed manipulation of the EE network. The basic concept here is the optimization of runoff production from crusted surface. Leaving a too-large crust surface will create a very sparse pattern of tree development. On the other hand, making the contour banks too close to each other will result in the generation of runoff that is too limited to support adequate development of trees. Therefore, the proper management should aim to optimize the runoff contributing area with the density of banks to create sustainable yet functioning afforestation.

Managing livestock activity within the context of the savannization project also introduces some optimization issues. On one hand, livestock activity may interfere with both runoff production (by heavy trampling) and the establishment of the planted trees (by browsing). However, in order to maintain maximum runoff production, herbaceous vegetation and shrubs should be kept at a low level. Only grazing and browsing by animals can assure this in a sustainable and environmentally friendly manner. If correctly managed grazing may have only a very light impact on the diversity of the annual plant community (Zaady et al. 2001).

17.4 ⬤ CONCLUDING REMARKS

The essence of management in the Negev Desert is related to the area and state of the two principal EEs—soil crust and shrubs—that control

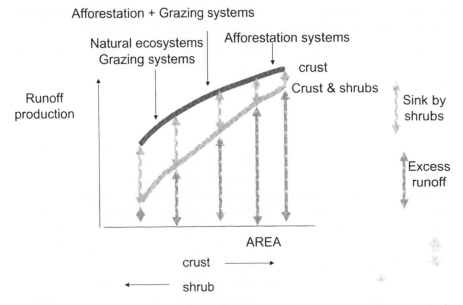

FIGURE 17.4 Ecosystem management in the Negev in relation to cyanobacteria–shrub engineering network. The four managed systems are located on a shrub-to-crust ratio gradient. Afforestation systems are managed for low shrub-to-crust ratio. Natural and grazing systems are managed for high shrub-to-crust ratio. For combined afforestation and grazing systems, an intermediate ratio is desirable. Brown line = runoff generation by crust; green line = runoff generation by crust and shrubs. (See color plate.)

the water regime and thus the primary production in the system. Figure 17.4 is a conceptual model summarizing the relationship between shrubs, crust, water regime, and management. When the system is dominated by only one EE (cyanobacteria), runoff production is mainly a function of the size of the crusted area (brown line, Figure 17.4). In a paired EE system, runoff is a negative function of area covered by shrubs (green line, Figure 17.4). The area between the two lines represents runoff absorbed by the shrubs. The area under the green line represents the leakage of water from the system. The various management options—natural area, rangeland, afforestation, and afforestation with livestock grazing—are defined by the crust:shrub ratio (Figure 17.4). For nature reserves and livestock production purposes, the management strategy should be to maintain high shrub:crust ratio so most water will remain in the system and the excess runoff will be small. For afforestation the opposite strategy should be employed, namely, low shrub:crust ratio, which yields more runoff for the tree usage. If a combined system of afforestation and livestock husbandry is desired, an intermediate ratio is recommended.

In summary, the various management options in the Negev are based on the optimization of the spatial ratio between the two patch types created by the principal EEs. In fact, the crux of the management in the Negev is the optimal landscape modulation according to desired exploitation.

REFERENCES

Abdel-Magid, A.H., Trilca, M.J., and Hart, R.H. (1987). Soil and vegetation response to simulated trampling. *Journal of Range Management* 40:303–306.

Abu-Rabia, A. (1994). *The Negev Bedouin and Livestock Rearing.* Oxford: Berg.

Alkon, P.U. (1999). Microhabitat to landscape impacts: Crusted porcupine digs in the Negev Desert highlands. *Journal of Arid Environments* 41(2):183–202.

Andrew, M.H. (1988). Grazing impact in relation to livestock watering points. *Trends in Ecology and Evolution* 3:336–339.

Avner, U., Carmi, I., and Segal, D. (1994). Neolithic to Bronze Age settlement of the Negev and Sinai in light of radiocarbon dating: A view from the southern Negev. In *Radiocarbon: Late Quaternary Chronology and Paleoclimate of the Eastern Mediterranean*, O. Bar-Yosef and R.S. Kra, Eds. Tucson, AZ: University of Arizona Press.

Bari, F.B., Wood, M.K., and Murray, L. (1993). Livestock grazing impacts on infiltration rates in a temperate range of Pakistan. *Journal of Range Management* 46: 367–372.

Bastrenta, B., Lebreton, J.D., and Thompson, J.D. (1995). Predicting demographic change in response to herbivory: A model of the effects of grazing and annual variation on the population dynamics of *Anthyllis vulneraria. Journal of Ecology* 83:603–611.

Berger, Y. (2006). *Linking landscape and species diversities: The case of woody vegetation patchiness and beetle species turnover.* Master of science thesis, Ben Gurion University, Israel.

Blackburn, W.H., (1975). Factors influencing infiltration and sediment production of semiarid rangelands in Nevada. *Water Resources Research* 11:929–937.

Boeken, B., and Shachak, M. (1998). Colonization by annual plants of an experimentally altered desert landscape: Source-sink relationships. *Journal of Ecology* 86:804–814.

——. (1998). The dynamics of abundance and incidence of annual plant species during colonization in a desert. *Ecography* 21:63–73.

Boeken, B., Shachak, M., Gutterman, Y., and Brand, S. (1995). Patchiness and disturbance: Plant community response to porcupine diggings in the central Negev. *Ecography* 18:410–422.

Burke, I.C. (1989). Control of nitrogen mineralization in a sagebrush steppe landscape. *Ecology* 70:1115–1126.

Cole, D.N. (1990). Trampling disturbance and recovery of cryptogamic soil crusts in Grand Canyon National Park. *Great Basin Naturalist* 50:321–325.

Cornet, A.F., Montaña, C., Delhoume, J.P., and Lopez-Portillo, J. (1992). Water flows and the dynamics of desert vegetation stripes. In *Landscape Boundaries*, A.J. Hansen and F. di Castri, Eds. New York: Springer-Verlag, pp. 327–345.

Crawley, M.J. (1989). Insect herbivores and plant population dynamics. *Annual Review of Entomology* 34:531–564.

Creda, A., and Lavee, H. (1999). The effect of grazing on soil and water losses under arid and Mediterranean climates: Implications for desertification. *Pirineos* 153-4:159–174.

Eldridge, D.J., and Greene, R.S.B. (1994). Microbiotic soil crusts: A review of their roles in soil and ecological processes in the rangelands of Australia. *Australian Journal of Soil Research* 32:389–415.

Eldridge, D.J., Zaady, E., and Shachak, M. (2002). Microphytic crusts, shrub patches and water harvesting in the Negev Desert: The Shikim system. *Landscape Ecology* 17:587–597.

Elkins, N.Z., Sabol, G.V., Ward, T.J., and Whitford, W.G. (1986). The influence of subterranean termites on the hydrological characteristics of a Chihuahuan desert ecosystem. *Oecologia* 68:521–528.

Finkelstein, I., and Perevolotsky, A. (1990). Process of sedentarization and nomadization in the history of Sinai and the Negev. *Bulletin of the American Schools of Oriental Research* 279:67–88.

Fuls, E.R. (1992). Ecosystem modification created by patch-overgrazing in semi-arid grassland. *Journal of Arid Environment* 23:59–69.

Gilad, E., von Hardenberg, J., Provenzale, A., Shachak, M., and Meron, E. (2004). Ecosystem engineers: From pattern formation to habitat creation. *Physical Review Letters* 93(098105):1–4.

Graetz, R.D., and Tongway, D.J. (1986). Influence of grazing management on vegetation, soil structure and nutrient distribution and the infiltration of applied rainfall in a semi-arid chenopod shrubland. *Australian Journal of Ecology* 11:347–360.

Greenwood, K.L., and McKenzie, B.M. (2001). Grazing effects on soil physical properties and the consequences for pastures: A review. *Australian Journal of Experimental Agriculture* 41:1231–1250.

Gutiérrez, J.R., Meserve, P.L., Contreras, L.C., Vásquez, H., and Jaksic, F.M. (1993). Spatial distribution of soil nutrients and ephemeral plants underneath and outside the canopy of *Porlieria chilensis* shrubs (Zygophyllaceae) in arid coastal Chile. *Oecologia* 95:347–352.

Halvorson, W.L., and Patten, D.T. (1975). Productivity and flowering of winter ephemerals in relation to Sonoran Desert shrubs. *American Midland Naturalist* 93:311–319.

Jones, A. (2000). Effects of cattle grazing on North American arid ecosystems: A quantitative review. *Western North American Naturalists* 60:155–164.

Lange, O.L., and Belnap, J. (2003). *Biological Soil Crusts: Structure, Function, and Management*, Ecological Studies 150. Berlin: Springer-Verlag.

Ludwig, J.A., and Tongway, D.J. (1995). Spatial organisation of landscapes and its function in semi-arid woodlands, Australia. *Landscape Ecology* 10:51–63.

Milchunas, D.G., Lauenroth, W., and Burke, I.C. (1998). Livestock grazing: Animal and plant biodiversity of shortgrass and the relationship to ecosystem function. *Oikos* 83:65–74.

Milchunas, D.G., Sala, O., and Lauenroth, W. (1988). A generalized model of the effects of grazing by large herbivores on grassland community structure. *American Naturalist* 132:87–106.

Moro, M.J., Pugnaire, F.I., Haase, P., and Puigdefábregas, J. (1997). Mechanisms of interaction between a leguminous shrub and its understory in a semi-arid environment. *Ecography* 20:175–184.

Mott, J.J., and McComb, A.J. (1974). Patterns in annual vegetation and soil microrelief in an arid region of western Australia. *Journal of Ecology* 62:115–126.

Mwendera, E.J., and Mohamed Saleem, M.A. (1997). Hydrologic response to cattle grazing in the Ethiopian highlands. *Agriculture, Ecosystems and Environment* 64:33–41.

Noy-Meir, I. (1973). Desert ecosystems: Environments and producers. *Annual Review of Ecology and Systematics* 4:25–51.

Noy-Meir, I., and Seligman, N.G. (1979). Management of semi-arid ecosystems in Israel. In *Management of Semi-Arid Ecosystems*, B.H. Walker, Ed. Amsterdam: Elsevier, pp. 113–160.

Oksanen, L., and Oksanen, T. (1989). Natural grazing as a factor shaping out barren landscape. *Journal of Arid Environments* 17:219–233.

Olff, H., and Ritchie, M.E. (1998). Effects of herbivores on grassland plant diversity. *Trends in Ecology and Evolution* 13:261–265.

Oren, Y. (2001). *Patchiness, disturbances, and the flow of matter and organisms in an arid landscape: A multi-scale experimental approach.* Ph.D. thesis, Ben Gurion University, Beer Sheva.

Osem, Y., Perevolotsky, A., and Kigel, J. (2002). Grazing effect on diversity of annual plant communities in a semi-arid rangeland: Interactions with small-scale spatial and temporal variation in primary productivity. *Journal of Ecology* 90:936–946.

Osem, Y., Perevolotsky, A., and Kigel, J. (2004). Site productivity and plant size explain the response of annual species to grazing exclusion in a Mediterranean semi-arid rangeland. *Journal of Ecology* 92:297–309.

Perevolotsky, A. (1987). Territoriality and resource sharing among the Bedouin of southern Sinai: A socio-ecological interpretation. *Journal of Arid Environments* 13:153–161.

Perevolotsky, A., and Seligman, N.G. (1998). Role of grazing in Mediterranean rangeland ecosystems. *BioScience* 48:1007–1017.

Perevolotsky, A., Perevolotsky A., and Noy-Meir, I. (1989). Environmental adaptation and economic change in a pastoral nomadic society: The case of the Jebaliyah Bedouin of the Mt. Sinai region. *Mountain Research and Development* 9:153–164.

Perevolotsky, A., Shachak, M., and Pickett, T.A. (2005). Management for biodiversity: Human and landscape effects on dry environments. In *Biodiversity in Drylands: Toward a Unified Framework*, M. Shachak, J.R. Gosz, S.T.A. Pickett, and A. Perevolotsky, Eds. New York: Oxford University Press.

Pickett, S.T.A., Shachak, M., Boeken, B., and Armesto, J.J. (1999). Management of ecological systems. In *Arid Lands Management—Toward Ecological Sustainability*, T.W. Hoekstra and M. Shachak, Eds. Urbana: Illinois University Press, pp. 8–17.

Proffitt, A.P.B., Bendotti, M., Howell, R., and Eastham, J. (1993). The effect of sheep trampling and grazing on soil physical properties and pasture growth for a red-brown earth. *Australian Journal of Agricultural Research* 44:317–331.

Pugnaire, F.I., Haase, P., Puigdefábregas, J., Cueto, M., Clark, S.C., and Incoll, L.D. (1996). Facilitation and succession under the canopy of a leguminous shrub, *Retama sphaerocarpa*, in a semi-arid environment in south-east Spain. *Oikos* 76:455–464.

Senft, R.L., Coughenour, M.B., Bailey, D.W., Rittenhouse, L.R., Sala, O.E., and Swift, D.M. (1987). Large herbivore foraging and ecological hierarchies. *BioScience* 37:789–799.

Skarpe, C. (1991). Impact of grazing in savanna ecosystems. *Ambio* 20:351–356.

Shachak, M., Jones, C.J., and Brand, S. (1995).The role of animals in arid ecosystems: Snails and isopods as controllers of soil formation, erosion and desalinization. *Advances in GeoEcology* 28:37–50.

Shachak, M., and Lovett, G.M. (1998). Atmospheric deposition to a desert ecosystem and its implication for management. *Ecological Applications* 8:455–463.

Shachak, M., Sachs, M., and Moshe, I. (1998). Ecosystem management of desertified shrublands in Israel. *Ecosystems* 1:475–483.

Shachak, M., and Yair, A. (1984). Population dynamics and the role of *Hemilepistus reaumuri* in a desert ecosystem. *Symposium of the Zoological Society of London* 53:295–314.

Shmida, A., Evenari, M., and Noy-Meir, I. (1986). Hot desert ecosystems: An integrated view. In *Hot Deserts and Arid Shrublands*, M. Evenari, I. Noy-Meir, and D.W. Goodall, Eds. Leiden: Elsevier, pp. 379–387.

Smith, J.L., Halvorson, J.J., and Bolton, H. Jr. (1994). Spatial relationships of soil microbial biomass and C and N mineralization in a semi-arid shrub-steppe ecosystem. *Soil Biology and Biochemistry* 26:1151–1159.

Turner, M.G., Romme, W.H., Gardner, R.H., O'Neill, R.V., and Kratz, T.K. (1993). A revised concept of landscape equilibrium: Disturbance and stability on scaled landscapes. *Landscape Ecology* 8:213–227.

Weltz, M., Wood, M.K., and Parker, E.E. (1989). Flash grazing and trampling: Effects on infiltration rates and sediment yield on a selected New Mexico range site. *Journal of Arid Environments* 16:95–100.

West, N.E. (1981). Nutrient cycling in desert ecosystems. In *Arid Land Ecosystems: Structure, Functioning and Management*, Vol. 2, D.W. Goodall and R.A. Perry, Eds. Cambridge, UK: Cambridge University Press, pp. 301–324.

West, N.E. (1989). Spatial pattern–functional interactions in shrub dominated plant communities. In *The Biology and Utilization of Shrubs*, C.M. McKell, Ed. London, UK: Academic Press, pp. 283–305.

West, N.E. (1990). Structure and function of microphytic soil crusts in wildland ecosystems of arid to semi-arid regions. *Advances in Ecological Research* 20:179–223.

Wilby, A., Shachak, M., and Boeken, B. (2001). Integration of ecosystem engineering and trophic effects of herbivores. *Oikos* 92:436–444.

Wright, J.P., and Jones, C.G. (2006). The concept of organisms as ecosystem engineers, ten years on: Processes, limitations, and challenges. *BioScience* 56(3):203–209.

Yair, A., and Shachak, M. (1987). Studies in watershed ecology of an arid area. In *Progress in Desert Research*, M. O. Wurtele and L. Berkofsky, Eds. Lanham, MD: Rowman and Littlefield Publishers, pp.146–193.

Zaady, E., Groffman, P.M., and Shachak, M. (1996). Litter as a regulator of N and C dynamics in macrophytic patches in Negev Desert soils. *Soil Biology and Biochemistry* 28:39–46.

Zaady, E., and Shachak, M. (1994). Microphytic soil crust and ecosystem leakage in the Negev Desert. *American Journal of Botany* 81:109.

Zaady, E., Yonatan, R., Shachak, M., and Perevolotsky, A. (2001). The effects of grazing on abiotic and biotic parameters in a semiarid ecosystem: A case study from the northern Negev Desert, Israel. *Arid Soil Research and Management* 15:245–261.

18

ECOSYSTEM ENGINEERS AND THE COMPLEX DYNAMICS OF NON-NATIVE SPECIES MANAGEMENT ON CALIFORNIA'S CHANNEL ISLANDS

Rob Klinger

18.1 ⬝ INTRODUCTION

Applying the concept of ecosystem engineer (EE, from here on) to restoration programs has recently been suggested as a way of increasing the likelihood of shifting a system to a more desired state (Byers et al. 2006). In this chapter I examine this suggestion as it relates to management of non-native animals and plants on California's Channel Islands. Feral goats, sheep, and pigs have not only had large-scale effects on abiotic and biotic elements of the islands, but responses to management of these species have been complex, varied, and sometimes surprising. Of perhaps greatest interest is that the responses resulted from interactions among the non-native species, some of which could potentially be considered ecosystem engineers (Byers et al. 2006).

Byers et al. (2006) suggest that, had ecosystem engineering been considered prior to removal of feral sheep and cattle from Santa Cruz Island, feral pig eradication would have preceded that of the grazers. This was because a rapid increase in non-native plants occurred after being released from grazing, and this hampered management of the pigs in the decade after the sheep and cattle were removed. The Byers et al. (2006) suggestion is not inaccurate or inappropriate, but management decisions on Santa Cruz Island had to be made in the context of considerable ecological uncertainty and a number of practical constraints. Because this is representative of the situation for many conservation programs, control and eradication of non-native species on the Channel Islands provide an excellent case study of the more general issue of how ecological concepts such as ecosystem engineering can be applied to management programs.

My approach will be to summarize some of the ecological and cultural aspects of the Channel Islands, describe non-native plant and animal impacts and outcomes from their management on Santa Cruz Island (SCI), and then address three main questions: (1) to what degree can non-native species on the Channel Islands be considered ecosystem engineers; (2) in hindsight, would application of the EE concept have altered management decisions or improved predictions of outcomes from non-native species management programs on the Channel Islands; and (3) how does the EE concept fit into future and ongoing non-native species management programs on the Channel Islands, as well as in other systems? While most of the emphasis will be on SCI, the examples are representative of issues common to the other Channel Islands, and likely many insular and mainland systems as well.

18.2 ● OVERVIEW OF CALIFORNIA'S CHANNEL ISLANDS

The Channel Islands consist of eight main islands in a 90-mile-long archipelago from 34°5″ to 32°48″ north latitude (Figure 18.1). They range in size from 2.6 km² to 249 km² and are 20 to 98 km from the mainland. The climate is a Mediterranean-type, but there is a strong north–south gradient in precipitation; rainfall on SCI averages 50 cm/year while San Clemente receives only 20 cm/year.

Scientific interest in the Channel Islands dates back to the nineteenth century, and a great deal is known about their natural history (Philbrick 1967, Power 1980, Hochberg 1993, Halvorson and Maender 1994, Browne et al. 2000, Garcelon and Schwemm 2005). As with most islands, their flora and fauna are depauperate when compared with similar sized areas on the mainland (Raven 1967). However, the moderate climate, topographic and geologic diversity, and range in size and isolation of the islands have resulted in vegetation that is both high in endemic species and characterized by a surprising diversity of plant communities (Junak et al. 1995). The total number of native plant species ranges from 88–480, with 7–13% of those being endemic. The most common vegetation communities are grasslands, coastal scrub, dune, oak woodland, chaparral, riparian, and remnant closed-cone conifer forests (Junak et al. 1995).

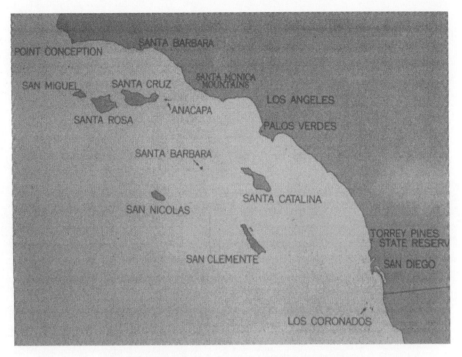

FIGURE 18.1 The California Channel Islands. (Adapted from Philbrick 1967.)

Humans have occurred on the islands for at least the last 9000 years (Glassow 1980). The extent and magnitude to which indigenous tribes altered the islands ecosystems are not well understood, but beginning in the nineteenth century changes associated with land use by Spanish and American settlers began to occur. The most widespread and serious of these changes were related to ranching and agricultural operations. Ungulates were introduced to the Channel Islands in the mid-nineteenth century, and by the 1880s there were established populations of feral sheep, goats, and pigs on virtually all of the islands (Halvorson 1994). Severe overgrazing and rooting resulted in massive erosion, altered hydrology, loss and fragmentation of vegetation communities (Figure 18.2), and extirpation of species (Brumbaugh 1980, Hobbs 1980, Minnich 1980, Junak et al. 1995). Exacerbating the effects of the feral animals were non-native plants that were introduced along with the animals. The heavy browsing and overgrazing of the native vegetation communities helped non-native plants to rapidly invade and eventually dominate the cover of many parts of the islands. This was most evident where grass-lands comprised predominantly of non-native annual grasses and forbs replaced areas that were once coastal scrub, chaparral, and pine forest (Hobbs 1980). Presently, estimates of the proportion of non-native plant species in the islands flora range from 20–48% (Junak et al. 1995).

FIGURE 18.2 Overgrazing by feral sheep on Santa Cruz Island resulted in many parts of the island where there was little if any vegetation cover. (See color plate.)

18.3 ➧ FERAL SHEEP AND PIGS ON SANTA CRUZ ISLAND

Abundance of feral sheep on SCI in the late nineteenth and early twentieth centuries was at least 50,000 (Van Vuren 1981). Ongoing attempts were made to control or contain the herd with roundups, hunting, and fencing, but these efforts had limited success. The sheep population remained largely unchecked until 1981, and anecdotal information indicates there were substantial fluctuations in their abundance (Van Vuren 1981). There are no data on pig abundance on SCI prior to the 1990s, but ranchers' memoirs provide clear evidence there were years when their population was very high. Rooting was extensive, especially in vineyards and agricultural areas, as well as oak woodlands and grasslands.

The Nature Conservancy (TNC) purchased an interest and a conservation easement in the western 90% of SCI in 1978, and soon began to consider eradication programs for the sheep and pigs (Schuyler 1993). As the initial step in the sheep eradication program they conducted a study of the sheep population ecology (Van Vuren 1981), as well as studies on the effects of overgrazing on the flora (Hochberg et al. 1980), fauna (Laughrin 1982), and soils (Brumbaugh 1980) of the island. TNC also conducted a brief survey of distribution and habitat use by the pigs (Baber 1982). All of the studies documented the enormous effects the sheep were having on the island, including loss of rare plants and vegetation communities, accelerated soil erosion, and reduced populations of some of the native animals (Van Vuren and Coblentz 1987). The strong conclusion from all of the studies was that TNC needed to implement a sheep eradication program as soon as possible. Although TNC felt it was desirable to attempt to eradicate both pigs and sheep simultaneously, the co-owner of the western 90% of the island would consent only to a program for the sheep. TNC began systematic hunting late in 1981 when there were an estimated 20,000 sheep on SCI (Schuyler 1993). By early 1989, 37,171 sheep had been shot off of the western 90% of the island (Schuyler 1993). Between 1500–5000 remained on the eastern 10% of the island through 2001, but systematic hunting and fencing prevented them from re-colonizing the western part of the island. The National Park Service acquired full ownership of the east end of SCI in 1997, and over the next 4 years rounded up and moved the remaining sheep to the mainland.

When the co-owner of the western part of SCI died unexpectedly in December 1987, full ownership of that part of the island passed to TNC. Although they were not prepared to begin a pig eradication program, TNC did decide to remove cattle from the island. Cattle ranching had been the primary land use on the island since the 1930s, with approximately 1800–1900 head occurring over 40–50% of the island during the 1980s. From 1988 to 1989 TNC rounded up all but six of the herd (they were kept for historical purposes) and shipped them to the mainland.

18.4 ● POST-ERADICATION FLORA AND FAUNA DYNAMICS

TNC recognized that there were many potential outcomes of the sheep eradication program (Schuyler 1993). The fundamental and most desirable one was that native biodiversity would be maintained or improved, but they were also aware that other outcomes could be undesirable or have unknown consequences. These included an increase in the pig population because of increased food and cover, increased fire frequency and extent resulting from increased vegetation biomass, and an increase in non-native plant species after being released from grazing pressure (Schuyler 1993). In 1984 a relatively small project was initiated to monitor vegetation change, primarily in grasslands. This was augmented with a broader photo-monitoring program. From 1990 to 1998 TNC implemented a conservation program that combined more extensive monitoring with field experiments to try to gain a better understanding of the dynamics of the pig population, rare plants, vegetation, selected vertebrate communities, and fire.

FERAL PIGS

Annual counts and systematic hunting were conducted from 1990–1999 to collect data on the density and population structure of the pigs. Estimates of their island-wide abundance ranged from 800–5400, with pronounced crashes following years with low rainfall and poor mast crops (Figure 18.3A; R. Klinger unpublished data). A population model indicated that over the 150 years the pigs had been on SCI they were characterized by large fluctuations in population growth, with a long-term geometric mean $\lambda = 1$ (Figure 18.3B; R. Klinger unpublished data).

Rooting occurred in all vegetation types, and in years when pig numbers were high an estimated 8–10% of the island was rooted. Impacts on rare plants could be severe. Populations of *Arabis hoffmannii* (Munz) Rollins and *Thysanocarpus conchuliferus* Greene, two of the rarest endemic plants on SCI, were either destroyed or seriously reduced by pig rooting. At the community level, shrub cover and density were reduced where rooting was frequent and intense, which favored both native and non-native annual forbs and non-native annual grasses.

VEGETATION AND RARE PLANTS

A rapid increase in vegetation cover occurred after a 5-year drought ended in 1991 (Klinger et al. 2002). Cover, density, and recruitment of trees and shrubs increased across the island, especially in chaparral and pine forests (Figure 18.4; Klinger, unpublished data). These were largely native species, including many endemics. However, non-native species comprised a significant proportion of the herbaceous layer in many vegetation types. Mean cover of non-native grasses and forbs was 87% in grasslands, 46% in oak woodlands, and 26% in coastal scrub. Only in chaparral and pine forests did non-native species comprise <10% of the ground cover (Klinger, unpublished data). Ironically, the richness and cover of native herbaceous species were greatest in sites where sheep grazing had been the most intense (Klinger et al. 2002).

(A)

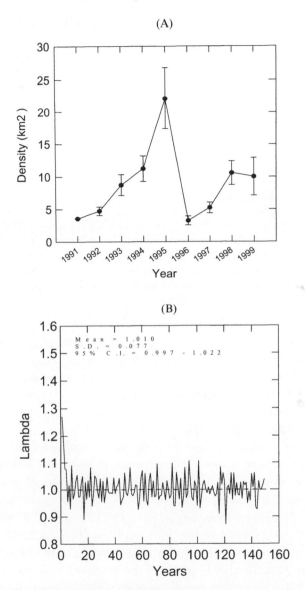

(B)

FIGURE 18.3 (A) Density estimates of feral pigs on Santa Cruz Island, California. (B) Stochastic population model of variation in finite population growth rate (Lambda) for the 150 years feral pigs were on Santa Cruz Island (Klinger, unpublished data).

The distribution of 77% of the island's 43 endemic species increased between 1991 and 1996 (Klinger et al. 2002). Of the five endemic species that were monitored most intensively, the abundance of *Berberis pinnata* Munz remained unchanged, *Malacothamnus fasciculatus* (B.L. Rob.) Kearney and the remaining populations of *Arabis hoffmannii* increased in population size, and *Dudleya nesiotica* Moran and *Thysanocarpus*

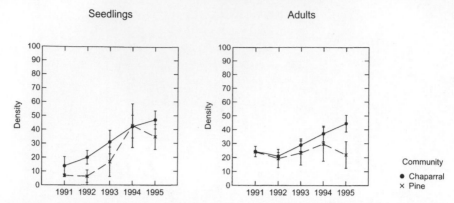

FIGURE 18.4 There was a significant increase in density (# stems/60 m²) of trees and shrubs in chaparral (solid line) and Bishop pine forest (dashed line) communities on Santa Cruz Island, California, in the decade after feral sheep were eradicated from the island. The decrease in density in pine forests in 1995 was due to sheep remaining on the east end of the island re-invading the eastern stand of pines (Klinger, unpublished data).

conchuliferus had serious declines. The decline in *Dudleya nesiotica* was strongly correlated with an increase in cover of alien grasses and forbs and a related buildup of organic litter. Besides the populations that had been rooted by pigs, the disappearance of six other populations of *Thysanocarpus conchuliferus* appeared to be associated with an increase in cover of alien annual grass (Klinger et al. 2002).

The most explosive increase by a single plant species was the perennial non-native forb fennel (*Foeniculum vulgare* Mill.) (Brenton and Klinger 1994). Fennel had occurred on the island since at least the 1880s, but was never particularly dense except in a few localized areas (Beatty and Licari 1992, Junak et al. 1995). Between 1991 and 1992 the percentage of plots with fennel more than doubled (12% to 27%) and mean fennel cover increased from 10% to over 25% (Figure 18.5). In 1995 it occurred in 32% of the plots and mean cover was >38%. By 1998 cover was 50%, including some areas that were virtual monocultures (Figure 18.5).

FIRE

TNC and the National Park Service conducted a fire management and research program on SCI from 1990–1999. The program is summarized in Klinger et al. (2004) and Klinger and Messer (2001), but an increase in fire frequency because of greater vegetation biomass following sheep eradication did not occur.

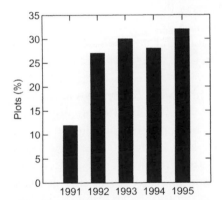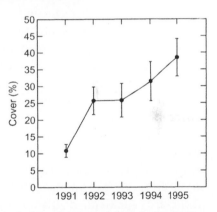

FIGURE 18.5 Primarily because of removal of cattle from Santa Cruz Island, the distribution and cover of the perennial non-native forb fennel (*Foeniculum vulgare*) doubled in a year and continued to increase. By 1998 mean cover was 50% and it occurred across 15% of the island (Klinger, unpublished data).

COMPLEX INTERACTIONS

As anticipated, there was a strong response at both species and community levels to removal of grazers. Some of these responses were not surprising, such as the increase in vegetation biomass. But others, such as changes in abundance of some species, were not expected. More important than responses by individual species though were the complex interactions that presented unanticipated management challenges.

More times than not these interactions involved multiple interacting species, both native and non-native, as well as abiotic factors such as rainfall (Klinger and Messer 2001, Klinger et al. 2002). Though there were a number of examples of these interactions, probably the most useful to examine from the perspective of ecosystem engineering is the expansion of fennel and the subsequent efforts to manage it (Brenton and Klinger 1994, Brenton and Klinger 2002, Ogden and Rejmanek 2005).

Several lines of evidence indicated it was actually the removal of the cattle more than the sheep that led to the increased distribution and abundance of fennel (Beatty 1991, Klinger et al. 2002). But while removal of cattle grazing was certainly a critical influence, Brenton and Klinger (1994) proposed three other contributing factors without which they believed fennel's rapid expansion would not have occurred. These were the species being pre-adapted to the Mediterranean-type climate, its phenology and growth form allowing it to out-compete other herbaceous species for light and moisture, and a 5–6 year drought ending in a series of 3 wet years. Removal of grazing pressure probably allowed fennel to begin storing nutrients in its deep, fleshy taproot. But its phenology, prolific seed production, and tall stature (Klinger 2000) likely gave it an important advantage over potential competitors. This was translated into rapid growth and spread, both vegetative and from seed, once the drought ended.

Questions arose whether pigs were enhancing the spread of fennel, either through dispersing their seeds or by disturbing the soil. Pig abundance was high in the fennel, possibly because the tall, dense stands provided them with more shade, cover, and food than grasslands and coastal scrub. But while there was a positive correlation between fennel cover and rooting intensity (Klinger, unpublished data), this did not necessarily mean pig disturbance created conditions favoring its spread. Fennel cover is greater in loose soils (Klinger, unpublished data), and it may have been that more intense rooting simply occurred in denser patches of fennel because it was easier for pigs to root in them. And while fennel seeds were often found on the hides of pig kills, they were also found on the pelage of rodents and feathers of mist-netted birds. So while pigs may have enhanced the spread of fennel, many native animals did as well.

The structure of the fennel stands had quite different effects on native plants and animals. Species richness of native grasses and forbs had a negative relationship with fennel cover (Figure 18.6A). Not all herbaceous species were shaded out though; the understory of the stands was comprised primarily of non-native annual grasses and forbs (Brenton and Klinger 1994, 2002; Ogden and Rejmanek 2005). In contrast, abundance of native vertebrates, including several endemic species, was

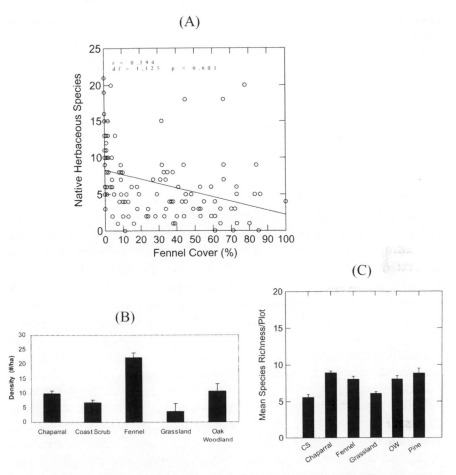

FIGURE 18.6 The spread of fennel between 1991 and 1998 had different effects on native plant and wildlife species on Santa Cruz Island, California: (A) the relationship between the number of native herbaceous species and fennel cover; (B) density of the endemic mouse *Peromyscus maniculatus insularis* in five vegetation types; and (C) mean avian species richness–plot in six vegetation types (Klinger, unpublished data).

greater than in some other vegetation types (Figure 18.6B, C). Presumably this was because of the structural complexity of the stands and higher levels of food.

18.5 ● NON-NATIVE SPECIES AS ECOSYSTEM ENGINEERS AND ECOSYSTEMS WITH MULTIPLE INVADERS

Alteration of ecosystem structure and processes on SCI occurred not just because of the presence of non-native species that had ecosystem-level effects, but also as a result of their removal. Although the EE concept did not exist prior to the removal of sheep and cattle from SCI, it is instructive to ask to what extent could non-native species such as feral sheep, feral pigs, and cattle, as well as fennel and annual grasses and forbs, be considered ecosystem engineers.

It could be argued that the feral sheep, pigs, and cattle do not fit the definition of an EE species. One part of the definition is that the EE process is independent of assimilation (Jones and Gutiérrez, Ch 1), and the denuding of the landscape by the sheep and the extensive rooting by the pigs were, in large part, a result of their feeding behavior. Whether they precisely meet the definition of an EE species though may not be as important as their effects; severe erosion, soil compaction, changes in hydrology, fragmentation and alteration of vegetation community structure and function, and at least local extirpation of species are serious ecosystem consequences by any standard. Another part of the definition of an EE is that they can be considered species that change the abiotic environment with subsequent consequences to biotic components of an ecosystem (Jones and Gutiérrez, Ch 1). The sheep and pigs clearly fit this part of the definition; their effects had profound consequences for both abiotic and biotic parts of the environment on SCI. Following from this are two important considerations though: (1) Some non-native species on SCI with effects consistent with the EE concept were clearly suppressing other potential non-native EE species; and (2) the relative magnitude of the effects of the sheep, pigs, and cattle depended in large part on their abundance.

SCI had been invaded by many non-native animals and plants, and these species had interacted for decades. The ecosystem engineering effects of the sheep, cattle, and pigs were extensive and severe, but they also suppressed (or masked) effects of other species that could also be considered ecosystem engineers. It wasn't until the sheep and cattle were removed that fennel, as well as the non-native annual grasses and forbs, could be thought of in an EE context. Physical features such as ecosystem structure (vegetation height and patchiness), light environment, and moisture availability become much different in stands of fennel and in grasslands dominated by non-native annuals than in the communities they replace (Gordon and Rice 1992, Dyer and Rice 1999, Brown and Rice 2000, Brenton and Klinger 2002, Ogden and Rejmanek 2005). On SCI this resulted in effects on the biotic features of the ecosystems, such as species composition and richness (Klinger and Messer 2001, Brenton and Klinger 2002). In essence, after the sheep and cattle were removed from SCI effects of one group of EE species (sheep and cattle) were replaced by others (fennel and, collectively, non-native annual plants), while effects from another species (pigs) remained largely unaltered.

In contrast with some other EE species (Wright and Jones 2006), none of the non-native species on SCI that could potentially be categorized as ecosystem engineers had intrinsically strong per capita effects. Rooting by pigs probably had a far greater per capita effect on ecosystem processes than grazing and browsing by sheep. But the abundance of feral sheep was at least an order of magnitude greater than that of feral pigs, so on an absolute basis their ecosystem engineering effects were far more extensive. In either case, though, trophic and nontrophic effects of both species on the island's ecosystems were proportional to their abundance. Similarly, the ecosystem effects of fennel and non-native annual herbaceous species did not become serious until they increased in abundance after the sheep and cattle were removed.

If management of one EE species in systems with multiple invaders could result in release of another EE species, then determining when a non-native species is most likely to begin having ecosystem engineering effects becomes very important. Biological invasion can be thought of as a process occurring in phases of colonization, establishment, spread, and equilibrium (Ricklefs 2005, Salo 2005). In addition, invasion effects tend to increase the longer a non-native species has persisted in a region (Rejmanek et al. 2005a). These two aspects of biological invasions may explain a great deal of the complex interactions among the sheep, pigs, cattle, fennel, and annual grasses on SCI. They may also provide a crucial link for evaluating when a non-native species could begin to have significant ecosystem engineer effects, either as a result of their being introduced to an area or as an outcome of management actions.

Because their per capita effects were relatively small when they were not abundant, thinking of species such as sheep, pigs, cattle, fennel, and non-native annual plants as ecosystem engineers when they are in the colonization and establishment phases of invasion is not particularly informative. But as they enter the spread phase and become much more widespread and abundant it does start to be useful to begin thinking of them as potential EE species. Whether they actually start altering ecosystem properties as they enter the equilibrium phase will depend on their long-term mean abundance and interactions with other species. The sheep, pigs, cattle, fennel, and non-native annual plants had co-occurred on SCI for many decades and each was clearly in the equilibrium phase of the invasion process (albeit a dynamic equilibrium). Interaction strengths among them were well established; grazing by the sheep and cattle prevented the plants from having strong ecosystem engineering affects. Removal of the sheep and cattle created a non-equilibrium situation though. As a result, interaction strengths were changed and the fennel and non-native annual plants rapidly increased in abundance and began to have major EE effects.

18.6 ◆ COMPLEXITY, UNCERTAINTY, AND THEIR ROLE IN SHAPING MANAGEMENT DECISIONS

From a conservation perspective, there were many desirable outcomes after the removal of the sheep and cattle from the western part of SCI. Aerial photographs indicated erosion rates had drastically declined, distribution and abundance of many endemic species increased (Schuyler 1993, Klinger et al. 2002), and recovery of woody species and shrub communities was occurring in many parts of the island (Wehjte 1994, Klinger et al. 2002). In these regards, few if any conservation practitioners would say sheep and cattle removal from SCI was not successful.

It was also clear though that the outcomes from the sheep eradication were more complex than anticipated and often depended on factors other than sheep. For example, relatively little attention had been given to outcomes from the cattle removal, but it was this action that contributed substantially to the expansion of fennel. Another example was pig abundance after the eradication. Densities were sometimes high, but dietary overlap between them and sheep is not extensive (Van Vuren 1981, Baber and Coblentz 1987) and variation in their abundance depended on factors other than sheep abundance (R. Klinger unpublished data). A final example was fire frequency. Plant biomass increased following the eradication of the sheep, but there was not an increase in fire frequency. The greater fuel loads certainly made this concern legitimate, but fire frequency on SCI appears to be related more to human ignition sources than fuel loads (Carroll et al. 1993, Keeley 2006, Klinger et al. 2006a), and human access to SCI has been highly regulated and controlled since at least the 1930s.

In part because of the complex and unpredictable outcomes and in part because of limited data, evaluation of the responses after the sheep were eradicated had to be done with an understanding that there was considerable uncertainty on *why* things happened, as opposed to *what* happened, following the eradication. Data were available on some aspects of post-eradication patterns, but they were primarily correlative so strong inferences could not be made.

Even with the successful outcomes of the sheep and cattle removal, TNC was clearly confronted with substantial management challenges. Despite the success of a trial pig eradication (1989–1991) in a 23 km^2

fenced area of SCI (Sterner and Barrett 1991) and complete eradication on Santa Rosa Island (1990–1992) by the National Park Service (Lombardo and Faulkner 2000), in 1990 TNC decided not to pursue an islandwide pig eradication program for an indefinite period of time. Consequently, pigs continued to threaten rare plant populations and rooting was severe and extensive in years when their numbers were high. The continued presence of sheep on the east end of SCI made ongoing hunts and fence repair a regular activity (Van Driesche and Van Driesche 2000). Fennel was spreading and non-native annual grasses and forbs dominated the herbaceous layer over a large proportion of the island, resulting in displacement of native plants, including some rare endemic species. These situations resulted in the bulk of the management effort by TNC in the 1990s being focused on phased, long-term studies of the pigs, fennel, and non-native annual plants (Brenton and Klinger 1994, Klinger 1997, Klinger and Messer 2001, Brenton and Klinger 2002, Klinger et al. 2002, Klinger et al. 2004, Ogden and Rejmanek 2005). In addition, experiments related to fennel and other non-native plants were conducted by researchers from the University of California (Dash and Gliessman 1994; Wenner and Thorp 1994; Barthell et al. 2001, 2005).

The same degree of complexity and uncertainty that characterized outcomes of the sheep eradication typified management experiments during the 1990s. In virtually all of the experiments, variation in many plant and animal responses was due more to rainfall patterns than treatment effects (Klinger and Messer 2001, Klinger et al. 2004, Ogden and Rejmanek 2005), a pattern that data from the monitoring program indicated was occurring across the island. It was also apparent that simple treatments alone would rarely if ever result in long-term reduction of non-native plant abundance (Brenton and Klinger 1994, Klinger and Messer 2001, Ogden and Rejmanek 2005). The management experiments and monitoring data provided a great deal of insight on what outcomes could realistically be expected to occur and how long they could be expected to persist. But they also reinforced the difficulty involved in shifting systems dominated by multiple non-native species to ones dominated by native species. This was especially apparent in situations where native animals and plants had different patterns of abundance in the non-native-dominated systems (Figure 18.6; Ogden and Rejmanek 2005). It became clear that management actions favorable to some native species could be unfavorable to others.

Given the complexity and uncertainties associated with outcomes of the sheep eradication program, would application of the EE concept have changed any of the management considerations or decisions related to the program? In all likelihood the answer is no. There were few if any precedents for an eradication on the scale of the sheep from SCI (Schuyler 1993), so it was difficult to precisely know what to expect. The impacts from the sheep were severe and ongoing, the recommendations from scientists and managers were consistently in favor of immediate eradication, and delay could potentially have resulted in extirpation of rare species. Although it wasn't articulated in EE terms, TNC clearly recognized that there was a good chance that populations of non-native species would increase after the eradication. But erosion rates were so severe that establishing vegetation cover, even if much of it was non-native, was imperative. From an EE perspective or not, there is little argument from anyone that it would have been optimal to try to eradicate the pigs at the same time as the sheep (Byers et al. 2006). But TNC was constrained by legal agreements from doing this. And, in hindsight, it is easy to suggest that greater attention should have been paid to potential outcomes from the cattle removal. But given the removal of the sheep it would have been difficult to justify the presence of a couple thousand head of cattle, and it is unlikely anyone would have been able to predict how rapidly fennel would expand. There is little that the EE concept could have done to change these considerations.

In 1993 the EE concept was informally suggested as a potential conceptual basis for some of the management activities on SCI (J. Crooks, personal communication). It was quite obvious that many non-native species on SCI could potentially have ecosystem-level effects, but at the time the concept was relatively new, controversial, and not particularly well developed. While the EE concept may have provided a mechanistic model of what effects were likely from a given species, it had little if any utility predicting what species were likely to become an EE as a result of management actions. Management goals and planning were already relatively specific (Klinger 1997), and the integration of experimentation and monitoring was being used to try to understand the numerous complex interactions and improve predictions of outcomes from large-scale management activities. In addition, the concept of a transformer species was considered a useful and well-accepted mechanistic and phenomenological model (Richardson et al. 2000) for describing effects of the fennel, pigs, and non-native annual herbaceous species. For these reasons, while it was believed that the EE concept was interesting from an ecological perspective, it would have added little in a practical sense to what was already in place for planning, implementation, or evaluation of management actions.

18.7 ◆ CONCLUSION: HOW DOES THE ECOSYSTEM ENGINEER CONCEPT FIT INTO ONGOING AND FUTURE NON-NATIVE SPECIES MANAGEMENT PROGRAMS ON THE CHANNEL ISLANDS?

Over the last 25 years there has been an upsurge in conservation activities on the Channel Islands and a substantial increase in research applied to management (Klinger and Van Vuren 2000). In large part this is because the islands have been invaded and impacted by so many non-native species. It is almost certain that management programs targeted at these species will be ongoing for many decades, and the EE concept does have the potential to make tangible contributions to these programs. In the last 10 years there has been considerable development of the concept (Wright and Jones 2006), including application to a number of invasive species problems (Crooks and Khim 1999, Byers 2002, Cuddington and Hastings 2004). There is also no reason the EE concept needs to be the only one underpinning management programs; it can be integrated with other ecological concepts such as trophic cascades or food web structure (Byers et al. 2006).

When thinking in terms of non-native species as ecosystem engineers it seems most appropriate to think of not just one species but multiple ones, and how they interact. Many systems have multiple invaders that have been present for long periods of time (Rejmanek et al. 2005b). The relationship between different phases of the invasion process, interactions among multiple invasive species, species abundance, and the strength of per capita effects may be the most helpful aspect of retrospectively applying the EE concept to management of non-native species on Santa Cruz Island. It is unlikely the EE concept in and of itself will be useful predicting what species will become problems on the Channel Islands, however this is an extremely difficult problem that is not inherent just to ecosystem engineering (Kolar and Lodge 2001, Underwood et al. 2004, Rejmanek et al. 2005a, Klinger et al. 2006b). But if linked to the invasion process and focused on species that either have known large per capita effects or are clearly in the spread phase of invasion, the EE concept may be extremely helpful in forecasting real and presumed effects of a species, and how their removal could change various ecosystem properties. Perhaps most important, it makes scientists and managers think about complex interactions and processes, which in and of itself is a crucial step in setting measures of success for management programs and designing potential follow-up programs.

The greatest challenge the EE or any ecological concept faces is the willingness of institutions to look at long-term benefits of applying them in management programs. For instance, 16 years after the end of the sheep eradication program, TNC and the National Park Service began a feral pig eradication program on SCI (Krajick 2005). The multi-million-dollar program has progressed very efficiently and effectively (Krajick 2005). Within 2 years the pigs have become functionally extinct (S. Morrison, The Nature Conservancy, personal communication), and persistent hunting of the remaining few will soon lead to their complete extinction. As with most eradication efforts, the majority of the resources have been targeted at the operational side of the program. Not without justification, success is primarily being measured as eradication (NPS 2003). Systematic monitoring designed specifically for outcomes from the eradication program has been limited to island fox (*Urocyon littoralis*) populations, mainly because of the purported relationship between pigs and golden eagle (*Aquila chrysaetos*) predation on the foxes (Roemer et al. 2002, Bakker et al. 2005). Despite strong recommendations, monitoring of broader community and ecosystem processes has largely been ignored.

In contrast, the approach of monitoring desired outcomes rather than simply eliminating a species is the central focus of non-native species management programs by the Santa Catalina Island Conservancy (J. and D. Knapp, Catalina Island Conservancy, personal communication). Although they are not using the EE concept, they have adopted an adaptive management approach (Holling 1978) because they have observed similar complex patterns following eradication programs for goats and pigs (Garcelon et al. 2005, Knapp 2005) on Catalina as were observed on Santa Cruz. This includes the dominance of many areas by non-native annual plants (Laughrin et al. 1994), the spread of fennel after a lag period of several years, and increased fire frequency as a result of greater fuel loads and human ignitions (Knapp 2005). Their goals are to improve the ability to predict potential undesirable outcomes from their management programs, and to prepare management actions before undesirable outcomes become too extensive to effectively deal with.

More generally, effects from control and eradication programs are clearly landscape-scale phenomena. Therefore, it will be important to think beyond simplistic notions of restoration when setting management goals, and recognize that trying to control or eradicate non-native species will often have very broad ecosystem effects with few predictable trajectories (Zavaleta et al. 2001, Zavaleta 2002). These programs are not ends in themselves, but likely only the first in a long series of ongoing management programs. Just where the EE species concept fits into this situation is difficult to say with certainty, but it is likely that there will be many circumstances where it can be used effectively to help plan and forecast outcomes from non-native species management programs.

REFERENCES

Baber, D.W. (1982). Report on a survey of feral pigs on Santa Cruz Island, California: Ecological implications and management recommendations. Unpublished report submitted to The Nature Conservancy, Santa Barbara, California.

Baber, D.W., and Coblentz, B.E. (1987). Diet, nutrition, and conception in feral pigs on Santa Catalina Island. *Journal of Wildlife Management* 51:306–317.

Bakker, V.J., Garcelon, D.K., Aschehoug, E.T., Crooks, K.R., Newman, C., Schmidt, G. A., Van Vuren, D.H., and Woodroffe, R. (2005). Current status of the Santa Cruz Island fox (*Urocyon littoralis santacruzae*). In *Proceedings of the Fifth California Islands Symposium*, National Park Service Technical Publications CHIS-05-01, D.K. Garcelon and C.A. Schwemm, Eds. Ventura, CA: Institute for Wildlife Studies, pp. 275–285.

Barthell, J.F., Randall, J.M., Thorp, R.W., and Wenner, A.M. (2001). Promotion of seed set in yellow star-thistle by honey bees: Evidence of an invasive mutualism. *Ecological Applications* 11:1870–1883.

Barthell, J.F., Thorp, R.W., Mason, C., Garvin, E., Johnson, E., Wells, H., and Wenner, A.M. (2005). An island ecosystem after honey bees: Is a noxious weed now competing for native pollinators? *Integrative and Comparative Biology* 45:961.

Beatty, S.W. (1991). *The Interaction of Grazing, Soil Disturbance, and Invasion Success of Fennel on Santa Cruz Island, CA*. Santa Barbara, CA: The Nature Conservancy.

Beatty, S.W., and Licari, D.L. (1992). Invasion of fennel into shrub communities on Santa Cruz Island, California. *Madrono* 39:54–66.

Brenton, B., and Klinger, R.C. (1994). Modeling the expansion and control of fennel (*Foeniculum vulgare*) on the Channel Islands. In *The Fourth California Islands Symposium: Update on the Status of Resources*, G.J. Maender and W.L. Halvorson, Eds. Santa Barbara, CA: Santa Barbara Museum of Natural History, pp. 497–504.

Brenton, R.K., and Klinger, R.C. (2002). Factors influencing the control of fennel (*Foeniculum vulgare* Miller) using Triclopyr on Santa Cruz Island, California, USA. *Natural Areas Journal* 22:135–147.

Brown, C.S., and Rice, K.J. (2000). The mark of Zorro: Effects of the exotic annual grass Vulpia myuros on California native perennial grasses. *Restoration Ecology* 8:10–17.

Browne, D.R., Mitchell, K.L., and Chaney, H.W. (2000). *Proceedings of the Fifth California Islands Symposium*. Santa Barbara, CA: Santa Barbara Museum of Natural History.

Brumbaugh, R.W. (1980). Recent geomorphic and vegetal dynamics on Santa Cruz Island, California. In *The California Islands: Proceedings of a Multidisciplinary Symposium*, D.M. Power, Ed. Santa Barbara, CA: Santa Barbara Museum of Natural History, pp. 139–158.

Byers, J.E. (2002). Impact of non-indigenous species on natives enhanced by alteration of selection regimes. *Oikos* 97:449–458.

Byers, J.E., Cuddington, K., Jones, C.G., Talley, T.S., Hastings, A., Lambrinos, J.G., Crooks, J.A., and Wilson, W.G. (2006). Using ecosystem engineers to restore ecological systems. *Trends in Ecology & Evolution* 21:493–500.

Carroll, M.C., Laughrin, L., and Bromfield, A. (1993). Fire on the California Islands: Does it play a role in chaparral and closed-cone pine habitats? In *The Third California Islands Symposium: Recent Advances in Research on the California Islands*, F.G. Hochberg, Ed. Santa Barbara, CA: Santa Barbara Museum of Natural History, pp. 73–88.

Crooks, J., and Khim, H.S. (1999). Architectural vs. biological effects of a habitat-altering exotic mussel, *Musculista senhousia*. *Journal of Experimental Marine Biology and Ecology* 240:53–75.

Cuddington, K., and Hastings, A. (2004). Invasive engineers. *Ecological Modelling* 178:335–347.

Dash, B.A., and Gliessman, S.R. (1994). Nonnative species eradication and native species enhancement: Fennel on Santa Cruz Island. In *The Fourth California Islands Symposium: Update on the Status of Resources*, W. Halvorson and G. Maender, Eds. Santa Barbara, CA: Santa Barbara Museum of Natural History, pp. 505–512.

Dyer, A.R., and Rice, K.J. (1999). Effects of competition on resource availability and growth of a California bunchgrass. *Ecology* (Washington, D.C.) 80:2697–2710.

Garcelon, D.K., Ryan, K.P., and Schuyler, P.T. (2005). Application of techniques for feral pig eradication on Santa Catalina Island, California. In *Proceedings of the Fifth California Islands Symposium*, National Park Service Technical Publications CHIS-05-01, C.A. Schwemm, Ed. Arcata, CA: Institute for Wildlife Studies, pp. 331–340.

Garcelon, D.K., and Schwemm, C.A. (Eds.). (2005). *Proceedings of the Fifth California Islands Symposium*, National Park Service Technical Publications CHIS-05-01. Arcata, CA: Institute for Wildlife Studies.

Glassow, M. (1980). Recent developments in the archaeology of the Channel Islands. In *The California Islands: Proceedings of a Multidisciplinary Symposium*, D.M. Power, Ed. Santa Barbara, CA: Santa Barbara Museum of Natural History, pp. 79–99.

Gordon, D.R., and Rice, K.J. (1992). Partitioning of space and water between two California annual grassland species. *American Journal of Botany* 79:967–976.

Halvorson, W.L. (1994). Ecosystem restoration on the Channel Islands. In *The Fourth California Islands Symposium: Update on the Status of Resources*, W.L. Halvorson and G.J. Maender, Eds. Santa Barbara, CA: Santa Barbara Museum of Natural History, pp. 567–571.

Halvorson, W.L., and Maender, G.J., (Eds.). (1994). *The Fourth California Islands Symposium: Update on the Status of Resources*. Santa Barbara, CA: Santa Barbara Museum of Natural History.

Hobbs, E. (1980). Effects of grazing on the northern populations of *Pinus muricata* on Santa Cruz Island, California. In *The California Islands: Proceedings of a Multidisciplinary Symposium*, D.M. Power, Ed. Santa Barbara, CA: Santa Barbara Museum of Natural History, pp. 159–166.

Hochberg, F.G. (Ed.). (1993). *Third California Islands Symposium: Recent Advances in Research on the California Islands.* Santa Barbara, CA: Santa Barbara Museum of Natural History.

Hochberg, M.S., Junak, S., and Philbrick, R. (1980). *Botanical Study of Santa Cruz Island for the Nature Conservancy.* Santa Barbara, CA: Santa Barbara Botanic Garden.

Holling, C.S. (1978). *Adaptive Environmental Management and Assessment.* New York: John Wiley & Sons.

Junak, S., Ayers, T., Scott, R., Wilken, D., and Young, D. (1995). *A Flora of Santa Cruz Island.* Santa Barbara, CA: Santa Barbara Botanic Garden.

Keeley, J.E. (2006). Fire in the South Coast Bioregion. In *Fire in California Ecosystems,* N. Sugihara, J.W. Van Wagtendonk, J. Fites, and A. Thode, Eds. Berkeley, CA: University of California Press, pp. 529–590.

Klinger, R.C. (1997). *Operating principles and ecological goals for Santa Cruz Island biological management* (unpublished report). Santa Barbara, CA: The Nature Conservancy.

Klinger, R.C. (2000). *Foeniculum vulgare.* In *Invasive Plants of California's Wildlands,* C.C. Bossard, J.M. Randall, and M.C. Hoshovsky, Eds. Berkeley, CA: University of California Press, pp. 198–202.

Klinger, R.C., Brooks, M.L., and Randall, J.A. (2006a). Fire and invasive plants. In *Fire in California Ecosystems,* N. Sugihara, J.W. Van Wagtendonk, J. Fites, and A. Thode, Eds. Berkeley, CA: University of California Press, pp. 726–755.

Klinger, R.C., and Messer, I. (2001). The interaction of prescribed burning and site characteristics on the diversity and composition of a grassland community on Santa Cruz Island, California. In *Proceedings of the Invasive Species Workshop: The Role of Fire in the Control and Spread of Invasive Species. Fire Conference 2000: The First National Congress on Fire Ecology, Prevention, and Management,* Tall Timber Research Station Miscellaneous Publication No. 11, K.E.M. Galley and T.P. Wilson, Eds. Tallahassee, FL: Tall Timbers Research Station, pp. 66–80.

Klinger, R.C., Schuyler, P., and Sterner, J.D. (2002). The response of herbaceous vegetation and endemic plant species to the removal of feral sheep from Santa Cruz Island, California. In *Turning the Tide: The Eradication of Invasive Species,* C.R. Veitch and M.N. Clout, Eds. Auckland, New Zealand: Invasive Species Specialist Group of The World Conservation Union (IUCN), pp. 163–176.

Klinger, R.C., Underwood, E.C., and Moore, P.E. (2006b). The role of environmental gradients in non-native plant invasions into burnt areas of Yosemite National Park, California. *Diversity and Distributions* 12:139–156.

Klinger, R.C., and Van Vuren, D. (2000). Thirty years of research on California's Channel Islands: An overview and suggestions for the next 30 years. In *Proceedings of the Fifth California Islands Symposium,* D.R. Browne, K.L. Mitchell, and H.W. Chaney, Eds. Santa Barbara, CA: Santa Barbara Museum of Natural History, pp. 323–329.

Klinger, R.C., Wills, R.D., and Messer, I. (2004). Design, implementation, and evaluation of a multi-scale prescribed burn program on Santa Cruz Island. In *Proceedings of the Symposium: Fire Management: Emerging Policies and New Paradigms,* Miscellaneous Publication No. 2, N.G. Sugihara, M.E. Morales, and T.J. Morales, Eds. San Diego, CA: Association for Fire Ecology, pp. 21–42.

Knapp, D.A. (2005). Rare plants in the Goat Harbor burn area, Santa Catalina Island, California. In *Proceedings of the Fifth California Islands Symposium*, National Park Service Technical Publications CHIS-05-01, C.A. Schwemm, Ed. Arcata, CA: Institute for Wildlife Studies, pp. 205–211.

Kolar, C.S., and Lodge, D.M. (2001). Progress in invasion biology: Predicting invaders. *Trends in Ecology & Evolution* 16:199–204.

Krajick, K. (2005). Winning the war against island invaders. *Science* 310:1410–1413.

Laughrin, L. (1982). *The vertebrates of Santa Cruz Island: Review, current status, and management recommendations.* Report to The Nature Conservancy, Santa Barbara, CA.

Laughrin, L., Carroll, M., Bromfield, A., and Carroll, J. (1994). Trends in vegetation changes with removal of feral animal grazing pressures on Santa Catalina Island. In *The Fourth California Islands Symposium: Update on the Status of Resources*, W.L. Halvorson and G.J. Maender, Eds. Santa Barbara, CA: Santa Barbara Museum of Natural History, pp. 523–530.

Lombardo, C.A., and Faulkner, K.R. (2000). Eradication of feral pigs (*Sus scrofa*) from Santa Rosa Island, Channel Islands National Park, California. In *Proceedings of the Fifth California Islands Symposium*, D.R. Browne, K.L. Mitchell, and H.W. Chaney, Eds. Santa Barbara, CA: Santa Barbara Museum of Natural History.

Minnich, R.A. (1980). Vegetation of Santa Cruz and Santa Catalina Islands. In *The California Islands: Proceedings of a Multidisciplinary Symposium*, D.M. Power, Ed. Santa Barbara, CA: Santa Barbara Museum of Natural History, pp. 123–128.

NPS. (2003). *Santa Cruz Island Primary Restoration Plan.* Ventura, CA: National Park Service, Channel Islands National Park.

Ogden, J.A.E., and Rejmanek, M. (2005). Recovery of native plant communities after the control of a dominant invasive plant species, *Foeniculum vulgare*: Implications for management. *Biological Conservation* 125:427–439.

Philbrick, R. (Ed.). (1967). *Proceedings of the Symposium on the Biology of the California Islands.* Santa Barbara, CA: Santa Barbara Botanic Garden.

Power, D.M. (Ed.). (1980). *The California Islands: Proceedings of a Multidisciplinary Symposium.* Santa Barbara, CA: Santa Barbara Museum of Natural History.

Raven, P.H. (1967). The floristics of the California islands. In *Proceedings of the Symposium on the Biology of the California Islands*, R. Philbrick, Ed. Santa Barbara, CA: Santa Barbara Botanic Garden, pp. 57–67.

Rejmanek, M., Richardson, D.M., Higgins, S.I., Pitcairn, M.J., and Grotkopp, E. (2005a). Ecology of invasive plants: State of the art. In *Invasive Alien Species: A New Synthesis*, SCOPE 63, H.A. Mooney, R.N. Mack, J.A. McNeely, L.E. Neville, P.J. Schei, and J.K. Waage, Eds. Washington, D.C.: Island Press, pp. 104–161.

Rejmanek, M., Richardson, D.M., and Pysek, P. (2005b). Plant invasions and invasibility of plant communities. In *Vegetation Ecology*, E. van der Maarel, Ed. Oxford, UK: Blackwell Publishing, pp. 332–355.

Richardson, D.M., Pysek, P., Rejmanek, M., Barbour, M.G., Pannetta, F.D., and West, C.J. (2000). Naturalization and invasion of alien plants: Concepts and definitions. *Diversity and Distributions* 6:93–107.

Ricklefs, R.E. (2005). Taxon cycles: Insights from invasive species. In *Species Invasions: Insights into Ecology, Evolution, and Biogeography*, D.F. Sax, J.J. Stachowicz, and S.D. Gaines, Eds. Sunderland, MA: Sinauer Associates, Inc., pp. 165–199.

Roemer, G.W., Donlon, C.J., and Courchamp, F. (2002). Golden eagles, feral pigs and insular carnivores: How exotic species turn native predators into prey. *Proceedings of the National Academy of Science* 99:791–796.

Salo, L.F. (2005). Red brome (Bromus rubens subsp. madritensis) in North America: Possible modes for early introductions, subsequent spread. *Biological Invasions* 7:165–180.

Schuyler, P. (1993). Control of feral sheep (*Ovis aries*) on Santa Cruz Island, California. In *The Third California Islands Symposium: Recent Advances in Research on the California Islands*, F.G. Hochberg, Ed. Santa Barbara, CA: Santa Barbara Museum of Natural History, pp. 443–452.

Sterner, J.D., and Barrett, R.H. (1991). Removing feral pigs from Santa Cruz Island, California. *Transactions of the Western Section of the Wildlife Society* 27:47–53.

Underwood, E.C., Klinger, R., and Moore, P.E. (2004). Predicting patterns of non-native plant invasion in Yosemite National Park, California, USA. *Diversity and Distributions* 10:447–459.

Van Driesche, J., and Van Driesche, R. (2000). Chapter 7: After all the sheep are gone: The recovery of Santa Cruz Island after 140 years of grazing. In *Nature Out of Place: Biological Invasions in the Global Age*. Washington, D.C.: Island Press, pp. 153–176.

Van Vuren, D. (1981). *The feral sheep of Santa Cruz Island: Status, impacts, and management recommendations*. Report to The Nature Conservancy, Santa Barbara, CA.

Van Vuren, D., and Coblentz, B.E. (1987). Some ecological effects of feral sheep on Santa Cruz Island, California. *Biological Conservation* 41:253–268.

Wehjte, W. (1994). Response of a Bishop pine (*Pinus muricata*) population to removal of feral sheep on Santa Cruz Island, California. In *The Fourth California Islands Symposium: Update on the Status of Resources*, G.J. Maender and W.L. Halvorson, Eds. Santa Barbara, CA: Santa Barbara Museum of Natural History, pp. 331–340.

Wenner, A.M., and Thorp, R.W. (1994). Removal of feral honey bee (*Apis mellifera*) colonies from Santa Cruz Island. In *The Fourth California Islands Symposium: Update on the Status of Resources*, W. Halvorson and G. Maender, Eds. Santa Barbara, CA: Santa Barbara Museum of Natural History, pp. 513–522.

Wright, J.P., and Jones, C.G. (2006). The concept of organisms as ecosystem engineers ten years on: Progress, limitations, and challenges. *BioScience* 56:203–209.

Zavaleta, E.S. (2002). It's often better to eradicate, but can we eradicate better? In *Turning the Tide: The Eradication of Invasive Species*, C.R. Veitch and M.N. Clout, Eds. Auckland, New Zealand: Invasive Species Specialist Group of The World Conservation Union (IUCN), pp. 393–405.

Zavaleta, E.S., Hobbs, R.J., and Mooney, H.A. (2001). Viewing invasive species removal in a whole-ecosystem context. *Trends in Ecology and Evolution* 16:454–459.

19

THE DIVERSE FACES OF ECOSYSTEM ENGINEERS IN AGROECOSYSTEMS

John Vandermeer and Ivette Perfecto

In its original intent, the ecosystem engineers definition was "those organisms that directly or indirectly modulate the availability of resources (other than themselves) to other species, by causing physical state changes in biotic and abiotic materials" (Berkenbusch and Rowden 2003). Such a definition would seem to allow all sorts of ecological interactions to be included under the theme of ecosystem engineers (both the trellis formed by the corn plant and the nitrogen supplied by the bean plant, as much as the beetle attracted to the field, would be part of an engineered ecosystem, as would almost any other "effect" on the environment). More recently the idea has been more restrictive "distinguishing it from trophic interactions" (Berkenbush and Rowden 2003, Jones et al. 1997, Wilby et al. 2001), from keystone species, and allowing it to take on both positive and negative values (Jones et al. 1997). Furthermore, the idea of "niche construction" (Olding-Smee et al. 2003, Sterelny 2005) is closely connected to the concept of ecological engineering—constructing a niche immediately conjures up images of engineering that niche.

While the intersection of these various lines of thought remains an interesting exercise that ought to be engaged, in the spirit of this volume we restrict our focus to ecological engineering as a subcomponent of species' effects on the environment, emphasizing the nontrophic effects (which, formally, as we understand it, constitute cases of ecological engineering), but perhaps straying into semitrophic effects when they seem particularly interesting and engineering-like. In the case of agroecosystems the decision as to what falls within the limits of ecosystem engineers is sometimes difficult. For example, as discussed further in the following text, if the main effect of one part of the decomposition cycle is to alter soil structure (e.g., forming conglomerates of soil organic matter and clay particles to alter the cation exchange capacity of the soil), does this effect fall within the rubric of ecosystem engineering? We believe many such examples exist in agroecosystems and take the position that at this point in development of the concept of ecosystem engineering inclusion is probably the wisest strategy. Consequently, although we exclude effects that are obviously only trophic, some of our examples are certainly on the margin between trophic and engineering (to say nothing of niche construction).

Agroecosystems, and other managed systems, present a unique situation for focusing the issue of ecosystem engineering. Even though the principal engineer is without doubt *Homo sapiens*, a focus on the particular activities of that one species would be of little interest, considering contemporary literature on ecosystem engineering. Yet, because of the normative behavior of that species, there is an important dichotomy that immediately comes to light. Much as with the more general topic of biodiversity, ecosystem engineers fall in two very general categories, planned and associated (Swift et al. 1996, Vandermeer et al. 1998)—some engineers are the direct consequences of the farmer's planning, and others are indirectly associated with the organisms introduced by the farmer.

Planned engineers are the plants and animals purposefully incorporated into the system by the farmer. They clearly "engineer" the entire system since the very definition of the system—cornfield, coffee plantation, pasture—is based on their overwhelming dominance. They create the habitat where other organisms live. However, once the planned elements are in place, a host of other organisms become associated with them. Many of those associated organisms are also engineers, equivalent to the engineers normally associated with ecosystem engineering in the standard literature (e.g., Berkenbush and Rowden 2003; Jones et al. 1994, 1997).

In addition to the planned–associated framework, we find it useful to cast the problem in another well-known framework, the response–effect conceptualization of Goldberg (1990). Multidimensional agroecosystems have already been characterized as falling into this classical response–effect framework (e.g., Vandermeer 1989). Thus, for example, in an intercropping system of corn and beans, the beans purportedly modify the environment of the corn by adding to the total nitrogen pool (being able to harvest nitrogen from the air as well as soil), while the corn modifies the environment of the bean with respect to physical structure (providing a trellis on which the bean vine can grow) (Vandermeer 1984, 1989). Yet it is evident that the corn and beans are competitive with one another when intercropped, since they both use some of the same nutrients, the same source of light, and the same space.

The response–effect framework nicely captures both of these aspects, the positive or facilitative (supplying more nitrogen or making a convenient trellis) and the negative (frequently, but not always, competition). These effects range from the purely physical (the corn supplies a trellis), through the questionably trophic (a rich nitrogen environment is thought by some workers to be a physical aspect of the environment and the shading of one plant by the other is competitive, but not resource competitive), to clearly trophic (resource competition). Yet the "effect" part of the response–effect framework is frequently thought of as ecosystem engineering (contingent, of course, on one's preferred definition of the concept to begin with). Competitive production (Vandermeer 1981, 1989) can be advantageous (a phenomenon most recently referred to as *complementarity*, Loreau et al. 2001) or not depending on conditions of response and effect. However, facilitation can be strong or weak (never, by definition, negative) depending on conditions (Vandermeer 1984, 1989). The balance between competition and facilitation (or, more generally, negative and positive) in organisms' effect on the environment and how it is related to the fundamental idea of ecosystem engineers has been previously acknowledged (Jones et al. 1997 citing Callaway and Walker 1997).

However, the corn and beans are planned components of the system, which, for most agroecosystems, represents only a small fraction of what is interesting ecologically. For example, in the Americas the corn attracts a beetle, the corn rootworm (*Diabrotica* spp.), whose larvae burrow into its roots and the beans attract the same beetle, whose adults eat bean foliage. The beetle is certainly not planned by the farmer, but is part of the associated biodiversity that normally comes with the agroecosystem. The bean, then, has another effect on the corn by attracting the beetle, which eventually has a negative affect on the corn, and the corn attracts the beetle, which has a negative effect on the bean. The corn and beans thus are "apparent competitors" with the beetle acting as the agent that has the proximate effect on the environment. The beetle itself is clearly part of the "associated" biodiversity.

Although this tale of corn and beans is obviously a "just so" story (Berkenbush and Rowden 2003), it does focus attention on the question of what should be included in a review of ecosystem engineering in agroecosystems. Ecosystem engineering is part of the more general "effect" an organism has on the environment, to which that organism and others must "respond." And in the case of managed ecosystems, some of those engineers are placed in the system by the planner, the "planned engineers," while others arrive by means unrelated to human planning, the "associated engineers."

19.1 ➤ PLANNED ECOSYSTEM ENGINEERS

In a trivial sense the planned elements of an agroecosystem are all ecological engineers. The farmer, being the ultimate engineer (but not the focus of this chapter) decides to transform the original ecosystem, but the crops he or she chooses inevitably have a major effect on the environment that is constructed. And frequently those effects are not directly trophic, so may be considered as ecological engineering. Perhaps the first acknowledgement of these effects was in the mid-nineteenth century by Liebig, in recognizing that some crops had different nutrient requirements than others and that the nutrient mix in the soil would be at least partially determined by the type of crop planted. It is likely that European farmers generations before him understood this principle, which is what led them to their various complicated rotational systems. However, it was Liebig who elaborated the idea in what might be the first formal scientific framework of ecological engineering.

A half century later, Albert Howard, an Englishman working in India, devised his famous Indore composting system, which was far more than simply a composting system. Howard's approach, which we can presume came from extensive observations of and conversations with Indian farmers, was focused on the health of the soil, arguing that a "healthy" soil, by which he meant one that contained a well-balanced mixture of worms, fungi, and bacteria and other microorganisms, would produce healthy food, while a soil devoid of those healthy elements would not produce such healthy food. Indeed, the connection between ecological health on the farm and the health-promoting qualities of food produced there was a key element of the early organic agriculture movement (Conford 2001). Howard thus noticed the immense effect of ecological engineers (i.e., worms, fungi, and bacteria and other microorganisms in the soil) in creating what he referred to as a healthy soil.

Modern soil science completely accepts the basic ideas of Howard (1940), acknowledging that ecosystem engineers operate in a variety of ways to maintain (or denigrate) soil fertility. For example, the practice of composting (Epstein 1997, Dreyfus 1990) is clearly an example of ecosystem engineering. Organic matter, a vital component of both physical and chemical processes in soils, is dramatically altered by the addition of compost, which is created by active treatment of organic residues with decomposing engineers (Insam et al. 1996, 2002). And the addition of worms as the major engineer has become commonplace, with vermiculture sometimes taking center stage in the production of compost (Frederickson et al. 1997, Berc et al. 2004).

In more conventional farming practices, the mechanical and chemical manipulation of soils is overwhelmingly dominant in forming the ecological base on which soil dynamics occur, and if we were to include the direct effects of *Homo sapiens* in this review, it would take center stage. However, in the spirit of nonhuman actors as engineers, we leave that topic to the already extensive literature easily accessible to the interested reader (Johnsen et al. 2001, Vandermeer 1995, Altieri 2000). Otherwise, excluding composting and related activities, most engineering effects on the soil are only indirect effects of the planned components, and we thus defer further discussion of them to the section on associated effects.

Crop rotations are ubiquitous in traditional farming systems, from the famous Norfolk rotation of industrializing England to corn–soybean rotations in the Midwestern United States today. Each crop in the cycle engineers some aspect of the environment for the other crops. Sometimes the critical factor is to eliminate a pest or pathogen that has built up (Altieri 1999, Liebman and Davis 2000, Johnson et al. 2001), other times to utilize a different mixture of nutrients from one year to the next (Douglas et al. 1998, Riedell et al. 1998). In either case, the crop in the one rotation clearly engineers part of the environment for the crop in subsequent rotations.

Intercropping systems are common in traditional tropical farming systems (Vandermeer 1989, 1995). Using the effect–response framework, it is easy to see how each of the crops may facilitate some aspect of environmental improvement for the other crops (Vandermeer 1984, Li 2003). The earlier example of corn and beans is only one in a large collection of examples that could be cited.

In many tropical agricultural systems trees are an integral part of a system that includes annual or semi-annual crops, in a system commonly referred to as *agroforestry*. Agroforestry systems must be regarded as just as paradigmatic for ecosystem engineering as trees in the forest. The question "What does a tree do in a forest?", the answer to which is thought to be a quintessential example of ecological engineering (Jones et al. 1997), applies equally to agroforestry systems. The literature on the effects of trees on the physical structure of microhabitats is huge as a glance at almost any issue of the journal *Agroforestry Systems* will attest.

Trees in agroforestry systems create windbreaks (Sturrock 1988; Bird 1991, 1998; Mayus et al. 1998), keep the understory cool (Rao et al. 1997, Rhoades 1996, Campanha et al. 2004), provision organic matter and nutrients (Palm 1995, Fassbender et al. 1991, Chander et al. 1998), provide refuge for natural enemies (Dix et al. 1995, Stamps and Linit 1998), disrupt the ability of pests to find crops (Altieri and Nicholls 2003, Nicholls and Altieri 2004), sequester carbon (Montagnini and Nair 2004), and likely produce other effects that have escaped our attention. Add to this any general engineering effect already attributed to trees in unmanaged systems (e.g., the paradigmatic example of Jones et al. 1997), and agroforestry emerges as a seemingly paradigmatic case of ecosystem engineering in general (Garcia-Barrios and Ong 2004).

In the modern industrial agricultural system, the planned ecosystem engineers have become less important as many of their functional effects have been taken over by the direct engineering activities of that one key species, *Homo sapiens*. Thus, for example, the important engineering activity of provisioning nitrogen, traditionally the job of legume engineers, has been replaced by the engineer located in the nitrate-manufacturing plant. The engineering activity of attracting parasitoids to control key pests has been taken over by the engineer located in the pesticide-manufacturing plant. Since we have chosen to ignore the direct effects of *Homo sapiens*, for the most part the conventional industrial system thus falls outside of the intended scope of this chapter.

Nevertheless, it is worth mentioning the overwhelming fact, that enormous elephant in the living room, that the human engineers in this case have had some major negative effects, not only on the practice of agriculture, but also on unintended environmental and health consequences (Raynolds 2004, Badgeley et al. 2007, Conway and Pretty 1991, Allen 1993, Lappé et al. 1998).

19.2 ⬤ ASSOCIATED ECOSYSTEM ENGINEERS

The effects of planned engineers are for the most part evident, as suggested earlier. Contrarily, the effects of associated engineers are largely indirect, subtle, and incompletely understood. To summarize them, it is convenient to think of the overall agroecosystem as having seven compartments into which matter enters and exits, and six processes whereby matter is transferred between compartments, some of which are strictly trophic, others of which are engineering. The seven compartments are plants, herbivores, carnivores, detritus, organic soup, humus, and nutrients, and the six processes are photosynthesis and nutrient absorption, primary consumption, secondary consumption, death, comminution, and catabolism. In this way the decomposition subsystem of ecosystems can be seen as fitting in with the trophic-based ideas of elementary ecology, and the main engineering aspects (i.e., comminution and catabolism) emerge as a simple subcomponent of the trophic hierarchy. We refer to these engineers as *decomposition engineers.* Their inclusion as engineers may be questionable in the spirit of the definition of engineers explicitly excluding trophic effects, but we feel they play such an important part in the modification of the environment that it is justified including them as a major category of ecosystem engineering (e.g., exodigestion clearly modifies the environment for organisms other than the exodigester; the soil organic matter is regarded by most researchers as a physical fact of soil chemistry, yet it is clearly engineered by these decomposition engineers).

However, this framework excludes one of the more important aspects of ecosystem engineering in the soil—those organisms whose activities directly affect the macrostructure of the soil. These are the digging and tunneling organisms, such as termites, ants, and earthworms, and the fungi whose mycelia act to bind particles together into peds as well as make physical connections among different peds, engineering a structural integrity beyond the simple ped structure. Termites are especially important in the tropical agroecosystems of Africa and South America (Black and Okwakol 1997, Jones 1990, Dangerfield et al. 1998, Martius 2001), earthworms are universally important (Lawton 1994, Fragoso et al. 1997, Edwards 1994, Edwards and Bohlen 1996), and ants are likely important but have not received the attention they deserve (Folgarait 1998). For clarity of presentation, we refer to them as *soil structure engineers.*

DECOMPOSITION ENGINEERS

Engineers in the soil reach their maximum importance in the process of decomposition. Decomposition of any piece of plant or animal matter is completed over a period of hundreds or thousands of years, yet the bulk of the decomposition occurs within one or a few years, depending on climatic conditions. The decomposition process thus has two components, the rapid cycle and the slow cycle. The rapid cycle is responsible for the mineralization of nutrients, while the slow cycle is responsible for the production of humus, and thus the contribution of organic matter to soil physical and chemical structure. Which materials go to which cycle depends on both the chemical makeup of the constituents and the activities of the microbial soil engineers.

In natural ecosystems the relevant process is the introduction of litter to the soil. This is almost always a seasonal phenomenon, even in seemingly nonseasonal tropical rain forests. In most forms of agriculture this seasonality is exaggerated because of planting and harvesting cycles, although many forms of permanent culture, especially in tropical regions, do not have such a strict cycle imposed by *Homo sapiens*, and can be expected to correspond more or less to the patterns of natural systems. In a non-managed system it is a reasonable expectation that the yearly seasonal cycle will result in something close to a complete cycling of nutrients through the fast cycle. But many agroecosystems are seemingly not at such an equilibrium state and have either rapid decomposition, such that most of the fast cycle is completed before the end of the seasonal cycle, or accumulate organic material because of slow decomposition. This timing of decomposition with the seasonal cycle is an important concept when it comes to the ecological management of decomposition, and is one of the reasons why various forms of composting activities have become part and parcel of many agroecosystems (Insam et al. 2002). Vermiculture, with its worm engineers, has become one of the most common forms (Hendrix and Edwards 1994, Frederickson et al. 1997, Berc et al. 2004).

At a mechanistic level, the process of decomposition can be thought of as involving three main transformations, all occurring simultaneously—leaching, catabolism, and comminution (Swift et al. 1979). Leaching is generally the transport of materials from one place to another, frequently through runoff or percolation of water. Catabolism is the chemical process whereby larger molecules are broken down into smaller ones. Comminution is the physical reduction in size of the organic material, a physical rather than a chemical process. All three transformations are strongly influenced by ecosystem engineers.

Soil ecosystem engineers are remarkably diverse, ranging in size from bacteria to snails. Furthermore these organisms do not fall into the same neat trophic categories as do the organisms in the herbivore–carnivore subsystem of the ecosystem. With the absence of either primary producer or herbivore in the decomposition subsystem, it is not surprising that trophic levels are not easily definable—all organisms are carnivores and top carnivores in a sense. Consequently a variety of alternative trophic classification schemes have been devised. One of the simplest is a two-way crossed classification, with the mode of nutrient acquisition (absorptive versus ingestive) crossed with the size of the organism (micro versus macro). This classification does not preclude the classic plant–herbivore–carnivore trophic system, but may be more useful when thinking of the decomposition subsystem, especially from the point of view of the nontrophic ideas of ecosystem engineering. We here discuss the absorptive versus ingestive categories.

Absorptive organisms involved in decomposition are mainly bacteria and fungi. They are the primary forces for decomposing highly commutated material, and are effectively the last stage in the decomposition process, engineering the catabolism that releases nutrients. For any given input into the cascading decomposition process, the microabsorptive organisms become more important as the system cascades through its full long-term cycle. Although special mechanisms may be important for various species and species groups, they all have a similar fundamental operation. The organism excretes exoenzymes, which create a digestive zone around its body, engineering a locally modified environment. The exoenzymes break down relatively large molecules outside of the body of the decomposing organism, and these smaller molecules then are absorbed and enter the normal intracellular metabolic process. Since a great part of digestion is external to the organism, digestion itself must be thought of as an engineering activity.

Absorptive organisms also share a strictly engineering characteristic the extent of which is not yet known, but the importance of which may be great. A number of studies report on complex interactions among various adsorptive microorganisms (e.g., Forlani et al. 1995, Nagarajah et al. 1970). For example, it is known that some bacteria concentrate near the surface of fungal hyphae, presumably for the purpose of scavenging on the products of the exodigestion of the fungi (Swift et al. 1979). In recognition of the basic processes involved in exodigestion, it would be surprising indeed if bacteria, and perhaps even fungi, did not regularly scavenge on the products of the exodigestion of other absorptive organisms. This process, if indeed it occurs and is common, has important implications for the long-term sustainability of decomposer ecosystems and places great emphasis on the potential role of biodiversity therein. Suppose, for example, the decomposition environment is composed of *Cellulomonas* (which produces cellulase as an exoenzyme) and *Bacillus* (which produces peptidases as exoenzymes). Is it possible that the *Cellulomonas* can obtain a fraction of its nitrogen nutrition from the small peptides produced near the body surface of *Bacillus* and the latter can obtain a fraction of its carbon from the carbohydrate products produced near the surface of *Cellulomonas* (from its digestion of cellulose)? Indeed, could *Bacillus* survive better in a cellulose-rich environment that contained *Cellulomonas* than one that didn't, through its ability to scavenge the products of exodigestion of *Cellulomonas*? If such a process actually operates to a significant extent in nature, bacterial engineers that have no obvious function in the actual decomposition process may in fact act as ecological buffers, providing exodigestive products on which other bacteria can scavenge when their preferred resource is temporarily in short supply. The significance of such complexity for questions of microbial biodiversity is obvious, and punctuates the strong connection between biodiversity and ecosystem engineering.

In contrast to absorptive organisms, ingesters are typically involved at a higher level in the decomposition process, usually more associated with comminution and less with catabolism—ingested resources are converted to constituent tissues, cells, and extracellular macromolecules and ejected as feces, returned to the soil organic material pool upon death, and, to a far lesser extent, released into the environment as metabolic products of catabolized molecules. The very idea of comminution is fundamentally one of engineering, thus implicating all ingesters as ecosystem engineers.

Ingestion in the case of macroingesters contributes mainly to the process of comminution and to a lesser extent to catabolism. Consider, for example, the earthworm, the great comminuter of temperate soils (Edwards and Bohlen 1996). Food is ingested through the mouth with the active participation of the pharynx, which operates as a pump. Food passes through the esophagus and arrives at the storage depot, the crop. From the crop, food is passed on to the gizzard, a muscular organ that grinds it, using small mineral particles that are also ingested. It is this physical grinding that has the main effect of comminution and is thus a major engineering process. The comminuted material is then passed to the intestine where digestion and absorption occurs. While earthworms produce a variety of enzymes, much of the digestion is accomplished through the use of symbiotic microorganisms, or simply by the microorganisms ingested with the organic matter.

Generally speaking, our present state of knowledge of the remarkably complex interactions that occur amongst all of these trophic-based engineers is slim. Yet, as a quick guide to the decomposition process, we can think of comminution and catabolism as working in sequence. Of course they do not in any real sense, but it is true that the first pass made by decomposing organisms at the organic material placed on or in the soil is dominated by comminution while the products of that comminution are acted on mainly by organisms that promote catabolism. Note that the products of comminution are not only a more finely dissected detritus but also the microorganisms that have incorporated the carbon from that detritus. So the engineering product of comminution might be most properly referred to as the soil organic "soup," since it includes both soil organic matter and soil microorganisms. It is thus not far off the track, as a heuristic device, to simply think of the organic detritus first being comminuted and then catabolized to form nutrients and humus. For the most part the initial comminution is accomplished by fungi and macroingestors, while the catabolism is accomplished mainly by fungi and bacteria. This view of the decomposition system is certainly too simplified to be useful in any active management plan, but it is a useful way to visualize the process of ecosystem engineering in the soil.

SOIL STRUCTURE ENGINEERS

The actions of earthworms, ants and termites, other arthropods, and fungi and other organisms may induce major structural modifications to the soil. One of the most obvious effects is the creation of more pore space by tunneling activity of earthworms. Sometimes this effect can be spectacular. Hoeksema and Jongerius (1959) report an increase of from 75 to 100% porosity in orchard soils that have earthworms when compared to those without earthworms. However, published results are highly variable with some authors finding insignificant effects of earthworms (Springett et al. 1992), and most others reporting figures on the order of 25% of total soil pore space being earthworm burrows. Edwards and Bohlen (1996) are of the opinion that, on average "earthworm burrows constitute only about 5% of total soil volume."

An important effect of increasing porosity is on water infiltration. Earthworm burrows contribute substantially to this process, especially those that are open to the surface, which is to say those constructed by anecic species. However, it is well documented that endogeic species also contribute to the infiltrateability of soils (Joschko et al. 1992). Burrows must be connected to one another to be effective at water infiltration, and tillage can significantly disrupt the network of burrows, thus reducing their function as water conduits (Chan and Heenan 1993). Other agricultural activities, such as pesticide applications, have been shown to reduce water infiltration by as much as 93% because of increased earthworm mortality.

On the other hand, earthworms can contribute to soil erosion, a fact first noted by Darwin (1881). Surface-deposited casts are susceptible to being carried away by water, and the bare spots created by large anecic species on the surface of the soil near their burrow entrances make patches of bare soil that then can be eroded. However, in most studies thus far reported (Edwards and Bohlen 1996), the beneficial effect of infiltration due to earthworm burrows outweighs the effects of water erosion. From the point of view of ecosystem engineering, both infiltration and erosion are consequences of ecosystem engineering.

The role of termites and ants would seem obvious (Whitford 1994). With respect to termites it has long been part of conventional wisdom that their effect on physical structure is enormous, given the evident size of their nests in African and South American savannas. As summarized by Black and Okwakol (1997; also see Lavelle 2002) termites have been linked with increasing aggregate stability, with increasing water penetration into the soil (i.e., porosity), with increasing hydraulic conductivity, and with pedogenesis itself. Unfortunately ants have not been as vigorously studied as termites with respect to soil physical structure, although their effect could be enormous. For example, Perfecto and Vandermeer estimated that *Atta cephalotes* could be responsible for complete soil turnover in as short a period as 200 years in a lowland rain forest in Costa Rica (Perfecto and Vandermeer 1993). Clearly more studies are warranted since this particular species is a very evident component of neotropical forests and moves a great amount of soil. Other less conspicuous species could have trivial effects, or the cumulative effect of all ants could be great.

19.3 ⬤ THE INTERACTION OF HUMAN ENGINEERS WITH ECOLOGICAL ENGINEERS: THE CASE OF PESTICIDES

Although this chapter for the most part ignores the human engineers that play so central a role in this particular ecosystem, it would seem perhaps too parochial not to at least mention some ways in which dramatic engineering feats have interacted with the ecological engineering that is the basis of this chapter. Indeed, some of these interactions have been quite spectacular, in a negative sense. When engineers make bridges whose span characteristics are such that harmonic vibrations cause their eventual collapse, we note the complexity of some design problems and the need to understand that complexity so as to avoid the problems in the future. Since the spectacular failure of the Tacoma Narrows Bridge in 1940, engineers take into account the subtle, but critical, aspects of bridge design, in particular the probability of harmonic coherence of oscillations causing expanding amplitudes. Human ecosystem engineers need to study their own failures and correct them in a similar fashion.

Pesticide-based agriculture emerged from the problems of overproduction generated by the end of World War II (WWII). The new agriculture was even referred to as "chemical agriculture" in some propaganda pieces. Problems with this new technology were evident from the start, but received massive public attention only after the publication of Rachael Carson's *Silent Spring* in 1962, which documented the fact that the massive use of pesticides was having a dramatic negative effect on the environment. Previously there had been much popular commentary about the human health effects of pesticides, a concern shared even by the pesticide manufacturers. But *Silent Spring* was the first popular account of environmental consequences, contributing not only to concern about environmental poisons, but perhaps providing the main springboard for the entire subsequent environmental movement.

What Carson said is now well known. Pesticides kill not only the targets, but also many species that are not targeted. Pesticides may concentrate in the higher trophic levels, thus making non-lethal doses at lower trophic levels quite dangerous at higher levels. Pests develop resistance to pesticides. The poisonous effects of pesticides and their residues may persist for a long time in the environment. These were the basic themes of her book, all of which implicate the ecosystem engineering aspects of pesticide effects.

A classic example of how Carson's principles worked themselves out, and how humans and pests can ultimately engineer an environment so hostile that certain species are effectively eliminated, is the case of cotton in Nicaragua (Falcon 1971). The beginnings of the story are in the same post-WWII climate referred to earlier. Landowners in Nicaragua saw cotton, the crop with an effectively undisruptable world demand, as a great potential due to the basic environmental conditions on Nicaragua's Pacific seaboard. The only problem was the cotton boll weevil, long known as a devastating pest in all of the Americas. But with the enthusiasm of the new focus on pesticides, boll weevils were simply another enemy to be vanquished and DDT was the armament that would do the job. Cotton was widely planted and DDT was sprayed a couple of times during the cotton growth cycle. The boll weevils died and the planters got rich. Yet only 5 years after its introduction DDT began losing its effectiveness. Furthermore the boll weevil reached densities in which it was an even worse pest than it had been earlier, and a new pest, the pink bollworm, had become even more important. Furthermore, several other pests now represented a pest suite that the planters had to deal with. But again there was an armament to solve the problem, this time the pesticide Aldrin. It had to be sprayed multiple times but was effective against both boll weevils and bollworms, along with a few other pests that had become important. Still, things got progressively worse. By the 1980s planters were spraying a cocktail of several kinds of pesticides over 27 times per growing season, to control 15 persistent and 9 occasional pests. The background environment had become saturated with a diversity of pest species that, by interacting with humans through the now well-known pesticide treadmill (van den Bosch 1989), engineered what became an impossible environment for the species we call cotton. It effectively became extinct because of these associated environmental engineers.

19.4 ● DISCUSSION

Agroecosystems are the most engineered of terrestrial ecosystems, by definition. However, by excluding *Homo sapiens*, the principal engineering species, we bring the agroecosystem into the more general categories of ecosystems that have already been analyzed from the point of view of ecosystem engineering. However, like other managed systems, because of the overwhelming importance of the human species, it is most convenient to look at the engineering consequences of those aspects that are directly planned by humans as separate from the aspects that indirectly arrive to the system associated with, but independent of, the direct actions of humans. Thus, many analysts of agroecosystems differentiate between planned and associated elements, and this chapter has continued with this practice.

Nevertheless, it is important to acknowledge the interaction between these two levels of engineering. A case study, the consequences of shade trees in coffee production, makes this interaction clear. Traditionally, coffee is grown under a forest-like canopy that varies from place to place and farm to farm, but is generally dominated by a few species of trees, frequently legumes. The shade cast by these trees is an obvious consequence of planned ecosystem engineering. However, there is a suite of associated organisms that are part of the engineering system, with important practical consequences for the farmer. Consider the particular case of southern Mexico.

Three agricultural pests are evident in the coffee farms in the mountains of southern Mexico, the green coffee scale (*Coccus viridis*), the coffee berry borer (*Hypothenemus hampei*), and the coffee rust (*Hemileia vastatrix*). The coffee berry borer is currently recognized as a devastating pest. The other two are occasional pests, but normally do not reach pest status, at least in more traditional modes of production. These three pests are evident to all coffee farmers in the region. However, subtle aspects of the engineers associated with the shade trees have only recently come to our attention.

Key to understanding the associated engineering system is a species of arboreally nesting ant, *Azteca instabilis*, which nests in the shade trees in the coffee farms. Typically it occupies less than 5% of the shade trees and is thus regarded as relatively rare over the entire landscape. However, in a classic mutualistic relationship, the ant tends the scale insect and protects it from its natural enemies—indeed, it is likely that the scale insect would disappear from the entire system if it did not receive protection from the ant, even though only about 3% of the area receives that attention (Vandermeer and Perfecto 2006). The 3% acts as a source for the scale population overall.

Since the ant tends the scale and the scale is a pest, it would seem logical to consider the ant as a pest also. However, the ant is also a predator on the coffee berry borer (Perfecto and Vandermeer 2006), making it a natural enemy of a more important pest. And a further examination of the system reveals a fungal disease that attacks the scale insects only after they have reached a local population density that is made possible exclusively through the engineering effects of the ants scaring away their potential predators and parasites. This fungal disease is known as the white halo fungus (*Verticillium lecanii*). Most recently it has come to our attention that this fungus is also a hyperparasite on the coffee rust. Thus the scale insect may act to engineer an environment that maintains the coffee rust under control.

In sum, the planned engineers, which are the various species of shade trees in the system, create conditions for an associated engineer, the arboreal nesting ant. That associated engineer (the ant) helps control an important pest (the coffee berry borer), but attracts an additional pest (the scale). However, that additional pest acts to engineer an environment in which a disease (the white halo fungus) controls an additional pest (the coffee rust). Thus the planned engineer generates a cascade of engineering consequences that reverberate throughout the system, an "engineering cascade," in which the tree engineers space for the ant, which engineers substrate for the fungus, and so on.

This sort of engineering cascade is likely to become more evident as ecologists unravel the various interactions that exist at all levels in agroecosystems. The cascades of organic matter decomposition, discussed earlier, contain within them a series of engineering cascades that, while elaborated by soil scientists in some cases, remain largely unstudied. And the cascades of pests and their regulating agencies present a similar picture, although most previous practical efforts rely, lamentably, on a more anachronistic mechanistic reductionism.

REFERENCES

Allen, P. (Ed.). (1993). *Food for the Future: Conditions and Contradictions of Sustainability.* New York: John Wiley and Sons.

Altieri, M.A. (1999). The ecological role of biodiversity in agroecosystems. *Agriculture Ecosystems and Environment* 74:19–31.

Altieri, M.A. (2000). The ecological impacts of transgenic crops on agroecosystem health. *Ecosystem Health* 6:13–23.

Altieri, M.A., and Nicholls, C.I. (2003). *Biodiversity and Pest Management in Agroecosystems.* Binghamton, NY: Food Product Press.

Badgley, C., Moghtader, J., Quinterok, E., Zakem, E., Chappell, M.J., Aviles-Vazquez, K., Samulon, A., and Perfecto, I. (2007). Organic agriculture and the global food supply. *Renewable Agriculture and Food Systems* (in press).

Berc, J.L., Muniz, O., and Calero, B. (2004). Vermiculture offers a new agricultural paradigm. *Biocycle* 45:56–57.

Berkenbusch, K., and Rowden, A.A. (2003). Ecosystem engineering—Moving away from "just-so" stories. *New Zealand Journal of Ecology* 27:67–73.

Bird, P.R. (1991). Tree and shelter effects on agricultural production in southern Australia. *Agricultural Systems* 4:37–39.

——. (1998). The windbreaks and shelter benefits to pasture in temperate grazing systems. *Agroforestry Systems* 41:35–54.

Black, H.I.J., and Okwakol, M.J.N. (1997). Agricultural intensification, soil biodiversity and agroecosystem function in the tropics: The role of termites. *Applied Soil Ecology* 6:37–53.

Callaway, R.M., and Walker, L.A.R. (1997). Competition and facilitation: A synthetic approach to interactions in plant communities. *Ecology* 78:1958–1965.

Campanha, M.M., Silva Santos, R.H., de Freitas, G.B., Prieto Martinez, H.E., Ribeiro Garcia, S.L., and Finger, F.L. (2004). Growth and yield of coffee plants in agroforestry and monoculture systems in Minas Gerais, Brazil. *Agroforestry Systems* 63:75–82.

Chan, K.Y., and Heenan, D.P. (1993). Surface hydraulic properties of a red earth under continuous cropping with different management practices. *Australian Journal of Soil Research* 31:13–24.

Chander, K., Goyal, S., Nadal, D.P., and Kapoor, K.K. (1998). Soil organic matter, microbial biomass, and enzyme activity in tropical agroforestry systems. *Biology and Fertility of Soils* 27:168–172.

Conford, P. (2001). *The Origins of the Organic Movement.* Edinburgh: Floris Books.

Conway, G.R., and Pretty, J.N. (1991). *Unwelcome Harvest: Agriculture and Pollution.* London: Earthscan Publications.

Dangerfield, J.M., McCarthy, T.S., and Ellery, W.N. (1998). The mound-building termite *Macrotermes michaelseni* as an ecosystem engineer. *Journal of Tropical Ecology* 14:507–520.

Darwin, C. (1881). *The Formation of Vegetable Mould Through the Action of Worms with Observations of Their Habits.* London: Murray.

Dix, M.E., Johnson, R.J., Harrell, M.O., Case, R.M., Wright, R.J., Hodges, L., Brandle, J.R., Schoeneberger, M.M., Sunderman, N.J., Fitzmaurice, R.L., Young, L.J., and Hubbard, K.G. (1995). Influences of trees on abundance of natural enemies of insect pests: A review. *Agroforestry Systems* 29:303–311.

Douglas, C.L., King, K.A., and Zuzel, J.F. (1998). Nitrogen and phosphorus in surface runoff and sediment from a wheat-pea rotation in Northeastern Oregon. *Journal of Environmental Quality* 27:1170–1177.

Dreyfus, D. (1990). *Feasibility of On-Farm Composting*. Rodale Research Center Kutztown, PA: Rodale Institute.

Edwards, C.A. (1994). *Earthworm Ecology*. Boca Raton, FL: CRC Press.

Edwards, C.A., and Bohlen, P.J. (1996). *Biology and Ecology of Earthworms*, 3rd ed. London: Chapman and Hall.

Epstein, E. (1997). *The Science of Composting*. Boca Raton, FL: CRC Press.

Falcon, L.A. (1971). Progreso del control integrado en el algodón de Nicaragua. *Revista Peruana de Entomología Agrícola* 14:376–378.

Fassbender, H.W., Beer, J., Heuveldop, J., Imbach, A., Enriquez, G., and Bonneman, A. (1991). The year balance of organic matter and nutrients in agroforestry systems at CATIE, Costa Rica. *Forest Ecology and Management* 45:173–183.

Folgarait, P.J. (1998). Ant biodiversity and its relationship to ecosystem functioning: A review. *Biodiversity and Conservation* 7:1221–1244.

Forlani, G., Mantelli, M., Branzoni, M., Nielsen, E., and Favilli, F. (1995). Differential sensitivity of plant-associated bacteria to sulfonylurea and imidazolinone herbicides. *Plant and Soil* 176:243–253.

Fragoso, C., Brown, G.G., Patron, J.C., Blanchart, E., Lavelle, P., Pachanasi, B., Senapati, B., and Kumar, T. (1997). Agricultural intensification, soil biodiversity and agroecosystem function in the tropics: The role of earthworms. *Applied Soil Ecology* 6:17–35.

Frederickson, J., Butt, K.R., Morris, R.M., and Danierl, C. (1997). Combining vermiculture with traditional green waste composting systems. *Soil Biology and Biochemistry* 29:725–730.

Garcia-Barrios, L., and Ong, C.K. (2004). Ecological interactions, management lessons and design tools in tropical agroforestry systems. *Agroforestry Systems* 61–62:221–236.

Goldberg, D.E. (1990). Components of resource competition in plant communities. In *Perspectives in Plant Competition*, J. Grace and D. Tilman, Eds. New York: Academic Press, pp. 27–49.

Hendrix, P.F., and Edwards, C.A. (1994). Earthworms in agroecosystems. In *Earthworm Ecology*, C.A. Edwards, Ed. Boca Raton, FL: CRC Press, pp. 287–296.

Hoeksema, K.J., and Jongerius, A. (1959). On the influence of earthworms on the soil structure in mulched orchards. In *Proceedings of the International Symposium of Soil Structure, Ghent 1958*, pp. 188–194.

Howard, A. (1940). *An Agricultural Testament*. Oxford: Oxford University Press.

Insam, H., Amor, K., Renner, M., and Crepaz, C. (1996). Changes in functional abilities of the microbial community during composting of manure. *Microbial Ecology* 31:77–87.

Insam, H., Riddech, N., and Klammer, S. (Eds.). (2002). *Microbiology of Composting*. Berlin: Springer-Verlag.

Johnsen, K., Jacobsen, C.S., Torsvik, V., and Sorensen, J. (2001). Pesticide effects on bacterial diversity in agricultural soils—a review. *Biology and Fertility of Soils* 33:443–453.

Johnson W.C. III, Brenneman, T.B., Baker, S.H., Johnson, A.W., Sumner, D.R., and Mullinix, B.G., Jr. (2001). Tillage and pest management considerations in a peanut-cotton rotation in the southeastern coastal plain. *Agronomy Journal* 93:570–576.

Jones, C.G., Lawton, J.H., and Shachak, M. (1994). Organisms as ecosystem engineers. *Oikos* 69:373–386.

——. (1997). Positive and negative effects of organisms as physical ecosystem engineers. *Ecology* 78:1946–1957.

Jones, J.A. (1990). Termites, soil fertility and carbon cycling in dry tropical Africa: A hypothesis. *Journal of Tropical Ecology* 6:291–305.

Joschko, M., Söchtig, W., and Larink, O. (1992). Functional relationships between earthworm burrows and soil water movement in column experiments. *Soil Biology and Biochemistry* 24:1545–1547.

Lappé, F.M., Colins, J., and Rosset, P. (1998). *World Hunger: Twelve Myths*. San Francisco, CA: Institute for Food and Development Policy.

Lavelle, P. (2002). Functional domains in soils. *Ecological Research* 17:441–450.

Lawton, J.H. (1994). What do species do in ecosystems? *Oikos* 71:367–374.

Li, L., Zhang, F., Li, X., Christie, P., Sun, J., Yang, S., and Tang, C. (2003). Interspecific facilitation of nutrient uptake by intercropped maize and faba bean. *Nutrient Cycling in Agroecosystems* 65:61–71.

Liebman, M., and Davis, A.S. (2000). Integration of soil, crop and weed management in low-external-input farming systems. *Weed Research* 40:27–47.

Loreau, M., Naeem, S., and Inchausti, P. (2002). Biodiversity and ecosystem functioning: Synthesis and perspectives. Oxford: Oxford University Press.

Martius, C. (2001). *The Effect of Termites as Ecosystem Engineers in the Humid Tropics*. San Francisco: American Geophysical Union, Fall Meeting.

Mayus, M., Van Keulen, H., and Stroosnijder, L. (1998). A model of tree-crop competition for windbreak systems in the Sahel: Description and evaluation. *Agroforestry Systems* 43:183–201.

Montagnini, F., and Nair, P.K.R. (2004). Carbon sequestration: An underexploited environmental benefit of agroforestry systems. *Agroforestry Systems* 61–62:281–295.

Nagarajah, S., Posner, A.M., and Quirk, J.P. (1970). Competitive adsorption of phosphate with polygalacturonate and other organic anions on kaolinite and oxide surfaces. *Nature* 228:83–85.

Nicholls, C.I., and Altieri, M.A. (2004). Designing species-rich, pest-suppressive agroecosystems through habitat management. In *Agroecosystems Analysis*, D. Ricker, R. Aiken, C. Wayne Honeycutt, F. Magdoff, and R. Salvador, Eds. Madison, WI: American Society of Agronomy, pp. 49–62.

Olding-Smee, K., Laland, N., and Feldman, M.W. (2003). *Niche Construction: The Neglected Process in Evolution*. Princeton, NJ: Princeton University Press.

Palm, C.A. (1995). Contribution of agroforestry trees to nutrient requirements in intercropped plants. *Agroforestry Systems* 30:105–124.

Perfecto, I., and Vandermeer, J. (1993). Distribution and turnover rate of a population of *Atta cephalotes* in a tropical rain forest in Costa Rica. *Biotropica* 25:316–321.

Perfecto, I., and Vandermeer, J. (2006). The effect of an ant/scale mutualism on the management of the coffee berry borer (*Hypothenemus hampei*) in southern Mexico. *Agriculture Ecosystems and Environment* 117:218–221.

Rao, M.R., Nair, P.K.R., and Ong, C.K. (1997). Biophysical interactions in tropical agroforestry systems. *Agroforestry Systems* 38:3–50.

Raynolds, L.T. (2000). Re-embedding global agriculture: The international organic and fair trade movements. *Agriculture and Human Values* 17:297–309.

Riedell, E.W., Schumacher, T.E., Clay, S.A., Ellsbury, M.M., Pravecek, M., and Evenson, P.D. (1998). Corn and soil fertility responses to crop rotation with low, medium, or high inputs. *Crop Science* 38:427–433.

Rhoades, C.C. (1996). Single-tree influences on soil properties in agroforestry: Lessons from natural forest and savanna ecosystems. *Agroforestry Systems* 35:71–94.

Springett, J.A., Gray, R.A.J., and Reid, J.B. (1992). Effect of introducing earthworms into horticultural land previously denuded of earthworms. *Soil Biology and Biochemistry* 24:1615–1622.

Stamps, W.T., and Linit, M.J. (1998). Plant diversity and arthropod communities: Implications for temperate agroforestry. *Agroforestry Systems* 39:73–89.

Sterelny, K. (2005). Made by each other: Organisms and their environment. *Biology and Philosophy* 20:21–36.

Sturrock, J.W. (1988). Shelter: Its management and promotion. *Agriculture, Ecosystem and Environment* 22/23:1–5.

Swift, M.J., Heal, O.W., and Anderson, J.M. (1979). *Decomposition in Terrestrial Ecosystems*, Studies in Ecology 5. Oxford: Blackwell Scientific Publications.

Swift, M., Vandermeer, J.H., Ramakrisnan, R., Anderson, J.M., Ong, C., and Hawkins, B. (1996). Biodiversity and agroecosystem function. In *Functional Roles of Biodiversity: A Global Perspective*, 1996 SCOPE, H.A. Mooney, J.H. Cushman, E. Medina, O.E. Sala, and E.D. Schulze, Eds. New York: John Wiley and Sons.

Van Breemen, N., and Finzi, A.C. (1998). Plant-soil interactions: Ecological aspects and evolutionary implications. *Biogeochemistry* 42:1–19.

van den Bosch, R. (1989). *The Pesticide Conspiracy*. Berkeley, CA: University of California Press.

Vandermeer, J.H. (1981). The interference production principle: An ecological theory for agriculture. *Bioscience* 31:361–364.

——. (1984). The interpretation and design of intercrop systems involving environmental modification by one of the components: A theoretical framework. *Journal of Biological Agriculture and Horticulture* 2:135–156.

——. (1989). *The Ecology of Intercropping*. Cambridge, UK: Cambridge University Press.

——. (1995). The ecological basis of alternative agriculture. *Ann Review of Ecology and Systematics* 26:201–224.

Vandermeer, J.H., van Noordwijk, M., Anderson, R.M., Ong, C., and Perfecto, I. (1998). Global change and multi-species agroecosystems: Concepts and issues. *Agriculture Ecosystems and Environment* 67:1–22.

Vandermeer, J.H., and Perfecto, I. (2006). A keystone mutualism drives pattern in a power function. *Science* 311:1000–1002.

Whitford, W.G. (1994). The importance of the biodiversity of soil biota in arid ecosystems. *Biodiversity and Conservation* 5:185–195.

Wilby, A., Shachak, M., and Boeken, B. (2001). Integration of ecosystem engineering and trophic effects of herbivores. *Oikos* 92:436–444.

20

MANAGEMENT AND ECOSYSTEM ENGINEERS: CURRENT KNOWLEDGE AND FUTURE CHALLENGES

Alan Hastings

20.1 — INTRODUCTION

The concept of ecosystem engineering may have particular impact in the area of applied ecology, as the five earlier chapters in this section demonstrate. The basic questions one would want to ask are when and how concepts from ecosystem engineering would or could improve management. Especially when a species is clearly an ecosystem engineer, the concept of ecosystem engineering is useful in an applied context, which is a theme that runs through all the chapters in this section. Among the many challenges in applied ecology is the problem of making management decisions, or providing management advice, in the face of limited information. Here, clearly, general rules need to be developed, and as suggested by Byers et al. (2006), the ecosystem engineering concept may prove useful in understanding the potential large impacts of removing species that have engineering effects. Another important aspect of applied ecology is that explicit timescales are often involved. Management decisions and options are often carried out over specific time frames, and consequently the timescale issues that arise with ecosystem engineers will be especially important. Here, as emphasized in Hastings et al. (2007), the recognition of the temporal and spatial scales over which ecosystem engineers operate can play an important role in recognizing some of the difficult issues that ecosystem engineers pose for management. For example, since the effects may be long term and also may in some ways be irreversible (Lambrinos 2007), challenges arise as to the relative costs and impacts of management decisions over the long term. Also, since the impacts of engineers may occur over larger spatial scales, management decisions at one location may effect the operation of ecosystems at other locations.

The challenges and opportunities provided by considering ecosystem engineers in management lead to problems in terms of the interplay between biological and physical aspects. However, they also raise issues that are particularly challenging within the human dimension. I will return to some of these human issues in following text, but first consider how the chapters in this section fit together to provide a larger insight into the interplay between management and ecosystem engineering.

20.2 ◆ EFFECTS AND IMPACTS OF SINGLE ENGINEERING SPECIES

One approach to employing this concept in management problems can be to look at a case study of essentially a single species that is unequivocally classified as an ecosystem engineer. This approach would be useful both for the case where the engineer is beneficial and for the case where the engineer is harmful to some aspect of ecosystem functioning. This is the approach taken in Grabowski and Peterson (2007) and Lambrinos (2007) respectively.

The easier case to understand may be the case of restoring a single species that has positive effects as an ecosystem engineer (Grabowski and Peterson 2007). In this case, the engineering concept provides important extra insights into the value of restoration, which fall into the general theme of the concept that the spatial and temporal impacts of engineers can be broader and longer than that of an individual's lifetime. In the case of oysters, there are both direct economic impacts since oysters themselves can be commercially harvested, and the impacts through engineering by the oysters that affect more general ecosystem services, like water quality. This recognition is first useful in a qualitative way, for emphasizing that the economic impact will in general be much greater than that calculated solely based on the market value of the oysters. Secondly, though much more difficult to carry out with limited data, this recognition of the engineering impact of oysters can be used in an explicit calculation of the true positive financial impacts of ecosystem engineering.

A second case is to recognize the impacts of an invasive engineer, and consider efforts for control and restoration (Lambrinos 2007). Once again, the recognition that ecosystem engineers have broader temporal and spatial impacts provides key insights into management of invasive species. As noted by Lambrinos (2007) these legacy effects of invasive ecosystem engineers provide special challenges for management, since the recovery of the ecosystem following removal of the engineer occurs in a very different fashion than recovery of species whose primary interactions are biotic and that do not produce long-lasting physical changes in the environment. The incorporation of the ecosystem engineering concept provides impetus for the development of models for management that take into account the explicit temporal and spatial effects. For the specific case reviewed by Lambrinos (2007) of *Spartina* invasions into Pacific estuaries, this means that control efforts that focus solely on the removal of living *Spartina* and do not directly confront the legacy effects of the engineer are unlikely to restore ecosystem function.

Both of these essentially single-species cases lead to new insights about active management of ecosystems. These are systems that are simple enough to understand quite completely in at least a heuristic way, and it is clear from both chapters (Grabowski and Peterson 2007, Lambrinos 2007) that explicit inclusion of ecosystem engineering concepts is essential for effective management.

20.3 ⬛ EFFECTS AND IMPACTS OF ENGINEERS IN THE CONTEXT OF ECOSYSTEMS

The ecosystem engineering concept can also be useful in the context of more complex systems, though of course analysis and prediction become substantially more difficult. The interaction among different ecosystem engineers is an important issue, and introduces new aspects not explicitly found in the single-species systems, and in the biotic interactions that have formed the core of much ecological thinking. Here, especially if several species can and do affect the physical environment in different ways, new kinds of dynamics and interactions emerge because species are interacting with each other potentially on the longer timescales and broader spatial scales that are typical of ecosystem engineers. As analyzed by Oren et al. (2007), one sees that the complications induced by interactions among ecosystem engineers can lead to much more complex behavior than in the somewhat simpler effectively single-species situations analyzed in the first two chapters of this section. Here the emphasis is both on active management, and also on the understanding of ecosystem processes. Oren et al. (2007) clearly demonstrate that the analysis of the ecosystem dynamics would be impossible without explicit inclusion of the engineering aspects.

Active management of ecological systems is difficult because the complexity of ecological systems makes responses difficult to predict. Even in relatively simple systems, removal or addition of species can have unforeseen consequences, as emphasized by Klinger (2007). Here the goal was to deal with non-native species on the Channel Islands off the coast of California. This is a system that has many non-native species, and it is simply not practical to simultaneously eradicate all of them. In addition, the native species have clearly been affected by the presence of the non-native species, so removal alone would not obviously lead to restoration. Klinger does concur with Byers et al. (2006) that applying the ecosystem engineering concept could be useful in understanding some of the large-scale impacts of removal of non-native species. However, Klinger also emphasizes a number of limitations related to complexity and the lack of predictability. The issue of whether the real impact of ecosystem engineer removal is apparent only after the impacts are seen, rather than being potentially predictable, is an important one that needs to be studied in many other systems. Another very important issue raised by Klinger of the constraints on management from political and social considerations is discussed further in following text.

In some ways, agricultural systems would seem to be easier to understand as the controls applied by humans are direct and well known. However, much of the response of these systems occurs on timescales that are real challenges to current ecological theory. Here is where the kinds of issues raised by Vandermeer and Perfecto (2007) about the role of ecosystem engineers come to the fore. Agricultural systems by their very nature are actively managed, and in many cases ecosystem engineering aspects play a major role. As they emphasize, soil clearly has a large physical component, and therefore understanding the dynamics of soil in the context of agriculture could profitably make use of concepts from ecosystem engineering. Other examples they raise that could be clearly understood as engineering include the role of shade provided by some plants for other plants. As Vandermeer and Perfecto (2007) suggest, the engineering concept could provide further insights into better agricultural processes.

What the three chapters here all have in common since they explicitly include multiple interacting species is a fundamental issue in the application of the engineering concept: What is engineering and what are biotic interactions? The different chapters focusing on species interactions reach somewhat different conclusions about the role of ecosystem engineering as the key concept. Nonetheless, all do reach one similar conclusion, namely that the ecosystem engineering concept potentially puts the focus on aspects that might otherwise not be as well studied, or included, such as interactions of organisms with the soil in Vandermeer and Perfecto (2007), or the kind of interaction among livestock, water, and vegetation (Oren et al. 2007).

20.4 ● CONCLUSIONS AND FURTHER DIRECTIONS

All of the chapters in this section carefully go over the importance of the inclusion of specific physical modification of the environment and how that would affect management decisions. In all cases, focusing only on the biotic interactions would be inefficient at best, and misleading at worst.

It is also important in discussing management and ecosystem engineering to consider aspects that are not included in these contributions and that are worthy of further detailed study. Although these chapters all take the careful first step toward understanding what are plausible solutions to management problems, or how ecosystem engineering fits into the context of existing management practices, explicit calculation of costs and benefits with proper economic aspects (like discounting and comparison of alternatives in the context of explicit optimization) is not emphasized. However, the field of bioeconomics has played a long role within fisheries (Clark 1990), and more explicit inclusion of economics in other areas of conservation is beginning (Naidoo et al. 2006). Yet, the problem of management with or by ecosystem engineers introduces new and important difficulties into the bioeconomic analysis. The kinds of hysteresis emphasized by Lambrinos (2007) lead to what are irreversible or nearly irreversible behaviors of ecosystems. These different forms of irreversibilities and cost structures can bias the optimal policy portfolio in different and counterintuitive ways. Furthermore, the spatial dimension adds another important form of interaction that can change results compared with the standard dynamic models that are typically used in bioeconomics.

And, as is known, and essentially discussed by Klinger (2007), economic considerations are not the only ones that come into play when making management decisions. Here, once again, the extended spatial and temporal impact of engineers naturally lead to complexities. The extended spatial impact means that actions by individuals, or individual agencies, affect, and are affected by, actions of others that are dealing with other parts of the spatial landscape. An important general question in environmental policy is the *collective-action* problem, where solutions require more than one individual or agency to act in concert. These problems are not easily understood using economic optimization procedures that do not take into account strategic interactions among actors. Another economic concept, *public good*, which essentially refers to a good (benefit) that can be enjoyed by a group that is not reduced by consumption of individuals, and where access is not controlled so all can draw the benefit, also clearly plays a key role in analysis of environmental policy.

Problems of management in the context of ecosystem engineers present interesting collective-action problems at two levels of analysis that have been the subject of previous social science research. In both cases, the collective-action problems stem from the public good character of eradication and restoration efforts, which require coordinated action and broad public support. The public good nature of the problems arises directly from the spatial interconnections among restoration sites, as facilitated by the spatial nature of engineering impacts. One actor's failure to deal with invasive engineering species on their land leaves spatially interconnected sites vulnerable, while restoration at one site with an engineer has positive spillovers for other sites. The literature on watershed management, including trust and policy networks (Sabatier et al. 2005), and the governance of common-pool resources provide several useful theories that can be applied to these questions (Ostrom 1990, Lubell 2004).

A second set of relevant studies that could be used to understand management issues associated with ecosystem engineers focuses on collective action and environmental behavior of citizens. Lubell (2002) has shown that the collective interest model of political participation and protest behavior can readily be applied to understanding environmental behavior. The collective interest model focuses on the perceived value of the public good, beliefs about efficacy in providing that public good, and the selective benefits and costs of action (Finkel and Opp 1991). These ideas could profitably be used to understand management issues associated with the large-scale impacts and controls needed when dealing with ecosystem engineers.

Thus the overall conclusion is that explicit consideration of ecosystem engineering has and will increasingly continue to provide vital new insights into management and applied ecology. As all the chapters in this section discuss, explicit consideration of engineers also leads to fascinating new problems that go beyond traditional ecology, and include new connections with social sciences in ways that will lead to better management and new insights into the functioning of ecological systems in an increasingly human-dominated world.

REFERENCES

Byers, J.E., Cuddington, K., Jones, C.G., Talley, T.S., Hastings, A., Lambrinos, J.G., Crooks, J.A., and Wilson, W.G. (2006). Using ecosystem engineers to restore ecological systems. *Trends in Ecology and Evolution* 21:493–500.

Clark, C.W. (1990). *Mathematical Bioeconomics: The Optimal Management of Renewable Resources*, 2nd ed. New York: John Wiley & Sons.

Finkel, S.E., and Opp, K.D. (1991). Party identification and participation in collective political-action. *Journal Of Politics* 53(2):339–371.

Grabowski, J., and Peterson, C. (2007). Restoring oyster reefs to recover ecosystem services, this volume.

Hastings, A., Byers, J.E., Crooks, J.A., Cuddington, K., Jones, C.G., Lambrinos, J.G., Talley, T.S., and Wilson, W.G. (2007). Ecosystem engineering in space and time. *Ecology Letters* 10:153–164.

Klinger, R. (2007). Ecosystem engineers and the complex dynamics of non-native species management on California's Channel islands, this volume.

Lambrinos, J. (2007). Managing invasive ecosystem engineers: The case of *Spartina* in Pacific estuaries, this volume.

Lubell, M. (2002). Environmental activism as collective action. *Environment and Behavior* 34:431–454.

——. (2004). Resolving conflict and building cooperation in the National Estuary Program. *Environmental Management* 33(5):677–691.

Naidoo, R., Balmford, A., Ferraro, P.J., Polasky, S., Ricketts, T.H., and Rouget, M. (2006). Integrating economic costs into conservation planning. *Trends in Ecology and Evolution* 21:681–687.

Oren, Y., Perevolotsky, A., Brand, S., and Shachak, M. (2007). Livestock and engineering network in the Israeli Negev: Implications for ecosystem management, this volume.

Ostrom, E. (1990). *Governing the Commons*. Cambridge, NY: Cambridge University Press.

Sabatier, P.A., Focht, W., Lubell, M., Trachtenburg, Z., Vedlitz, A., and Matlock, M. (Eds.). (2005). *Swimming Upstream: Collaborative Approaches to Watershed Management*. Cambridge: MIT Press.

Vandermeer, J., and Perfecto, I. (2007). The diverse faces of ecosystem engineers in agroecosystems, this volume.

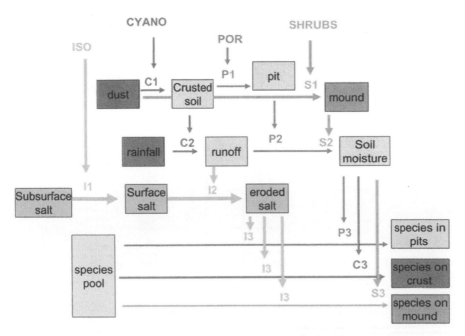

FIGURE 17.1 The engineering network in the Negev Desert and its effects on species richness. Three ecosystem engineers—cyanobacteria (CYANO, C), porcupines (POR, P), and shrubs (S)—modulate the landscape by creating three patch types: crusted soil, pits, and mounds. Landscape-level engineering controls state changes in an ecosystem process of water flow (C2, P2, S2). Water flow (P3, C3, S3) affects species assemblages on the three patch types. An additional engineer-isopod (ISO, I) modulates another ecosystem process-desalinization, by transferring salty soil from subsurface to the surface (I1). The salty soil is eroded from the system by runoff water (I2), a desalinization process that affects plant assemblages (I3). Notice that the engineering network is a network that links processes at various levels of organization (horizontal boxes and arrows) induced by the engineers (vertical arrows).

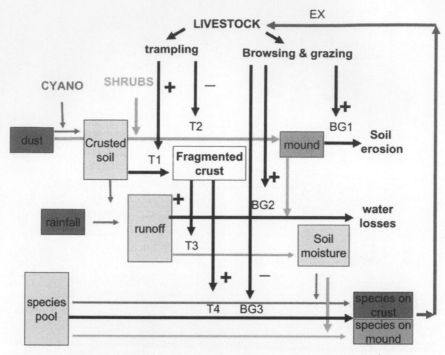

FIGURE 17.2 The joint network of natural ecosystem engineers and livestock. Livestock introduces a new state variable (fragmented crust) to the natural engineering network by trampling on the crust (T1). Crust trampling increases (+) soil moisture (T3) and species richness on crust (T4). Trampling on shrubs decreases (–) soil mound (T2). Browsing and grazing increase soil erosion (BG1) and water losses (BG2) and decrease species diversity (BG3). EX refers to plant species exploitation by grazing.

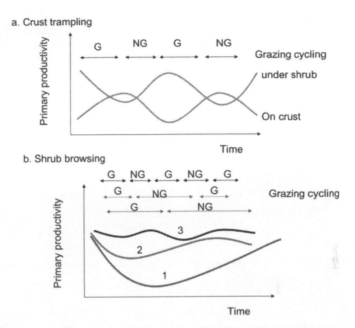

FIGURE 17.3 A conceptual model of the effects of crust trampling (a) and shrub browsing (b) on primary productivity related to shrub and crust patch types under shifting of grazing (G) and non-grazing (NG) periods. 1, 2, and 3 refer to shrub traits: 1 = high palatability and low recovery; 2 = intermediate palatability and recovery; 3 = low palatability and high recovery.

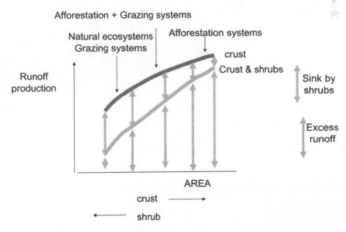

FIGURE 17.4 Ecosystem management in the Negev in relation to cyanobacteria–shrub engineering network. The four managed systems are located on a shrub-to-crust ratio gradient. Afforestation systems are managed for low shrub-to-crust ratio. Natural and grazing systems are managed for high shrub-to-crust ratio. For combined afforestation and grazing systems, an intermediate ratio is desirable. Brown line = runoff generation by crust; green line = runoff generation by crust and shrubs.

FIGURE 18.2 Overgrazing by feral sheep on Santa Cruz Island resulted in many parts of the island where there was little if any vegetation cover.

INDEX